DR. M. HENINI

Diode Laser Materials & Devices

A Worldwide Market & Technology Overview to 2005

First Edition

ELSEVIER
ADVANCED
TECHNOLOGY

UK	Elsevier Science Ltd, The Boulevard, Langford Lane, Kidlington, Oxford OX5 1GB, UK
USA	Elsevier Science Inc, 665 Avenue of the Americas, New York, NY 10010, USA
JAPAN	Elsevier Science Japan, Tsunashima Building Annex, 3-20-12 Yushima, Bunkyo-ku, Tokyo 113, Japan

Author: Roy Szweda

Programme Editor: Roisin Reidy

First edition, 2001

Library Cataloguing in publication Data

ISBN 1 85617 386 0

Published by
Elsevier Advanced Technology
The Boulevard, Langford Lane, Kidlington, Oxford OX5 1GB, UK
Tel: +44 1865 843695
Fax: +44 1865 843971

Typeset by Variorum Electronic Publishing Ltd, Lancaster
Printed in the United Kingdom by Biddles Ltd, King's Lynn

Contents

List of Tables

Figures

1 Introduction

This report examines the development of the diode laser industry over a six-year period, 2000 to 2005, incorporating analysis of trends in markets, technologies and industry structure. It is designed to provide key information to users and manufacturers of substrates, epitaxial wafers (epiwafers) and devices.

The coverage includes components, laser diodes and the semiconducting (SC) wafers and epiwafers on which most of these devices are made.

The geographical coverage of the report includes North America, Japan and Europe, which together will account for over 90% of the production and consumption of diode laser materials and devices over the next five years.

However, many other regions and countries have activities in this field including South-East Asia (Taiwan, South Korea, Singapore, Malaysia, etc), China, India, Australia and Eastern Europe (Russia, Poland, Hungary, Czech Republic). Activities in these countries are commented on where relevant, but are not quantified in the market data.

1.1 Report Structure

Chapter 2 summarizes the main conclusions and market data.

The size, quality, and particularly the price, of substrates and wafers are key factors in determining the ability of companies to produce competitive laser products. Chapter 3 examines trends in materials technologies for laser diodes, the impact of the device markets on wafer demand and the main suppliers. The chapter introduces the semiconductor materials that are presently, or will likely become, important to the fabrication of diode laser devices. The principal distinguishing properties of these materials are explained with reference to their application.

Chapter 4 examines the basic application sectors (many of which overlap) for laser diode devices as well the basic commercial opportunities, changes and forces acting within each sector. The chapter also examines the market for the basic types of device as well as the promising newer types. For each type of device, market data and forecasts are provided and future prospects described. The chapter gives an overview of the applications markets for semiconductor optoelectronic components, describing each market sector: telecommunications, computer, automotive, aerospace, consumer and industrial. The application data are presented for the following industrial groupings:

- Automotive.
- Computers.
- Consumer.
- Industrial.
- Military and aerospace.
- Telecommunications.
- Others.

A full five-year forecast and analysis is provided by application and region.

Chapter 5 gives a background and overview of developments in the principal technological research and development and production processes for devices. The main focus is on the most important enabling technology for the production of the present and future generations of laser diodes and related devices. This process is crystal growth and involves the following sequence:

- Bulk growth of single crystals;
- Epitaxial growth of semiconductor single-crystal layers;
- Ion implantation;
- Device fabrication, i.e. gate and contact formation, etc.; and
- Packaging and testing.

Chapter 6 presents profiles of substrate suppliers, epiwafer suppliers and merchant and captive producers of GaAs devices.

Chapter 7 gives an alphabetical list (by country) of universities and leading industrial laboratories involved in the areas of diode laser research.

A directory of suppliers is given in Chapter 8, and Chapter 9 contains appendices listing acronyms and exchange rates.

1.2 Products

The products considered in this report are those that have an established market presence. Coverage is also included for materials and devices that are not yet in widespread use but which are expected to exert an impact during the forecast period, as are longer-term trends and research and development activities.

Coverage includes materials and devices based on the following materials: gallium arsenide, indium phosphide, gallium nitride, indium antimonide, silicon carbide, silicongermanium and organic polymers.

1.3 Markets

Market figures were compiled during 2001, and data are provided for 2000, the base year, through to 2005, the final forecast year. Where relevant markets are expressed both in volume and value terms. Values are expressed in constant 2001 US dollars. Please note, table totals may not add up exactly, due to rounding up/down.

Markets and forecasts are considered according to the following product categories:

- Substrates.
- Epiwafers.
- Discrete devices.

1.4 Regions

This study considers the topic on a global basis. Markets and forecasts are broken down according to the following four regions:

- North America.
- Western Europe.
- Japan.
- Rest of the world (ROW).

1.5 Research Background

The information in this report was derived from a combination of sources: extensive literature and internet research and a comprehensive interview programme. The latter was conducted with key personnel from leading laser materials and optoelectronics device manufacturers, researchers and other organizations. Other secondary sources employed included:

- Manufacturers' product literature;
- Company reports and profiles;
- Company accounts and other financial data;
- Technical articles and research publications;
- Conference proceedings;
- Press releases and other promotional material; and
- Industry, trade association and government statistics.

This report was produced in association with the industry's premier publication, III-Vs Review. Readers are also referred to the companion reports in this series, *Gallium Nitride & Related Wide Bandgap Materials & Devices: A*

Market & Technology Overview 1998-2003, Second Edition and *Gallium Arsenide Electronic Materials & Devices: A Strategic Study of Markets, Technologies & Companies Worldwide 1999-2004*, Third Edition. These publications contain additional research and development information on the status of GaAs, GaN, SiC and other materials as applied to opto- and micro-electronic devices.

2 Executive Summary

The key findings and forecast results of the report are presented in summary form. Reflecting the structure of the report it is broken down according to the following levels:

- Diode laser substrate markets;
- Diode laser epitaxial wafer markets; and
- Diode laser device market by application.

The structure of the GaAs electronic materials and devices industry as covered in this report is shown in Figure 2.1.

The family of devices that come under the title of diode lasers is one of the most important in today's electronics industry. The diode laser not only permits the high-speed processing but also the storage of data, i.e. information, audio and video. It thus underpins a huge range of products from fibre optic communications through computers to home entertainment systems. It is an industry that is continually developing, with new variants of existing types along with new devices appearing on a regular basis. Moreover, it is an industry that is experiencing a period of heightened interest. This is for two reasons:

- After a period of some of the highest growth rates in its history, the industry has seen a major downturn in fortunes. It is expected to pick up again in due course, but the timing of the upturn is a matter of considerable industry-wide preoccupation.
- In the past few years the market has seen the commercial introduction of several new diode laser types. The most important of these include violet diode lasers, vertical cavity diode lasers and high-power diode lasers.

The market is thus diverging and extending its market value. A most notable transition is from signal processing, i.e. fibre optics and data storage which make up the bulk of the market, to the diode laser as a compact source of light *energy*. This has already resulted in an increase of market share compared with traditional lasers such as gas or crystal lasers. The application has only just

Figure 2.1 Structure of the Diode Laser and Devices Industry as Covered in this Report

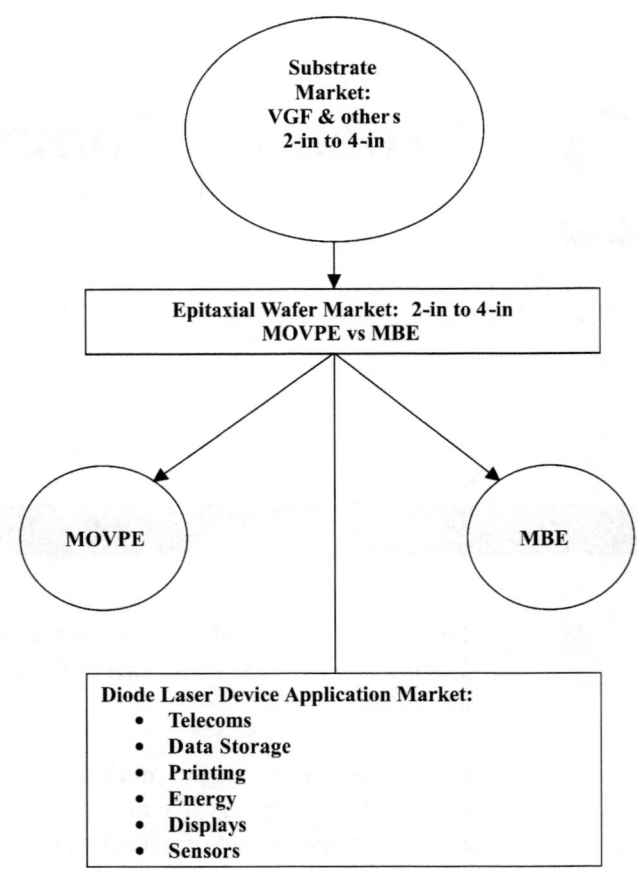

begun, however, with the diode laser expected to find considerable take-up in a wide variety of applications such as welding and cutting as well as in the general lighting market replacing filament and other lamps.

Today the diode laser market is established as a multi-billion-dollar business with many players and new participants being added annually. Not all of the start-ups and spin-out companies described in this first edition will be still active by the end of the forecast period, but for many a great deal of business will be accrued.

The diode laser business is an attractive but competitive segment of the electronics industry. It was characterized by having a higher degree of stability compared with the cyclic mainstream business dominated by silicon-based ICs. However, recently commercial success has faltered largely as a result of over-expectations from the telecommunications industry. Offsetting what is expected to be a temporary setback is the continuing success from a relatively new market, that of DVDs for movies and software.

This report is timely because of the industry's huge expectations of full recovery. This period will be characterized by the adoption of new materials, wafers and process equipment as suppliers strive to meet the need for improved economics while adding performance. This will primarily be accomplished via a transition from what has been the industry standard for over a decade, i.e. the 2-in wafer, to 3- or 4-in diameter wafers. The report thus arrives at a watershed period for the diode laser industry. The market information provided here is intended to assist with business planning for the next five years as the market extends, diversifies and builds on its strengths to become a multi-billion-dollar per year business.

The actual value of the market is the inverse of this structure and this is reflected in the market data provided here. The highest-value tier, that of devices, is followed by epiwafers and finally substrates. It is the business of participating companies to add value at each stage as material is processed step by step through to packaging and installation in the final product.

At the lowest tier is the source materials sector, i.e. gallium and arsenic, which is not covered here. Some attention is given to the device processing equipment sector that was worth around US$800 million in 2000 for all III-V applications, i.e. opto- and microelectronic devices.

Systems represent the highest-value tier but since diode lasers represent only a small fraction, i.e. one or two devices in a fibre optic system, this amounts to less than 10% of the total manufacturing value of the product. There is also pressure to reduce this still further even though the market is growing strongly. The market is also growing through the introduction of new products and not just replacement markets.

The worldwide merchant markets for diode laser materials are presented here. This chapter is basically split into two parts: substrates and epiwafers. The term 'substrate' refers to the unprocessed slice, while epitaxial wafer (epiwafer) refers to the substrate after epitaxial growth but prior to device fabrication.

Secondly, the market for semiconducting epiwafers, e.g. those based on GaAs and InP and their alloys, is estimated and forecast.

Coverage includes an examination of the two principal epiwafer types: metallorganic vapour-phase epitaxy (MOVPE) and molecular beam epitaxy (MBE). These will remain the principal growth techniques for the fabrication of diode lasers and related devices for the period covered by this report.

At the time of writing, the events of 11 September 2001, and subsequent military actions were having an effect on the worldwide marketplace. On the one hand, the terrorist attack served to further depress the total worldwide stock market, hindering its recovery. On the other hand, the military actions in Afghanistan triggered increases in defence spending within the world's largest market, North America. In that region the largest single defence contract in its history was won by Lockheed Martin Corp. The US$200 billion contract was for the Joint Strike Fighter (JSF), with anticipated international sales reaching as much as double that over the programme's life, which could be nearly half a century. Analysts were warning that such defence spending increases could last only a couple of years. This could therefore just be a 'blip' on the overall downward trend.

In certain quarters, industry analysts were predicting an increase in spending on equipment in several key areas as a result of the attacks. These would not only boost offensive systems, such as aircraft weapons and on-board systems, but also intelligence gathering and processing. As a result, developments can be expected to include:

- Increased security across the board in the transportation industry not only at airports but also in aircraft;
- Replacement equipment for that lost in the World Trade Center with the possibility of more distributed information networks so as to minimize damage in future incidents; and
- Increased interest in detection and monitoring systems for weapons of mass destruction as a result of the anthrax attacks in the USA.

However, to confirm the industry consensus that the optoelectronic components business had not been picking up in the later months of 2001, one of the world leaders in the market place, JDS Uniphase, reported a wider net loss for its first quarter and predicted weakness ahead as telecommunications capital spending and the economy continued to slump. It could still not predict the bottom of the depression and stated that it may come sometime between December 2001 and March 2002.

2.2 Diode Laser Substrate Markets

In the first part of this chapter, the market for semiconducting (SC) substrates, e.g. gallium arsenide (GaAs) and indium phosphide (InP), is estimated and forecast. Coverage includes an examination of the principal substrate types by size, price, etc.

Vertical gradient freeze (VGF) will remain the principal crystal growth technique for the fabrication of diode lasers and related devices for the period covered by this report.

- Other substrates used to fabricate diode lasers include sapphire substrates, which are used for the fabrication of violet diode lasers, but this is presently only a very small market.

In 2000 the world market for substrates for diode lasers was worth US$113 million (see Table 2.1). However, in the period 2000–01 the optoelectronics industry suffered a serious downturn which produced many cancellations of orders for associated substrates and other materials. This resulted in a market downturn where the industry fell by as much as 20% to US$90 million in 2001. Positive growth was expected to resume in 2002, of the order of 27%, to reach a market value of US$115 million. This strong return to growth reflects harder unit prices as demand firms up.

As a result of continuing good growth, the total substrate market will grow to US$290 million in 2005. It is expected to be sustained in value in terms of sales, but there are likely to be significant shifts in production emphasis by product type, e.g. substrate diameter should shift from 2 to 3 in, with 4-in diameter also coming into play over the longer term.

The total world market for substrates by region is summarized in Figure 2.2 and Table 2.1.

Table 2.1 Worldwide Diode Laser Substrate Market 2000–05 (US$ million)

	2000	2001	2002	2003	2004	2005
North America	33	26	33	44	59	83
Japan	36	29	37	48	65	92
Europe	24	19	24	32	43	62
Rest of world	20	16	21	27	38	54
Total	113	90	115	151	205	290

The world substrate market is analysed and forecasted by:

- Substrate size — 2- and 3-in and others, i.e. larger or smaller wafers;
- Crystal growth type, i.e. VGF;
- Substrate size; and
- Geographic regions.

In 2001 the diode laser substrate market continued to be dominated by 2-in substrates. However, the 3-in substrate market is a solidly growing sector and will have matched the 2-in market by 2004–05 (see Table 2.2).

Table 2.2 Total World Diode Laser Substrate Market by Size 2000–05 (US$ million)

	2000	2001	2002	2003	2004	2005
2 in	82	66	74	84	95	107
3 in	22	18	30	50	85	144
Other	9	7	11	17	26	39
Total	113	90	115	151	205	290

In the long term, substrate prices will always be in decline. This is the general rule for this tier of the industry irrespective of the material type. This accepted fact of business life is due to a combination of intense pricing pressure coupled with increasing demand and improved yields, year-on-year. Nevertheless, prices

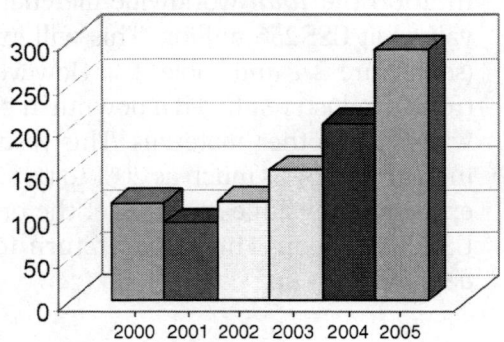

Figure 2.2 Worldwide Diode Laser Substrate Market 2000–05 (US$ million)

Figure 2.3 Total World Diode Laser Substrate Market by Size 2000–05 (US$ million)

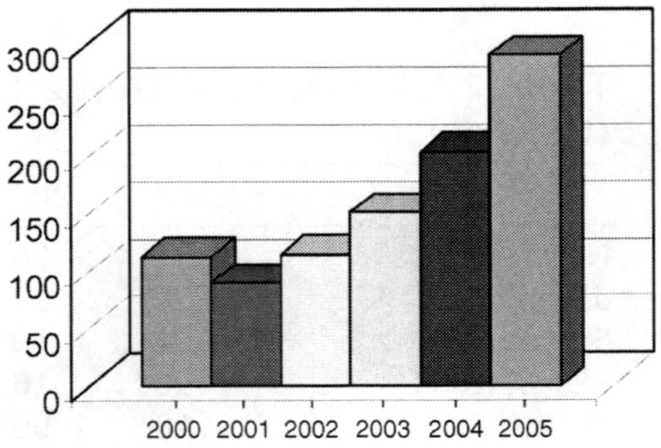

still vary considerably depending on the volume, geographical region, etc. This is irrespective of the type of substrate.

Over the period 2000–02, all substrates will be subject to a fairly serious decline in prices. This is due to the industry downturn which caused cutbacks in device orders for both substrates and epiwafers. From a solid US$20 per square inch region average value, the price will have fallen to a low of about US$10 by early 2002. This is likely to be an industry record for a high-to-low period.

Other key technical and business trends that are likely to have some impact on the diode laser substrate market for the forecast period are discussed. These include:

- Alternative crystal growth methods;
- Other III-V semiconductors;
- Substrates for violet diode lasers; and
- GaAs-on-silicon epitaxial growth.

There is every chance that once the optoelectronics market resumes its upward progress the market will see substantial divergence in materials and process technologies. This will be in direct response to the market demand for ever-more performance at ever-lower prices.

2.3 Diode Laser Epitaxial Wafer Markets

In 2000 the *total* worldwide merchant market for diode laser epiwafers was valued at US$256 million. This will exceed US$525 million by the year 2005 (see Figure 2.4 and Table 2.3). However, in the period 2000–01 the optoelectronics industry suffered a downturn producing cancellations of orders for epiwafers and other materials. This resulted in a market downturn where the industry fell by as much as 20% to US$205 million in 2001. Positive growth was expected to resume in 2002, of the order of 26%, to reach a market value of US$254 million. This strong return to growth reflects harder unit prices as demand firms up.

With continuing good growth the total epiwafer market will grow to US$ 525 million by 2005. It is expected to be sustained in value in terms of sales but there are likely to be significant shifts in production emphasis by product type, e.g. diameter should shift from 2 to 3 in, with 4-in diameter also being important over the longer term. Pricing and application issues should also see changes in the next five years.

Table 2.3 Worldwide Diode Laser Epiwafer Market 2000–05 (US$ million)

	2000	2001	2002	2003	2004	2005
North America	78	62	78	98	126	164
Japan	90	72	88	110	140	180
Europe	51	41	51	64	81	104
Rest of world	37	30	37	47	59	77
Total	256	205	254	318	406	525

The principal requirement for the fabrication of diode lasers is multilayers of compound semiconductors on a semiconducting or other substrate. In the past five years many of the technological difficulties that characterize these materials have been overcome. Today the industry is on the threshold of a new era of mass production capability. This has arisen from the commercial availability of much larger, i.e. 3- and 4-in diameter substrates. Several companies are already installing fabrication facilities to process these wafers largely for the mass production of diode lasers for telecommunications applications.

In contrast, the predominant wafer size for silicon is 8 in (200 mm) with only a few fabs currently operational with 12-in (300-mm) wafers. However, for the foreseeable future, the majority of diode laser manufacturers will make do with smaller-diameter (2 in) wafers until the volume that their business handles mandates re-equipping and scaling up. For the next couple of years its exclusive reliance on discrete devices will dictate that it remains based largely on 2-in diameter wafers. However, by the end of the forecast period the larger wafers will become the dominant types.

Today's merchant epiwafer marketplace is roughly equally split between MBE and MOVPE as the epitaxy technique of choice. In this market estimation MOVPE has a slightly larger market share than MBE, as shown in Table 2.4. However, MBE is making strides as companies are equipping themselves with

Figure 2.4 Worldwide Diode Laser Epiwafer Market 2000–05 (US$ million)

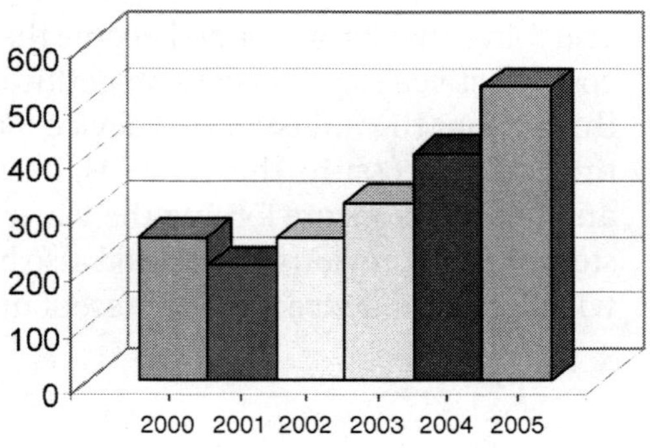

special phosphorus cracker accessories so as to produce multiwafer P-alloys necessary for visible lasers and so on. Another factor in favour of MOVPE is the imminent higher-volume production of violet diode lasers, which so far are made exclusively by this method.

Table 2.4 World Epiwafer Market: MBE *vs* MOVPE 2000–05 (%)

	2000	2005
MBE	48	47
MOVPE	52	53

In recent years the epiwafer market has experienced something of an unsettled period. This is a direct result of increased demand for these materials over the period 1999–2000 followed by a downturn in the market, a situation that prevails at the time of writing. There is every confidence that the market will resume good growth and it is more a matter of when rather than if. When that market growth resumes, the leanest and fittest epiwafer suppliers will be those that reap the most benefit. Much on the minds of these companies at this time is ensuring that they are best equipped with optimal equipment capable of handling the largest wafers, which will provide them with the optimal economics in what will be a very price-sensitive marketplace.

There has been a considerable shift in emphasis within epi-based materials over the years. This reflects the importance of the more modern techniques such as MOVPE and MBE over older techniques such as VPE and LPE. The former represent more advanced epiwafers and hence constitute the higher value segment, whereas the older techniques represent the more basic low-cost product. The latter is more likely to be captive rather than merchant, however.

A new set of dynamics has therefore arisen in the epiwafer market. On the one hand, demand had been high keeping prices buoyant. On the other hand, yields have increased through new generations of multiwafer epitaxy systems and allowed cost reductions to be enjoyed for the first time.

In the past year or so, the 3-in epiwafer has started to become the more important wafer size. Volumes of such wafers have now reached many thousands per annum worldwide. The 2-in wafer is still the most important commercially but its use will begin to decline over the next 3–5 years. This has the effect of pushing up prices and is exacerbated by the reduced emphasis on these products by major epiwafer providers as they wish to obtain the maximum economic advantage that accrues from larger wafer sizes.

The 3-in epiwafer will soon become the benchmark epiwafer size. The improved cost-effectiveness and cost-competitiveness are becoming more widely known throughout the industry. In a device marketplace that is experiencing unremitting price pressure the move is likely to see accelerated schedules. These and other factors are forcing the wafer down a pricing curve that is likely to be steeper than previously. It should also be noted that this standardization on 3 in will also be influenced by the present mixed reception to any larger sizes.

2.4 Diode Laser Device Market by Application

This report also examines the basic application sectors (many of which overlap) for diode lasers as well the basic commercial opportunities, changes and forces acting within each sector. The report also examines the market for the basic types of device according to wavelength. For each type of device, market data and forecasts are provided and future prospects described.

After a period of spectacular growth during 1999–2000, the industry suffered a downturn brought about in large part by over-ordering of components for the telecommunications industry so as to support the exponentially increasing demand for higher bandwidth communications. The period late 1999 to early 2002 was characterized by users of diode lasers and related products working through inventories and as such the industry consensus was for one of the lowest market valuations for the first year of the new millennium.

This sudden downturn was superimposed onto a general industry recessionary trend. Manufacturing worldwide was going through a period of readjustment as sales of computer, consumer, automotive and other products were growing at much reduced rates. Thus it was necessary to impose low or even negative growth rates on several, but not all, application sectors covered in this report.

The highest-value market (light-emitting diodes (LEDs) lead in terms of volume of units shipped) in semiconductor optoelectronics is now the laser diode family of devices. Today the wavelength spectrum is now complete with the following widely available:

- Infrared (IR) telecommunications, data and power;
- Visible I red (DVD), green (laser pointers); and
- Visible II violet lasers for instrumentation and analysis.

At present the most commercially important part of the spectrum remains the IR region. This is due to the matching requirements for the optical characteristics of fibre optics. It is also important because the data storage industry is still dominated by compact disc for computers, music and games.

Up to fairly recently the diode laser industry has been concerned with that device's ability to handle data streams. Whether it is sending signals down a fibre or reading a DVD, power levels of a few milliwatts are sufficient. However, the advent of diode lasers with output powers of 1 W and above have brought into being several interesting new applications arising from the use of the diode laser as a compact energy source. These high-power diode lasers (HPDLs) have become successful as light-pumping sources for solid-state lasers. However, their biggest opportunity is only just emerging — as compact power sources of coherent light. The wavelength range of interest is from 635 nm up to 2 μm, where they are becoming increasingly attractive for functions such as soldering and welding, as well as in microsurgery.

Today's diode laser market is constantly expanding with new types being introduced on a fairly frequent basis. These types were, until recently, largely low-power devices for the processing of data or sending signals (or storing them). The laser industry is, however, experiencing a change in emphasis. Semiconductor lasers are becoming important as compact energy sources. This presently represents less than 10% of the semiconductor laser market but has great potential for future growth.

For the purposes of this forecast, the semiconductor laser diode market is broken down according to various sub-types. IR diode lasers continue to dominate today's market. This is due, for the most part, to fibre optic telecommunications but most of the new laser types such as HPDLs and vertical cavity surface emitting lasers (VCSELs) also operate in the IR region.

Laser applications can be said to fall into four categories:

- Signal transmission — fibre optics.
- Optical data storage CD — and DVD-R.
- Directed energy — materials processing, welding, etc.
- Sensing — pollution monitoring, analysis, etc.

The industry is experiencing a shift to other wavelengths in several important respects:

- To longer wavelengths for telecommunications and sensor applications;
- To shorter wavelengths for optical data storage applications; and
- To very short wavelengths for instrumentation and higher data density.

While the bulk of all these application areas involve IR wavelengths (fibre optics, DWDM, VCSELs, HPDL, and so on), visible (red) wavelengths are becoming very important, especially for higher data storage capacity for movies and archiving via read-only DVD and recordable DVD.

- These non-telecommunications applications, especially DVD data storage, are helping sustain the total market during the period of the telecommunications downturn.

Coverage of the following sub-types of diode laser is included:

- Infrared diode lasers.
- Optical fibre systems.
- Free-space optics.
- Bragg gratings.
- Distributed feedback (DFB) lasers.
- Fabry–Perot (FP) lasers.
- Quantum cascade (QC) lasers.
- Fibre amplifiers.
- VCSELs.
- Microchip diode lasers.
- Violet diode lasers.
- Laser pointers.
- CD lasers.

- DVD lasers.
- Erbium-doped fibre amplifiers (EDFAs).
- Raman lasers.

Laser diodes have much in common with LEDs from which they were developed. Laser diodes have become by far the most important type of laser in use today and in many respects the diode laser is replacing the traditional crystal laser in many application areas.

The evolution of the laser diode is by no means over. Not only is the generation of new market applications an on-going activity through the evolution of existing devices but also through the creation of new types of laser diode. Owing to further refinement of its efficiency, operating wavelength and power, the laser diode continues to displace successfully traditional solid-state and gas lasers.

VCSELs stand to become one of the key drivers of the semiconductor optoelectronics market mostly via telecommunications, but as these low-cost lasers become more understood by designers they will compete with other light sources. For example, some industry observers are impressed with progress in the development of VCSEL-based white light sources.

The strong growth areas within this business are expected to be principally in diode lasers especially for telecommunications, data communications and optical data storage applications. These sectors have the potential to almost double in size by the end of the forecast period.

The dynamics of the market over the past year has, however, seen an abrupt downturn. But this is expected to return to positive growth at some point in next five years. A strong upward trend will then be likely to continue to the end of the forecast period. The timing of this resurgence had yet to become clear at the time of writing but the consensus — which is used here — was that it will resume in 2002 but not return to the 1999–2000 level until at least one year later.

Over the period 2000–01, the laser diode market was affected worst of all in the semiconductor optoelectronic components marketplace. This is due to two factors: major order cutbacks leading to low demand for IR telecommunications devices and a general downturn across all the electronic goods sector. The laser diode segment lost one-fifth of its value over the period 2000–01. It will not return to this value for at least another two years.

While this downturn was severe it could have been worse. Several growth markets have buoyed up the segment. These include:

- A rise in optical data storage via DVD success — DVD-ROM and DVD players that include IR lasers for retro-compatibility; and
- A growth in high-power diode lasers which are mainly IR devices.

Contrary to these positive factors, all aspects of the diode lasers market are under strong price pressures. This leads to amelioration of otherwise good growths in terms of volumes of components shipped.

2.5 Diode Laser Application Market Forecast

In 2000 the *total* worldwide merchant market for diode lasers was valued at US$ 5.6 billion. This will exceed US$ 8.7 billion by the year 2005. The market estimate for the year 2000 and the five-year forecast from 2001 to 2005 are shown in Figure 2.5 and Tables 2.5, 2.6 and 2.7. However, in the period 2000–01 the optoelectronics industry suffered a downturn producing cancellations of orders for devices. This resulted in a market downturn where the industry fell by as much as 12% to US$ 5.0 billion in 2001. Positive growth was expected to resume in 2002, of the order of 15%. This strong return to growth reflects harder unit prices as demand firms up and latterly increased unit shipments.

With continuing good overall growth, the total diode laser device market will grow to US$ 8.7 billion by 2005. This is expected to be sustained in value in terms of sales, but there are likely to be significant shifts in production emphasis by product type. Telecommunication lasers will resume good growth even though pricing may be harsher than in the past. This and other sectors such as HPDLs and VCSELs will expand at the cost of optical data storage such as CDs and DVDs. These markets will see continued growth but pricing will be under severe pressure forcing down the overall market value and hence the contribution to the total market. Other interesting areas for growth include violet diode lasers, which will break out from the high-value, low-volume instrumentation market and into printing and eventually the consumer and data mass storage market for the all-solid-state digital video cassette recorder.

As a result of the different growth rates for the various types of diode laser over the forecast period, some types will improve their market share at the expense of others. Once again this is partly due to their position in the product lifecycle. As a result, it is expected that data storage devices, i.e. the CD and DVD diode laser families, will see further pricing pressure and hence reduced market share even though unit shipments will still increase. By contrast the violet diode laser, which has yet to approach volume production, will see a near doubling of market share by 2005.

Table 2.5 Worldwide Diode Laser Market by Application 2000–05 (US$ million)

	2000	2001	2002	2003	2004	2005
Automotive	126	118	138	162	192	228
Computer	645	598	672	758	857	972
Consumer	770	715	811	921	1050	1199
Industrial	441	420	494	582	688	815
Military/aerospace	240	221	253	290	335	388
Telecommunication	3343	2813	3226	3701	4248	4879
Other	130	122	142	168	198	234
Total	5696	5006	5736	6583	7568	8715

Table 2.6 Worldwide Diode Laser Market by Region 2000–05 (US$ million)

	2000	2001	2002	2003	2004	2005
North America	1811	1628	1873	2159	2492	2883
Japan	1525	1289	1472	1685	1932	2220
Europe	1448	1271	1458	1676	1928	2222
Rest of world	911	818	932	1063	1215	1390
Total	5696	5006	5736	6583	7568	8715

Table 2.7 Worldwide Diode Laser Market by Device Type 2000–05 (US$ million)

	2000	2001	2002	2003	2004	2005
FOL	3588	3025	3470	3983	4571	5247
CD	689	620	671	726	785	850
VCSEL	342	308	378	465	572	704
HPDL	285	271	335	416	516	640
Red	621	590	644	703	768	838
Violet	171	193	237	290	356	436
Total	5696	5006	5736	6583	7568	8715

While there will continue to be good growth for all types of diode laser over the longer term, some types will perform substantially better than others. This is partly due to their position in the product lifecycle, e.g. the CD diode laser family is mature and unit prices are now very low. By contrast the violet diode laser has yet to approach the middle of its product cycle and so growth rates are more robust. The relative growth rates for all of the diode laser types covered in this report are summarized in Table 2.8.

Figure 2.5 Worldwide Diode Laser Market by Application 2000-05 (US$ million)

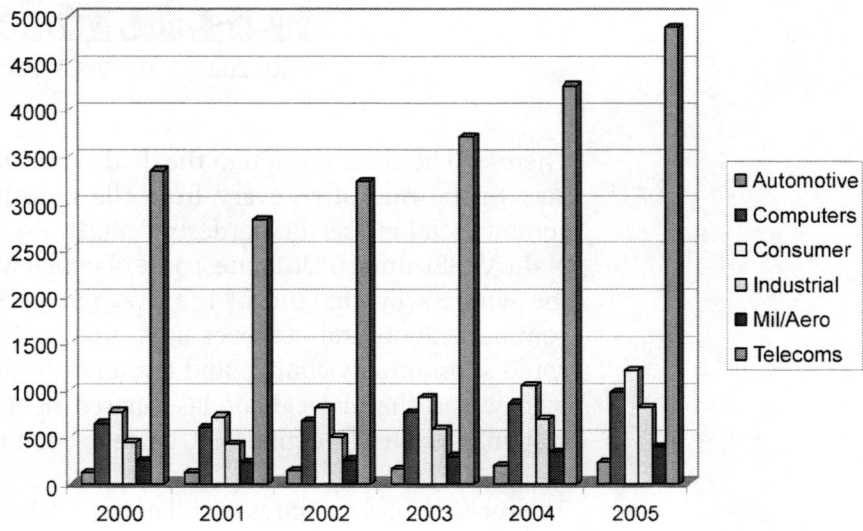

Table 2.8 Growth Rates for Diode Laser Devices by Type (%)

	2000	2002–05
FOL	−15.7	14.7
CD	−10.0	8.2
VCSEL	−10.0	23.0
HPDL	−5.0	24.0
Red	−5.0	9.2
Violet	13.0	22.5
Average	−12.1	14.6

Figure 2.6 Worldwide Diode Laser Market by Region 2000–05 (US$ million)

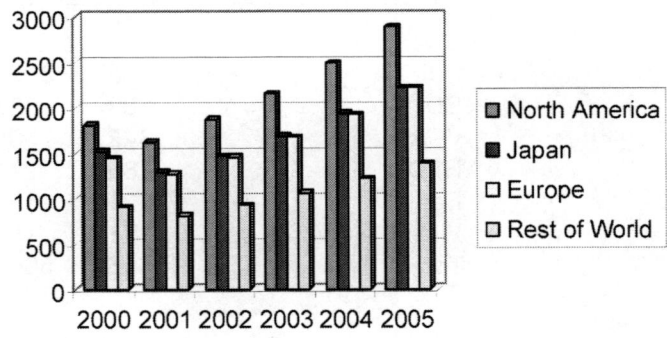

Figure 2.7 Worldwide Diode Laser Market by Device Type 2000–05 (US$ million)

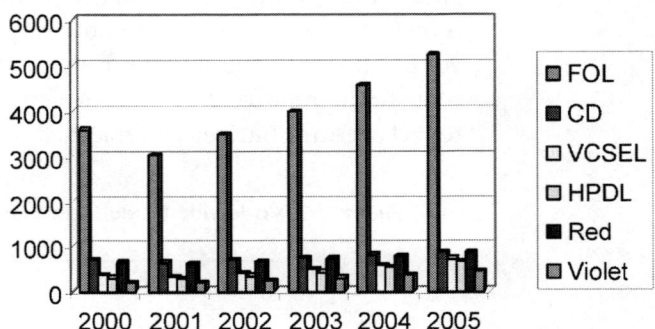

There will be some erosion of the diode laser fibre optic segment. This is due in part to the rate of recovery from the industry downturn that hit the telecommunications sector hardest of all. It also reflects the likely improvement of the VCSEL and HPDL share, some of which will also be fibre related. Basically, the winners by the end of the five-year period will be the fibre optic telecommunication and compact light energy (i.e. HPDL) sectors. However, this could substantially change and the data storage sector could see an improvement when the violet diode laser-based next-generation optical data storage systems, such as the digital VCR, come to commercial fruition.

The relative market shares for all of the diode laser types covered in this report are summarized in Table 2.9.

Table 2.9 Market Share for Diode Laser Devices by Type (%)

	2000	2005
FOL	63	60.2
CDL	12.1	9.8
VCSEL	6	8.1
HPDL	5	7.3
Red	10.9	9.6
Violet	3	5.0
Total	100	100.0

2.6 Market by Application Sector

2.6.1 Telecommunications Markets for Diode Lasers

The market value of devices in the telecommunications sector was estimated to be more than US$3.3 billion in 2000. An industry downturn — which affected the telecommunications sector, and hence the diode laser sector, particularly badly to the tune of 16% — was responsible for the lower 2001 market value of US$2.8 billion. However, the market is expected to see positive growth restored in the period 2001–02. This will boost the market growth from 2002 onwards and increase by an annual average rate of 15% up to a 2005 value of US$4.9 billion. The market estimation and five-year forecast are summarized in Figure 2.8 and Table 2.10.

Table 2.10 Diode Laser Telecommunications Market by Region 2000–05 (US$ million)

	2000	2001	2002	2003	2004	2005
North America	1039	909	1048	1210	1396	1612
Japan	843	652	744	849	969	1107
Europe	934	790	908	1043	1199	1379
Rest of world	528	461	526	599	683	780
Total	3343	2813	3226	3701	4248	4879

Figure 2.8 Diode Laser Telecommunications Market by Region 2000-05 (US$ million)

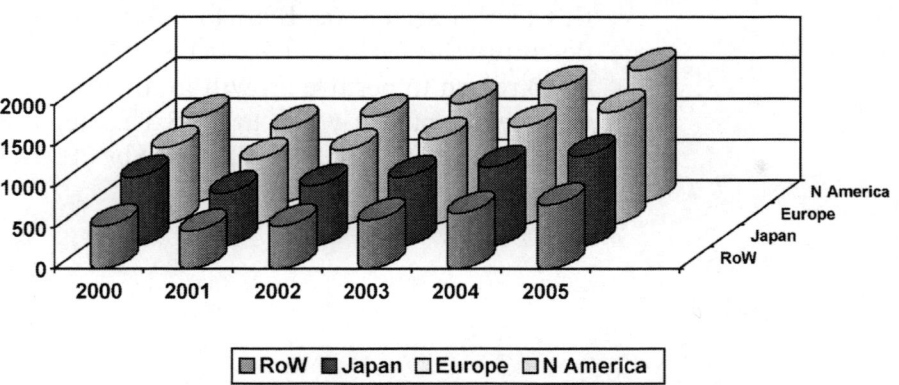

2.6.2 Consumer Markets for Diode Lasers

The consumer sector is amongst the most important in today's marketplace. The value of this market sector was US$770 million in 2000 but because of the industry downturn this fell by 7% to US$715 million in 2001. The market is expected to return to positive growth from 2002 onwards, increasing by an annual average rate of nearly 14% up to 2005. This important device family will therefore increase to US$1.2 billion by 2005. This market data is summarized in Figure 2.9 and Table 2.11.

Table 2.11 Worldwide Diode Laser Consumer Market by Region 2000–05 (US$ million)

	2000	2001	2002	2003	2004	2005
North America	257	238	270	307	350	401
Japan	236	219	249	284	325	372
Europe	135	125	142	161	183	209
Rest of world	143	132	149	169	192	218
Total	770	715	811	921	1050	1199

Figure 2.9 Worldwide Diode Laser Consumer Market by Region 2000-05 (US$ million)

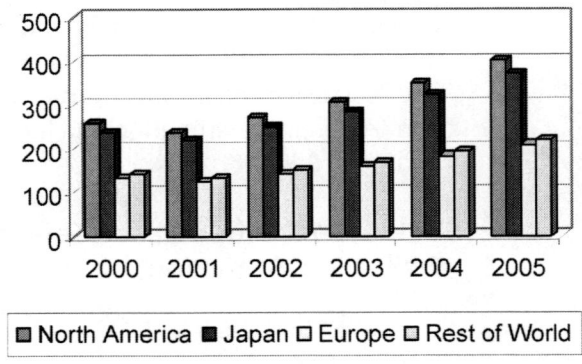

2.6.3 Computer Markets for Diode Lasers

The computer sector is a diverse application marketplace for diode lasers but it is dominated by the need for diode lasers for optical data storage. The value of this market sector was US$645 million in 2000 but because of the industry downturn this fell by over 7% to US$598 million in 2001. This market is expected to return to positive growth from 2002 onwards increasing by an annual average rate of nearly 13% to 2005. This important device family will therefore increase to more than US$972 billion by 2005. This market data is summarized in Figure 2.10 and Table 2.13.

Figure 2.10 Worldwide Diode Laser Computer Market by Region 2000-05 (US$ million)

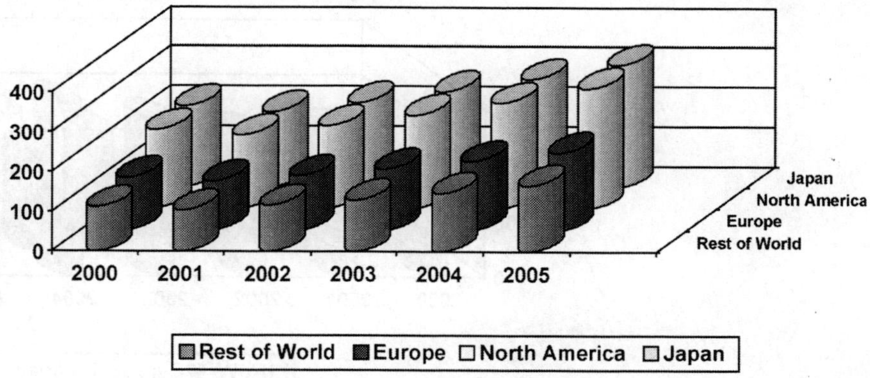

Table 2.12 Worldwide Diode Laser Computer Market by Region 2000–05 (US$ million)

	2000	2001	2002	2003	2004	2005
North America	198	184	208	235	266	303
Japan	204	189	213	240	272	308
Europe	131	122	136	153	173	195
Rest of world	111	103	116	130	147	166
Total	645	598	672	758	857	972

2.6.4 Industrial Markets for Diode Lasers

The market value of diode laser devices in the industrial sector is estimated to be US$441 million in 2000. This will drop to US$420 million by 2002. The growth in the period 2001–02 is likely to fall by 4.6% as a result of the industry downturn. From 2002 to 2005 this will transform into positive growth of 17.8% per annum reaching US$815 million by 2005. The market estimate and forecast are summarized in Figure 2.11 and Table 2.13.

Table 2.13 Worldwide Diode Laser Industrial Market 2000–05 (US$ million)

	2000	2001	2002	2003	2004	2005
North America	140	133	157	186	221	264
Japan	134	128	150	176	208	246
Europe	114	109	128	150	177	210
Rest of world	53	51	59	69	81	96
Total	441	420	494	582	688	815

2.6.5 Military/Aerospace Markets for Diode Lasers

The market value of devices in the military/aerospace sector was estimated to be US$240 million in 2000. This will fall to US$221 million in 2002. The period 2001–02 will fall by 8% as a result of the industry downturn. From 2002 to

Figure 2.11 Worldwide Diode Laser Industrial Market by Region 2000–05 (US$ million)

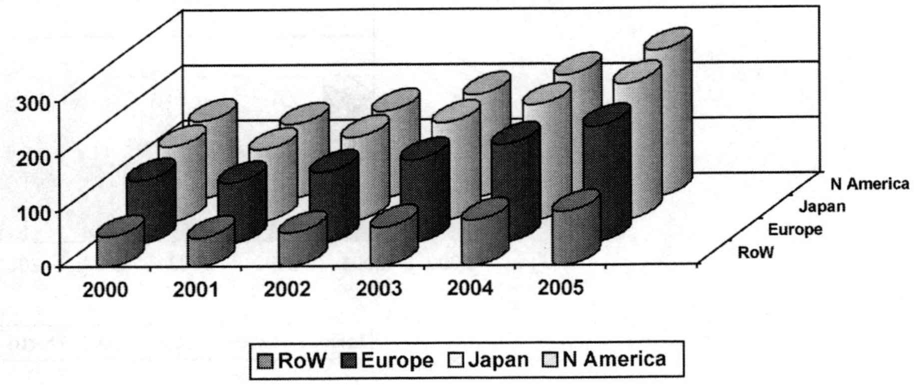

2005 this will transform into positive growth of more than 15% per annum. As a result by 2005 the market will have grown to a value of US$388 million. The market estimate and forecast are summarized in Figure 2.12 and Table 2.14.

Table 2.14 Worldwide Diode Laser Military/Aerospace Market by Region 2000–05 (US$ million)

	2000	2001	2002	2003	2004	2005
North America	101	93	106	122	141	163
Japan	38	35	40	46	54	62
Europe	74	68	78	90	104	120
Rest of world	26	24	28	32	37	42
Total	240	221	253	290	335	388

2.6.6 Automotive Markets for Diode Lasers

The market value of devices in the automotive sector was estimated to be US$126 million in 2000. This market has suffered from the general industry downturn and fell by nearly 7% to US$118 million in 2001. This will increase to U$138 million in 2002 representing an average annual growth rate of 18%, reaching US$228 million in 2005. The market estimate and forecast are summarized in Figure 2.13 and Table 2.15.

Figure 2.12 Worldwide Diode Laser Military/Aerospace Market by Region 2000-05 (US$ million)

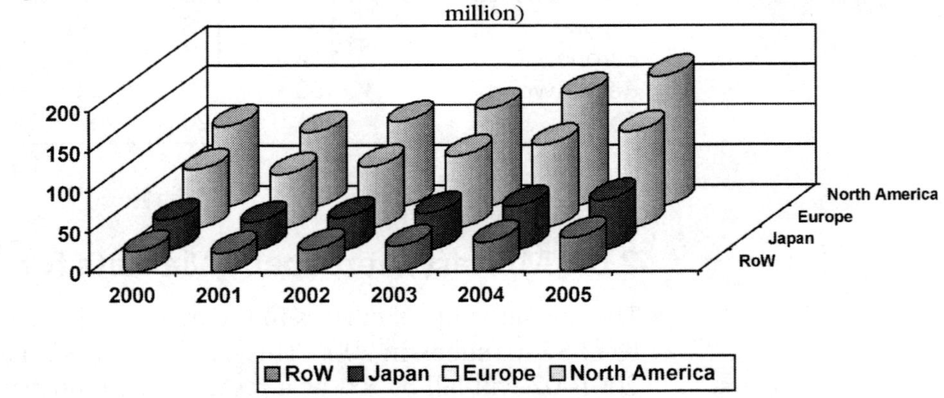

Figure 2.13 Worldwide Diode Laser Automotive Market by Region 2000–05 (US$ million)

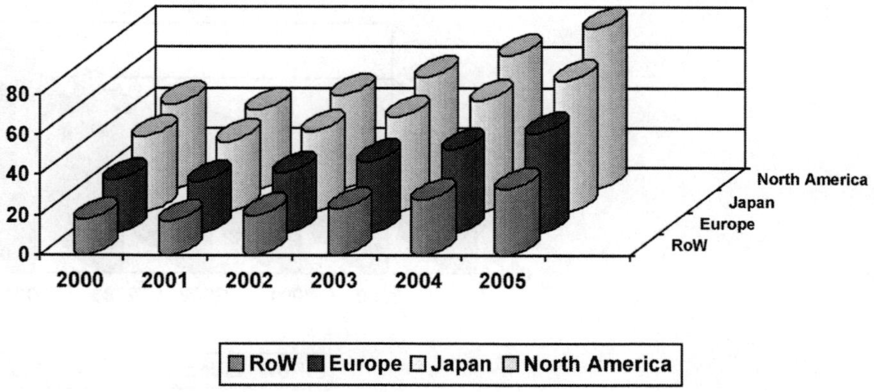

Table 2.15 Worldwide Diode Laser Automotive Market by Region 2000–05 (US$ million)

	2000	2001	2002	2003	2004	2005
North America	43	40	47	56	67	80
Japan	37	35	40	47	55	65
Europe	27	26	30	36	42	50
Rest of world	18	17	20	23	28	33
Total	126	118	138	162	192	228

2.6.7 Other Markets for Diode Lasers

The market value of diode laser devices in the category under the generic title of 'other' was estimated to be worth US$130 million in 2000. This will decrease to US$122 million in 2001 as a result of the industry downturn, a fall of just over 6%. From 2002 to 2005 growth will recover, and proceed at nearly 18% average annual growth, reaching US$234 million by 2005. This forecast information is presented in Figure 2.14 and Table 2.16.

Table 2.16 Worldwide Diode Laser Other Market by Region 2000–05 (US$ million)

	2000	2001	2002	2003	2004	2005
North America	33	31	37	43	51	60
Japan	32	30	36	42	50	59
Europe	33	31	36	42	50	59
Rest of world	31	29	34	40	47	56
Total	130	122	142	168	198	234

Figure 2.14 Worldwide Diode Laser Other Market by Region 2000-05 (US$ million)

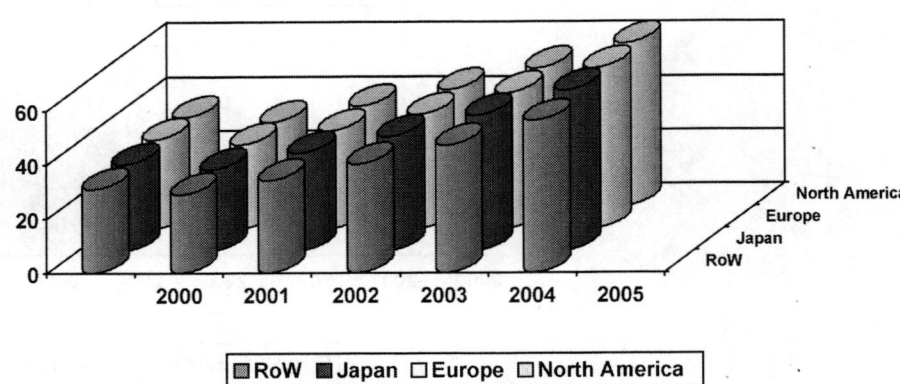

3 Diode Laser Materials Markets

Table 3.1 Worldwide Diode Laser Substrate Market 2000–05 (US$ million)

	2000	2001	2002	2003	2004	2005
North America	33	26	33	44	59	83
Japan	36	29	37	48	65	92
Europe	24	19	24	32	43	62
Rest of world	20	16	21	27	38	54
Total	113	90	115	151	205	290

Table 3.2 Worldwide Diode Laser Epiwafer Market 2000–05 (US$ million)

	2000	2001	2002	2003	2004	2005
North America	78	62	78	98	126	164
Japan	90	72	88	110	140	180
Europe	51	41	51	64	81	104
Rest of world	37	30	37	47	59	77
Total	256	205	254	318	406	525

3.2 Introduction

In this chapter the worldwide merchant markets for diode laser materials are presented. The chapter is basically split into two parts: substrates and epiwafers. The term 'substrate' refers to the unprocessed slice, while epitaxial wafer (epiwafer) refers to the substrate after epitaxial growth but prior to device fabrication.

It should also be noted that this report is concerned mainly with the *merchant* market for diode laser materials and components, i.e. those sold to commercial customers and not for in-house requirements. The combination of merchant

Figure 3.1 Worldwide Diode Laser Materials — Substrates and Epiwafers — Market 2000–05 (US$ million)

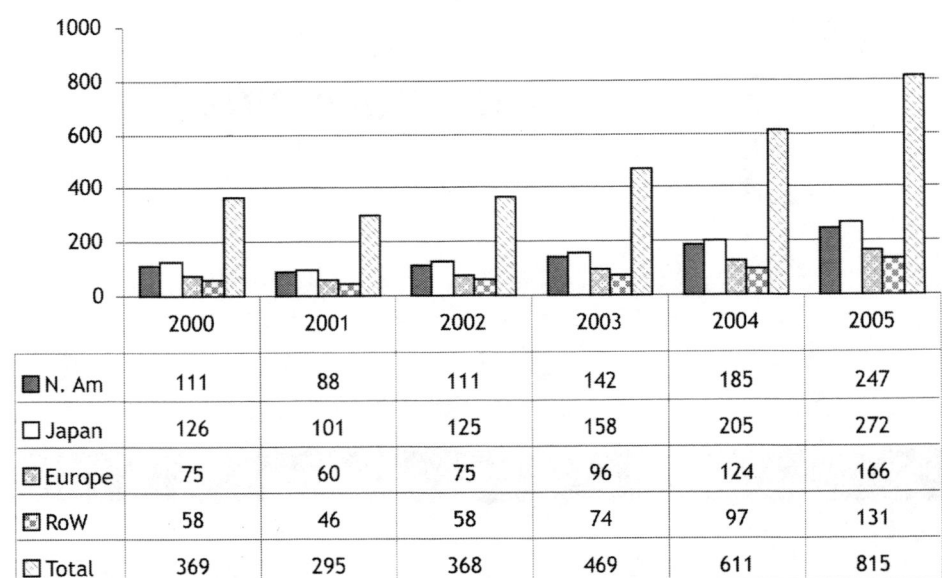

	2000	2001	2002	2003	2004	2005
■ N. Am	111	88	111	142	185	247
□ Japan	126	101	125	158	205	272
▨ Europe	75	60	75	96	124	166
▨ RoW	58	46	58	74	97	131
□ Total	369	295	368	469	611	815

and captive represent the total available market (TAM) but it is unlikely that the market will ever move entirely over to merchant.

- Substrates are almost exclusively merchant with the exception of Japanese suppliers which feed both the in-house epiwafer requirements, i.e. *captive*, and those for merchant market customers.
- Epiwafers, by contrast, are fairly evenly split into merchant and captive markets with device companies making use of both approaches according to market demand.
- Semiconducting substrates are also required for other opto- and micro-electronic devices and so the estimations here do not represent the TAM for these materials.
- The total substrate and epiwafer markets must also include the substantial contribution from semi-insulating (SI) III-V materials as used to fabricate microwave devices, etc. (See *Gallium Arsenide Electronic Materials & Devices, A Strategic Study of Markets, Technologies & Companies Worldwide*.)

The structure and dynamics of these markets are therefore complex and to some extent the market analysis presented is simplified somewhat. Further-more, the market has been rendered less readily understood by the appearance of multi-product companies. Examples include:

- EMCORE, a former MOVPE equipment maker, now also manufactures epi-wafers as well as devices;
- IQE, a former MOVPE epiwafer specialist, now manufactures MBE epiwa-fers as well as substrates via its acquisition of Wafer Technology Ltd;

- Vertical gradient freeze (VGF) substrate specialist AXT also manufactures diode laser devices; and
- MBE epiwafer pioneer Picogiga acquired an optoelectronic component maker (Modulight) as well as wafer recycler company (Picopolish).

The analysis has therefore used published company information as a guideline in concert with an empirical approach derived from the diode laser device market forecast in the following chapter. Prices of diode laser substrates and epiwafers are not readily available but useful ballpark figures can be attributed to the basic products and a market estimation can then be generated.

3.2.1 Substrates

In the first part of this chapter the market for semiconducting (SC) substrates, gallium arsenide (GaAs) and indium phosphide (InP), will be estimated and forecast. Coverage includes an examination of the principal substrate types by size, price, etc.

VGF will remain the principal crystal growth techniques for the fabrication of diode lasers and related devices for the period covered by this report.

- Other substrates used to fabricate diode lasers include sapphire substrates, which are used for the fabrication of violet diode lasers, but this is presently only a very small market.

Therefore the principal subject for this market data and forecast will be semiconducting materials.

3.2.2 Epiwafers

In the second part of this chapter the market for semiconducting epiwafers, e.g. those based on GaAs and InP and alloys, will be estimated and forecast. Coverage includes an examination of the two principal epiwafer types: metallorganic vapour-phase epitaxy (MOVPE) and molecular beam epitaxy (MBE). These will remain the principal growth techniques for the fabrication of diode lasers and related devices for the period covered by this report.

- MBE is taken to include all variants of this technique, e.g. gas source MBE.
- Other, older techniques such as vapour-phase epitaxy (VPE) and liquid-phase epitaxy (LPE) which are still used for the fabrication of some diode lasers are largely captive and therefore not included in this market forecast for the merchant market. For both types of epiwafer, market data and forecasts are given and future prospects assessed.

3.3 The World Market for Substrates 2000–05

3.3.1 Introduction

In 2000 the *total* world market for substrates for diode lasers was worth US$113 million. However, in the period 2000–01 the optoelectronics industry suffered a serious downturn which produced many cancellations of orders for associated substrates and other materials. This resulted in a market downturn by as much as 20% to US$90 million in 2001. Positive growth was expected to resume in 2002, of the order of 27%, to reach a market value of US$115 million. This strong return to growth reflects harder unit prices as demand firms up.

As a result of continuing good growth, the total substrate market will grow to US$290 million in 2005. It is expected to be sustained in value in terms of sales but there are likely to be significant shifts in production emphasis by product type, e.g. substrate diameter should shift from 2- to 3-in, with 4-in diameter also coming into play over the longer term.

The total world market for substrates by region is summarized in Figure 3.2.

The world substrate market is analysed and forecasted by:

- Substrate size — 2- and 3-in and others, i.e. larger or smaller wafers;
- Crystal growth type, i.e. VGF;
- Substrate size; and
- Geographic regions.

The market for substrates by diameter from < 2-in and larger is summarized in Tables 3.2, 3.3 and 3.4.

Figure 3.2 Total Worldwide Diode Laser Substrate Market 2000–05 (US$ million)

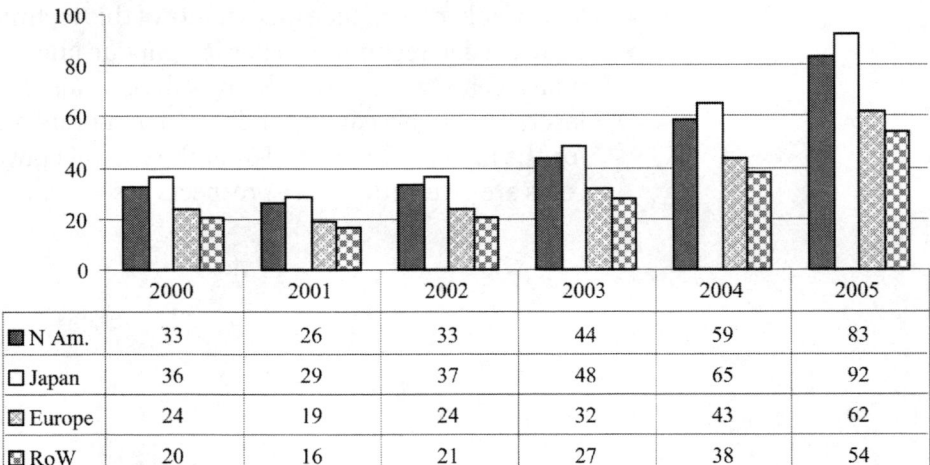

	2000	2001	2002	2003	2004	2005
N Am.	33	26	33	44	59	83
Japan	36	29	37	48	65	92
Europe	24	19	24	32	43	62
RoW	20	16	21	27	38	54

Table 3.3 Total World Diode Laser Substrate Market Summary: 2-in (US$ million)

	2000	2001	2002	2003	2004	2005
North America	24	19	21	24	27	31
Japan	26	21	24	27	30	34
Europe	17	14	16	18	20	22
Rest of world	15	12	13	15	17	19
Total	82	66	74	84	95	107

Table 3.4 Total World Diode Laser Substrate Market Summary: 3-in (US$ million)

	2000	2001	2002	2003	2004	2005
North America	6	5	9	15	25	42
Japan	7	6	10	16	27	46
Europe	5	4	6	11	18	30
Rest of world	4	3	5	9	15	26
Total	22	18	30	50	85	144

Table 3.5 Total World Diode Laser Substrate Summary: Others (US$ million)

	2000	2001	2002	2003	2004	2005
North America	3	2	3	5	7	10
Japan	3	2	4	5	8	12
Europe	2	2	2	4	6	9
Rest of world	2	1	2	3	5	8
Total	9	7	11	17	26	39

3.3.2 Size Trends for Diode Laser Substrates 2000–05

In 2001 the diode laser substrate market continued to be dominated by 2-in substrates. However, the 3-in substrate market is a solidly growing sector and will have matched the 2-in market by 2004–05.

Table 3.6 Total World Diode Laser Substrate Market by Size 2000–05 (US$ million)

	2000	2001	2002	2003	2004	2005
2-in	82	66	74	84	95	107
3-in	22	18	30	50	85	144
Other	9	7	11	17	26	39
Total	113	90	115	151	205	290

'Other' comprises research and development (R&D) and 4-in substrates so it should grow from under 10% of the market in 2000 to over 10% by 2005. This reflects the importance not only of R&D but also the growing importance of 4-in substrates, of GaAs as well as InP.

Figure 3.3 Diode Laser Substrate Market by Diameter 2000–05 (US$ million)

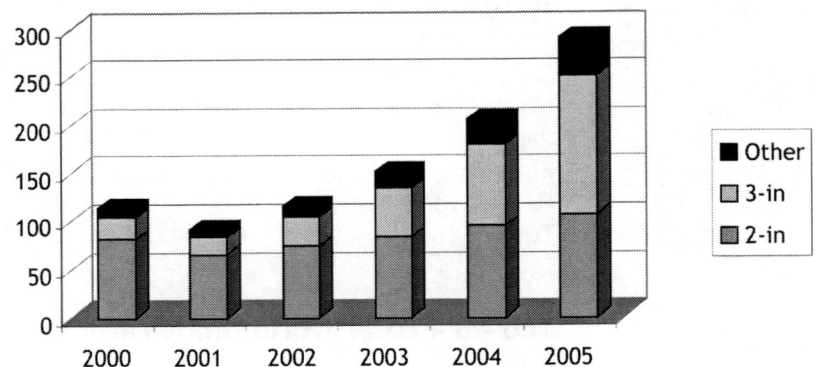

Figure 3.4 Worldwide Diode Laser Substrate Market by Size 2000

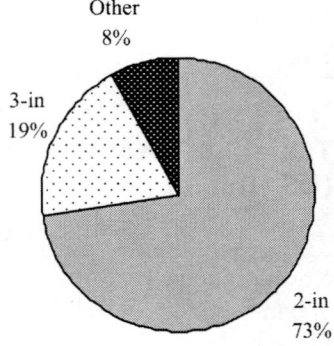

As shown in Table 3.6 and graphically in Figures 3.4 and 3.5, the relative market share by diameter will see the 3-in surpass the 2-in market in importance by 2005. However, at the time of writing companies were considering moving to larger sizes such as the 4-in for even greater economic gains. These substrates require further work to perfect production technologies from epitaxial growth to device fabrication using larger sizes and so will only come into play when the market resumes firm upward growth in the next two years.

VGF is the dominant crystal growth technology, accounting for over 90% of wafers consumed in 2000. Substrates grown by all other techniques combined accounted for less than 10% of shipments. However, the dominant crystal growth method for III-Vs, LEC, is suitable for microelectronic applications and has a large-scale installed base worldwide. Consequently, it is expected to expand its position only slightly by 2005 unless a major step forward is made with respect to improvements in crystal quality.

- The benefits offered by VGF include a low etch pit density (EPD), reduced mechanical stress and potentially lower cost.
- In particular, VGF has found favour amongst optoelectronics companies because of its complementarity ('epi-ready') to epi-based processes. However, so far it has achieved only a limited penetration of microwave device suppliers.

Figure 3.5 Worldwide Diode Laser Substrate Market by Size 2005 (%)

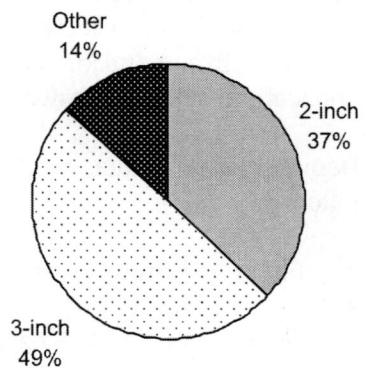

More information on crystal growth technologies is given in Chapter 5.

Today's substrate marketplace is characterized by the broadest range of available diameters in its history. The principal diameters are 2-, 3-, 4- and 6-in. The larger-diameter wafers have been demonstrated for optoelectronics applications but are not likely to be commercially significant outside the microwave sector for at least another three years owing to the need to scale up the back-end device processing capabilities of existing plant.

● This description is, however, superficial and hides the fact that in the market substrate specifications remain very much dependent on specific user requirements.

In 2000 the dominant substrate size was 2-in accounting for three-quarters of the market. New fabs coming on line or planned are likely to be based on 3- or even 4-in wafer and this will result in market share peaking and beginning a decline thereafter.

At present the majority of leading producers of devices manufacture on smaller-diameter substrates, i.e. 2- or 3-in. Use of 2-in substrates will also continue in R&D but will suffer the disadvantage of higher prices as volume users exit this market segment.

● Recently, leading substrate manufacturers such as Freiberger Compound Materials (FCM) have already announced cessation of 2-in SI substrate manufacture. However, these are still available via other lower tier players such as Atomergic, which sources its products in part from the former Soviet Union.
● The most important size for volume production of certain types of GaAs microelectronic devices is 6-in which already accounted for 11% of the market in 2000. This market share reflects the price premium of this size that is expected to drop significantly as mass production volumes gear up.

The strong growth areas within the business are expected to be principally in VGF and, to a much lesser extent, in other methods.

Within this area the product types have seen good growth over the period 1998–2000. However, in the period 2000–01, and for the short term, overall growth for the optoelectronics sector has reduced. As a consequence of this downturn in demand for diode lasers — particularly for the telecommunications application sector — the demand for diode laser materials has also diminished.

Demand for substrates of all kinds comes from the principal device sectors as follows:

- Integrated circuits (primarily for microwave applications).
- Discrete devices I — diode lasers.
- Discrete devices II — light-emitting diodes, detectors, etc.
- R&D.

Previously, the industry had been experiencing a shortfall in capacity. Substrate vendors could for the first time sell everything that they could make. Historically, however, supply has always exceeded demand. This situation has reasserted itself and so from 2000 to perhaps as long as 2002, it will continue to be the case that supply will more than meet demand. New additional capacity will be needed when the market resurgence takes effect. During this period the substrate market may thus not be able to fulfil potential. However, it will benefit from the resultant hardening of prices.

- While the bulk of orders come from companies that have production facilities, a small percentage, under 10%, of the demand comes from research institutes and organizations.

South-East Asia (including Japan) is the biggest regional *consumer* and *producer* of diode laser substrates, followed by North America. Over the next five years, demand from the rest of the world is expected to increase further as more and more diode laser manufacturing is carried out in countries such as Taiwan.

- There are over 30 volume diode laser device production facilities around the world (see Chapter 6). Merchant diode laser epiwafer suppliers are also fairly large consumers of substrates — there are around 20 of such companies worldwide.

Home to two of the largest merchant epiwafer houses, Europe's substrate demand is substantial. It is also home to several device manufacturers. As a whole, European substrate demand is expected to not grow quite as strongly as that of North America. However, it will see an increase driven by demand from key companies such as Infineon, IQE and Picogiga.

- The situation is made complicated by the fact that a US or Japanese company might well use a European merchant epiwafer supplier.

Rest of the world countries have previously provided a smaller demand for wafers. But new fabs have started to appear especially in Taiwan and are likely to contribute to the region's growing appetite for substrates. The question is less about *demand* than about *supply*. All of the top substrate vendors have strong supply agreements with North American, European and Japanese customers. For key substrate products all capacity is already allocated into a year ahead.

Rest of the world companies, especially those new fabs in Taiwan, were finding it difficult to source sufficient substrates. Now with the market having receded this is likely to become less of a problem. An additional factor is their competition with Japanese vendors which conflict at upper tiers, especially epiwafers and devices. When Japanese demand accelerates, Taiwanese companies (and others) may experience supply difficulties as Japanese vendors strive to meet their customers' demands.

- The Japanese share of the overall market has continued to fall over the past few years, due to increased competition from Western suppliers such as AXT in the USA, and FCM in Germany. In addition, currency exchange rates, including the strong yen, have a negative impact on foreign imports and Japanese exports.

Conversely, the Japanese home market is still to a large extent closed to non-Japan-based substrate suppliers. Some penetration was expected but has so far been limited to sampling of VGF material. This reflects the traditional close working relationship between companies in Japan. Western suppliers are presently less interested in this potential market than fulfilling existing orders. The expertise that these companies have accumulated in substrates could become attractive to Japanese device houses in due course.

However, Japan's wafer industry still has strengths and despite substantial domestic over-capacity, it has not yet experienced the shakeout of suppliers seen in North America and Europe. But with market share increasingly concentrated among the top few suppliers, some Japanese companies have quietly withdrawn from the market to focus on other 'value-added' sectors of the industry, such as epitaxial wafers.

Perhaps surprisingly, little progress has been noted for suppliers originating in the rest of the world. There are very few important suppliers of wafers in countries otherwise active in microelectronics, such as the former Soviet Union, South Korea, Taiwan, China and Eastern Europe. The few companies that currently operate outside the principal geographic regions have not yet become commercially competitive. This is not expected to change significantly over the forecast period given the present industry set-up and competitive supply situation. There would appear to be little opportunity for any new entrant to this business sector for the foreseeable future.

In 2000 over 80% of the substrates used by foundries were for device production compared with less than 20% for R&D/process development. Although R&D will fall as a percentage of total substrate consumption, it will remain an important demand driver.

Most foundries support substantial R&D programmes, either from internal resources or from government. Many larger corporations have pilot lines for research, in addition to the main production line. Some Japanese companies have research in four or five separate locations. Together, these R&D activities have added up to considerable substrate consumption.

In terms of device production, discrete devices will continue to be the main driver, accounting for over 90% of substrates used today. Quite a few companies are involved in microwave discrete devices where overall yields are higher than in the digital industry, for example. Discrete devices will therefore continue to be important for the period covered by this report.

Only a few companies manufacture very high volumes of devices and their contribution to substrate demand is expected to be considerable.

Discrete devices are the most mature sector of the device market. Their manufacture is a high-yielding process overall, delivering up to several thousand units off a 2-in substrate (the discrete device industry has yet to move to the very largest available substrates). Once the price differential has moved in favour of the larger sizes then there is every likelihood that discrete device manufacturers will also move up. By then process equipment will have been written off and new investments will have been made. Of course, by then it will be less attractive to buy equipment to handle smaller substrates. Within five years such equipment will either be unavailable or be prohibitively expensive to own and run.

3.3.3 Merchant and Captive Production

Very few device manufacturers have captive crystal growth facilities. Nor do many of the independent merchant epiwafer suppliers have any captive crystal growth manufacturing.

Quite a number of Japanese substrate suppliers are vertically integrated, i.e. they not only manufacture substrate products but also epiwafers and in some cases devices. However, virtually none of the large device makers, e.g. Mitsubishi and Sony, have continued with captive substrate supply, relying exclusively on close working relationships with a few external sources.

- Amongst the very few industry changes to have appeared in the past three years is the availability of epiwafers and other products from an equipment supplier, e.g. EMCORE Corp (see company profile). However, this does not as yet include substrates.

Captive production of substrates accounted for less than 10% of world demand in 2000. No additional investments in captive crystal growth have been announced nor are any anticipated in the near future.

Most Japanese epiwafer vendors are supplied with substrates from their own wafer divisions, including Furukawa Electric, Mitsubishi Chemical Corp, Hitachi Cable and Sumitomo Electric Industries. Conversely, previously a substrate vendor, Japan Energy now outsources its substrate requirements for its epiwafer business.

- In Europe, there have been no captive crystal growth production capabilities for some time.
- Recently, merchant epiwafer company IQE became the first of its kind to acquire a substrate supplier, Wafer Technology.

A high proportion of demand for substrates will continue to come from North American companies.

The key companies that presently supply substrates are (in alphabetical order):

- AXT.
- Dowa Mining.
- Freiberger Compound Materials.
- Hitachi Cable.
- M/A-COM.
- Sumitomo Electric.
- Wafer Technology.

Long-established substrate suppliers such as Litton Airtron and Japan Energy have recently exited the substrate business for a variety of reasons.

Large-volume substrate users have at least two, and usually three, qualified sources of supply. They buy complete boules of wafers, rather than sourcing from many different boules, a sign of the relative immaturity of the industry compared with that of silicon. As has long been the case, qualifying a substrate vendor is expensive and time-consuming. Typically it may take over a year to complete and involves a close interaction between the various parties. The use of larger substrates is not a sign that this era is drawing closer at least as far as the modus operandi of the substrate manufacturing tier is concerned. More hands-on feedback during the qualifying process is mandatory for the first stages of qualification of larger substrates. With each generation of substrate sizes this becomes more involved and time-consuming.

3.3.4 Impact of Device Demand on Substrate Consumption

Demand for substrates is mainly driven by production of devices (although R&D is also a consumer). However, it is a tradition of the industry that the growth rate of the substrate markets has been proportionately lower than the growth rate of the device markets.

At present this scenario is less certain while the larger diameter substrates enjoy exceptional growth. However, this situation is likely to be only temporary. This is because of the relationship between device production and substrate demand. This depends mainly on three variables: yields, device count per wafer and wafer prices.

- Yields. As yields increase, fewer substrates are required to produce a given number of devices. Yields for the industry as a whole are generally fairly good compared with the IC industry. Even so, improvements are occurring for devices in volume production.
- Die sizes. Since cost is related to chip size, optoelectronic devices are much less vulnerable in this respect compared with ICs, which are continually refined to achieve the same level of performance in a smaller area, resulting in fewer substrates being required for a given production volume.
- Substrate prices. As explained above, in the period 1999–2000 substrate prices had been hardening in contrast to the industry trend for the past decade, but prices will continue to erode over the longer term.

Not all product lines follow this pattern. For example, in some sectors there is a trend towards increasing die sizes owing to the incorporation of more on-chip functions, i.e. the move towards optoelectronic integration. For the bulk of the market discrete devices tend to be less area dependent. However, the industry is in a transition phase at present with a fairly high degree of design evolution underway. Designs are seldom frozen for long and until they reach that point, die size creep is an inevitable problem for all fabs to contend with. This will remain a general trend of the market.

3.3.5 Substrate Pricing Trends

Table 3.7 Average Substrate Price (US$ per sq in)

	2000	2001	2002	2003	2004	2005
2-in	17	14	12	11	13	18
3-in	19	14	11	15	12	11
Other	22	19	17	20	17	14
Total	58	48	40	47	42	42

Over the long term, substrate prices will always be in decline. This is the general rule for this tier of the industry irrespective of the material type. This accepted fact of business life is due to a combination of intense pricing pressure coupled with increasing demand and improved yields, year-on-year. Nevertheless, prices still vary considerably depending on the volume, geographical region, etc. This is irrespective of which type of substrate, LEC, VGF, etc.

Over the next three years it is likely that the price for larger substrates will continue to show relatively slow price erosion until at least the end of 2001. At the same time prices will rise for the smaller-diameter, i.e. 2- and 3-in, substrates. In the longer term, pricing decline will also be a feature of the 4-in market as this diameter continues through its product lifecycle.

Referring to Figure 3.6, the substrate pricing trends can be summarized as follows. Over the period 2000–02, all substrates will be subject to a fairly serious

Figure 3.6 Semiconductor Substrate Pricing Trends 2000–05 (US$ per sq in)

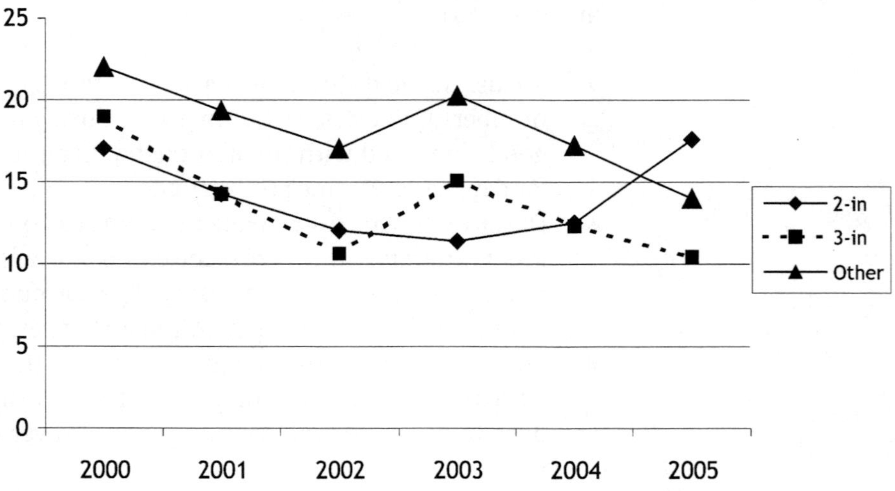

decline in prices. This is due to the industry downturn, which caused cutbacks in device orders and also for substrates and epiwafers. From a solid US$20 per square inch region average value, the price will have fallen to a low of around US$10 per square inch by early 2002. This is likely to be an industry record for a high-to-low period.

- The consensus is that substrate prices will see the start of a recovery in 2002. By then all inventory will have been used up. Because vendors have scaled back production, there may well be a shortage of substrates in mid-2002.

By the end of 2002 prices will all begin to firm and an upward trend will be a possibility. For the current industry standard, the 2-in substrate, this period will be likely to mark the beginning of the end. By the end of 2002 the larger sizes will become the preferred choice for the industry leaders.

The period 2003–05 will be one of increased competition and some hardening of prices for some substrate products cannot be ruled out. Two-inch substrates will begin to command higher prices as the emphasis shifts to 3- and 4-in substrates. Manufacturers of substrates have previously shifted their production emphasis away from less profitable products. For example, not long ago FCM ceased all 2-in substrate supply in favour of meeting demand for 6-in substrates for microwave markets.

At this time 3- and 4-in substrates will begin their downward market pricing curve. This is due to two factors:

- Increased demand from device manufacturers wishing to accrue improved economics; and
- Substrate suppliers making these more attractive and wanting to enlarge market share.

This market was also under considerable pricing pressure.

Previously, substrate price erosion was helped by government subsidies such as the Title III Program in the USA. However, the differential with silicon substrates will remain substantial. While new crystal growth technologies might provide lower price substrates due to various factors such as higher yields or longer ingots, this has yet to be proved in the field by any major manufacturer. There is also an in-built long timeframe before any such new substrate product gains widespread acceptance with any of the major substrate purchasers who prefer to stick with the tried-and-trusted materials that underpin their manufacturing operations.

It cannot be ruled out that a new start-up may begin to offer such innovative products within the timeframe considered by this report. This is, however, unlikely. This is because of the very close — almost symbiotic — relationship that is favoured between users and suppliers today. But there are precedents in other materials such as silicon carbide (SiC), but these are rare. The technology exists for gallium nitride (GaN) crystal products but this has yet to move on from small-scale commercial status despite an apparent consensus of the desirability of such products within the huge market for GaN optoelectronic devices.

- Currently, GaN substrates are amongst the smallest on the market, about 1 in in diameter. The bulk of the blue emitter market is therefore based on alternative substrates such as sapphire and SiC.

For other types of substrate applications such as microwave devices, 6-in substrates were priced at around US$500 in 2000, depending on size of order, specification, etc. Alternatively, 4-in substrates were priced at around US$175 in volume in 2000. This market was also under considerable price pressure. A fair number of these companies operate in both areas of substrate manufacture so have been hit particularly hard in the past year.

The increasing availability of cheaper raw materials and substrates has as yet not turned out to have an impact on the world substrate market. Such materials were becoming available from suppliers in the former Soviet Union and elsewhere and in the longer term they may become established in commercial markets. The unreliability of supply and generally lower quality are factors that serve to negate their successful penetration of the very tight market for substrate products.

The industry could see further escalation of gallium prices. At the time of writing gallium was going through a period of being in short supply. This was hardening prices, and for some grades the price was climbing. This was effectively squeezing the substrate manufacturers that were under pressure from their customers to keep prices down. The situation is being exacerbated by the comparable success being enjoyed by the microelectronics industry, which is a volume consumer of GaAs substrates and epilayer growth products.

New gallium ore mines take some time to come into play. But it is believed that adequate supplies will exist for at least another decade. Also, the industry is now recycling scrap III-Vs in order to meet the environmental and economic requirements of future markets and to ensure long-term supply. It should therefore have the effect of squeezing substrate suppliers who are forced by their customers to lower prices.

It is difficult to predict the overall impact of changing gallium prices over the longer term because they have in the past come down, as new production capacity has come on stream. Gallium suppliers — some of which, such like Dowa Mining, are themselves GaAs product suppliers — are likely to have to add capacity and this should ease the situation over the medium to longer term. It should therefore have the effect of lowering prices and easing the situation for substrate suppliers so they can lower their prices.

Substrate prices will continue to fall, due to a combination of intense pricing pressure, increasing demand and improving yields in crystal growth. However, prices will still vary considerably by volume and geographical region.

3.3.6 Other Substrate Trends

3.3.6.1 Introduction

The key technical and business trends that are likely to have some impact on the diode laser substrate market for the forecast period are discussed in this section. These include:

- Alternative crystal growth methods.
- Other III-V semiconductors.
- Substrates for violet diode lasers.
- GaAs-on-silicon epitaxial growth.

There is every chance that once the optoelectronics market resumes its upward progress the market will see substantial divergence in materials and process technologies. This will be in direct response to the market demand for ever-more performance at ever-lower prices.

In prospect is the possibility of a transformation of the substrate industry via the use of alternative substrate technologies; in particular, the GaAs-on-silicon process recently announced by Motorola. This may prove acceptable for microelectronic devices but less useful for defect-sensitive devices such as diode lasers. However, it was once considered impossible to produce violet diode lasers by growing GaN on sapphire because defects would prevent useful operation. Technology is continually solving such problems and given the enormous market potential should it succeed there are likely to be several companies interested in carrying out the requisite R&D via licensing deals with Motorola.

3.3.6.2 Substrate Growth and Materials

Today's market is represented almost entirely by VGF for diode lasers. Other methods have been tried but as yet have not made a serious commercial impact. For example, horizontal gradient freeze (HGF) may become important if further R&D is applied. At present it is difficult for HGF to reach the low EPD levels required for lasers. The same applies to LEC. For example, the EPD in silicon-doped VGF semiconducting GaAs is less than 500 per sq cm or better, whereas for HGF it is around 5000 and for LEC as high as 50 000.

Semiconducting InP is also important in the diode laser market. However, quantitative market research is complex because it is difficult to differentiate between diode lasers and detectors, the two complementary market applications for InP semiconducting substrates.

Other indium-containing compound semiconductors also have some promise for diode laser applications. These include indium arsenide (InAs) and indium antimonide (InSb) which are becoming a market for optoelectronics. At present this is mostly for detectors but R&D is underway for diode lasers. Also, there is some R&D concerning InAs for long-wavelength (3–5 µm) diode lasers and detectors. Over the long term gallium antimonide may also become important but this is currently making only a small contribution to the R&D market segment. This amounts to a total (i.e. all device applications and not just diode lasers) of around 3000 wafers per year, which are priced at the US$ 300 level. It therefore constitutes a million-dollar-per-year market.

More important are the issues surrounding larger-size substrates (which are included in the 'other' segment in this estimate). This particularly includes the 4-in substrates made from either GaAs or InP. At present only six customers are using some 4-in GaAs substrates. There had been more discussions as to the use of 4-in InP but plans have been delayed due to the market downturn.

Metamorphic epitaxial growth technology whereby, for example, layers of InP alloys are grown via a special intermediate buffer layer on top of GaAs, is proving to be of considerable interest to makers of high-frequency microwave devices. However, its future for diode lasers is less certain owing to possible high levels of defects.

3.3.6.3 GaN and SiC vs Sapphire

In the early 1990s Nichia Chemical Industries became the first company to commercialize devices based on the heteroepitaxial growth of gallium nitride (GaN) on sapphire. Similarly, Cree built a strong business based on the heteroepitaxial growth of GaN on silicon carbide (SiC) to produce LEDs but not as yet commercial diode lasers. Combined, these have permitted a new industry sector that did not exist 10 years ago. However, over 90% of this business is based on LEDs. Violet diode lasers are currently a small, high-unit-value market that has yet to fulfil market expectations. Current industry thinking is that violet diode lasers will become a higher volume business some time in the next 2–3 years. This will be as a result of the commercial development of the digital video cassette recorder (VCR) based on optical data storage.

When this market and related business takes off there will be a step function in the demand for substrates and epiwafers. It is likely that these will be sapphire and the devices will be constructed from MOVPE epiwafers. Moreover, they will most likely be mass produced by large captive Japanese manufacturers such as Sony, Pioneer and Matsushita. These have a long history in longer-wavelength diode lasers, i.e. IR for CD-ROM and red for DVD.

Since sapphire substrates are not semiconducting (they are in fact very highly resistive) they are not included in this market forecast. It cannot be completely ruled out that these violet diode lasers might use other substrates such as GaAs or SiC in due course. Also, no type of laser has been demonstrated by the new GaAs-on-silicon technology described below. The consensus is that the crystal defect density will be too high for useful diode lasers to be produced by this method so the present situation will continue for at least another three years.

3.3.6.4 GaAs-on-Si Substrates

Another area, which is enjoying revitalized interest, is that of GaAs grown on silicon substrates. Over the past two decades many routes to accomplish this have been tried but have so far failed to capture market share. For years, optoelectronics manufacturers have focused on ways to replace expensive GaAs, InP, etc., substrate materials with silicon substrates, which are economical, easy to manufacture and available in diameters as large as 300 mm.

In the summer of 2001 Motorola and IQE announced the world's first demonstration 8- and 12-in GaAs-on-Si wafers. In essence, the new technology is the first to combine successfully the best properties of workhorse silicon technology with the speed and optical capabilities of high-performance compound semiconductors.

The materials technology involves the molecular beam epitaxial growth of strontium titanate (STO) on silicon wafers which is a combination popular with developers of advanced memory devices. The key to the process is that STO is well matched to the crystalline properties of silicon but also to those of GaAs, making it possible to grow GaAs structures on top of the coated silicon wafer. The work was initially carried out at the Phoenix, Arizona, laboratories of Motorola and later in the Bethlehem, Pennsylvania, production facility of IQE. Subsequently, some of these wafers were shipped to IQE's facility. Here, engineers deposited multiple layers of III-V semiconductors on the substrates, fabricating LEDs and other optoelectronic devices.

The STO makes the Motorola–IQE approach unique. Previous attempts at mating GaAs and silicon involved intermediate layers such as those made from germanium or III-V semiconductor materials. These yielded encouraging but far from perfect results. Devices made from these combinations were functional but performance and/or lifetime were inferior to devices made of GaAs on GaAs.

- As GaAs crystal quality and wafer pricing improved, the need for alternative substrates became less pressing. Still, the economics of III-V device processing have seldom been out of the spotlight. Today, device makers are keener than ever to reduce manufacturing costs and improve their products for a competitive marketplace.

The STO on silicon approach solves a problem that has been vexing the semiconductor industry for several decades. It opens the door to significantly less expensive optical communications and other devices by potentially eliminating the current cost barriers holding back many advanced applications.

Analysts do not expect the technology to hurt III-V substrate markets in the short term. Current fabs are fitted for 2-in wafers and will take time to tool-up for bigger wafers. Discrete die size is fairly constant for optoelectronic devices, so there is less pressure to move to bigger wafers. Nevertheless, in due course price pressure will force this change.

Recently, the potential of the technology was demonstrated with the fabrication of several light-emitting devices such as a LED. Working mobile telephone components have also been constructed. While the recent results have been based on GaAs-on-silicon, the method is not restricted to this combination. Motorola is now working on developing the optimum intermediate layer for other important optoelectronic materials such as InP. This technology should support chip clock speeds of more than 70 GHz and long-wavelength lasers that are critical to fibre optic communications.

3.4 World Substrate Supplier Situation

Production of substrates is dominated by a handful of major players that together account for over 80% of the market.

Worldwide there are more than 15 main suppliers of compound semiconductor and related substrates. Despite the shakeouts that have forced six companies out of business, there was until recently substantial over-capacity, especially in Japan.

A further reduction in suppliers may occur as the business further consolidates into the hands of the majors. Despite the upturn, others may choose to exit the business altogether. Should some products become in shorter supply there may be opportunity for improved market success but this is unlikely. Substrate R&D is expensive and profits are notoriously narrow. Hence the necessary commercial development of the larger sizes of semiconducting substrate has been slow in the past even in the hands of the major players.

There are around 10 Japanese companies (see Chapters 6 and 8) with compound semiconductor substrate capability. Most Japanese vendors are located within large diversified companies involved in many aspects of compound semiconductors. The two dominant suppliers are Sumitomo Electric Industries (SEI) and Hitachi Cable, which are also leading vendors of other III-V products such as epiwafers.

The Japanese market has so far been largely closed to Western suppliers, due to a combination of slow growth in demand, and massive domestic over-capacity. With price becoming an increasingly important factor, currency factors have proved unfavourable to importers. Penetration of the Japanese market by Western suppliers is not expected to improve substantially over the next few years. It is likely that other companies exiting the business will come from this region. They may choose to favour higher-margin products such as epiwafers or chemicals.

In the USA, the dominant players are M/A-COM and AXT. The pioneering AXT is no longer the only commercial supplier of VGF, but has emerged as the major US force. These companies were the beneficiaries of the now-concluded Title III program to develop improved GaAs materials.

Europe has three major suppliers of compound semiconductor substrate materials.

- Freiberger Compound Materials (FCM) in Germany has made further progress in the market focusing on 6-in substrates and latterly expanding production to triple capacity. In late 1999 it also introduced 4-in SI VGF substrates.
- Wafer Technology (UK), the second most important overall European vendor, is a more diversified supplier of III-V materials, including both LEC and VGF, as well as InAs, InSb and GaN substrates.

- Crismatech InPact, the French substrate supplier, is focused principally on InP for which it has an international marketing arrangement with FCM.

In North America and Europe, Litton Airtron had established a lead primarily due to its large share of the 4-in market for semi-insulating materials for microelectronics. It was followed closely by M/A-COM. In 2001 Litton exited the business leaving the market to M/A-COM and AXT, which was steadily increasing its presence and is estimated to have now captured over 20% of the world market.

Of the Japanese vendors, Hitachi Cable and Sumitomo Electric have the most significant presence in North America and Europe, with other companies just barely visible. The market share of all the Japanese players has dropped significantly over the past year or two, due to increasing competition from Western suppliers on price and quality.

Within the rest of the world collective region, there are few other significant manufacturers of substrates. A steady but small business is obtained by newer companies via the former Soviet Union that might become more significant over the period covered by this report but this has yet to materialize.

While there would appear to be substantial commercial opportunity for a substrate supplier based in Taiwan, no such operator presently exists there. There are now many strong device companies in Taiwan including more than a dozen optoelectronic fabs. This represents a huge single market for III-V substrates and related products but at present this market is entirely served by foreign vendors.

3.5 Merchant GaAs Epiwafer Markets

3.5.1 Introduction

After substrates, the second tier in the diode laser market is that of *merchant* epiwafer manufacturing. Currently it is one of the stronger growth areas in the industry. In addition, the potential for future growth is also one of the best in the marketplace. However, it is also technically one of the most challenging activities and is becoming a very competitive business.

This section concentrates on the MOVPE- and MBE-based merchant epitaxy business. The major players worldwide are in the main dedicated to one or another of these approaches. More recently, a few companies have emerged having a broader base which includes more than one epitaxial growth type. This is a direct result of several factors not the least being the needs of the market. One of the key trends in the market is less emphasis on the growth technique from the customer viewpoint, i.e. customers are less interested in the 'how' as 'will it do the job' and if so 'when' the product will arrive.

It is worth reiterating that this report is concerned only with the *merchant* market for diode laser materials and components, i.e. those sold to commercial customers and not for in-house requirements. The combination of merchant and captive markets represents the TAM, but it is unlikely that the market will ever move entirely over to merchant.

Nevertheless, there is an industry trend to increase the percentage of manufacturing that is outsourced. Such a trend is clearest at the upper tiers of the manufacturing chain where, for example, telephone handset or VCR makers are closing obsolete factories and handing over manufacturing responsibility to subcontractors. The latter are able to take advantage of new plant located in lower-cost regions such as Eastern Europe or South-East Asia.

3.5.2 The World Market for Epiwafers 2000–05

In 2000 the *total* worldwide merchant market for diode laser epiwafers was valued at US$256 million. This will exceed US$525 million by the year 2005. However, in the period 2000–01 the optoelectronics industry suffered a downturn which resulted in cancellations of orders for epiwafers and other materials. This resulted in a market downturn where the industry fell by as much as 20% to US$205 million in 2001. Positive growth, on the order of 26%, was expected to resume in 2002 to reach a market value of US$254 million. This strong return to growth reflects harder unit prices as demand firms up.

With continuing good growth the total epiwafer market will increase to US$525 million by 2005. It is expected to be sustained in value in terms of sales but there are likely to be significant shifts in production emphasis by product type, e.g. diameter should shift from 2- to 3-in, with 4-in diameter also being important over the longer term. Pricing and application issues should also see changes in the next five years.

The total worldwide market for diode laser epiwafers by region is summarized in Table 3.8 and Figure 3.7.

Figure 3.7 Total Diode Laser Merchant Market for Epiwafers 2000–05 (US$ million)

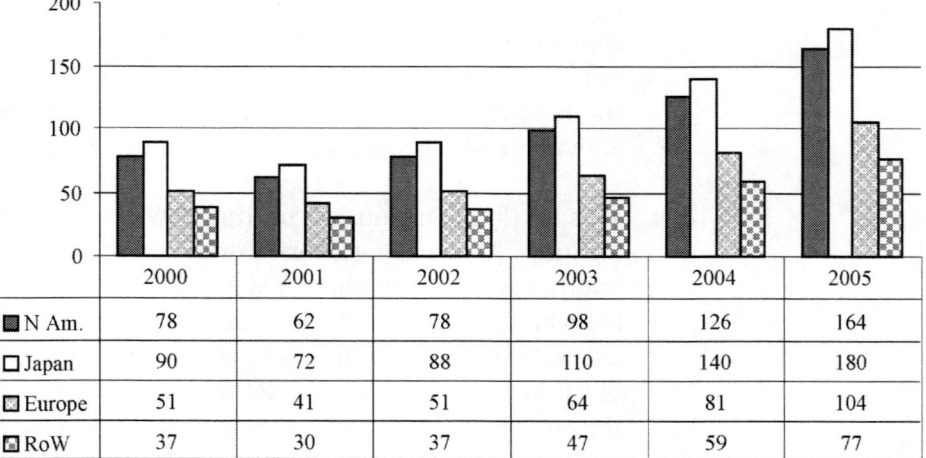

	2000	2001	2002	2003	2004	2005
■ N Am.	78	62	78	98	126	164
□ Japan	90	72	88	110	140	180
▨ Europe	51	41	51	64	81	104
▧ RoW	37	30	37	47	59	77

Table 3.8 Total Diode Laser Merchant Market for Diode Laser Epi-wafers 2000–05 (US$ million)

	2000	2001	2002	2003	2004	2005
North America	78	62	78	98	126	164
Japan	90	72	88	110	140	180
Europe	51	41	51	64	81	104
Rest of world	37	30	37	47	59	77
Total	256	205	254	318	406	525

The analysis of the diode laser epiwafer market is described here. This addresses quantitative differences in production types, e.g. MBE *vs* MOVPE, as well as markets split by region and application. The five-year forecast is also given. This covers the same areas as the 2001 market analysis up to 2005. The issues of market split by epitaxy type, captive *vs* merchant are fully assessed. Prospects for the future development of these and related issues are also discussed.

The technical overview of epiwafer manufacture is given in Chapter 5 which covers the present and future principal growth technologies used in diode lasers. Only brief details will be given here where relevant.

Key technology points for epiwafers are that virtually all next-generation devices rely on epitaxy. These devices are not practical to make, as they use older techniques. The latest diode lasers and related components, for example, can only be made using epitaxial growth. Nevertheless, in common with the ion implantation method popular for the manufacture of transistors, new devices still rely on a high-quality substrate. Rather than create the device within the substrate, epitaxy adds new surface layers from which the device is then made.

However, epitaxy is not a new technology and has already been popular for a wide range of opto- and microelectronic devices and ICs. The earlier methods will remain in use for the lowest cost devices, for example. They are inadequate for new heterostructure-type devices. Such devices have been made possible by new techniques, particularly MOVPE and MBE.

The beginning of the twenty-first century represents a key transition period for the business. The fortunes of virtually all players in the diode laser industry will depend on the decisions made now.

Unlike the earlier analysis of the substrate marketplace, analysis for the epiwafer market has to address the split between merchant and captive. The reason for this is that there are many signs of a continuing reliance on captive epiwafer production over the next five years. Worldwide, the industry approach is reliant on the significant manufacture of epiwafers and related products by independent as well as vertically integrated device companies.

Accordingly, the market forecast is divided up according to basic subdivisions such as wafer diameter: 2- and 3-in plus 'other' sizes which includes wafers smaller and larger than these basic sizes. It is also broken down by epitaxy type, i.e. MBE *vs* MOVPE, and so on.

The total epiwafer market is summarized in Tables 3.9, 3.10 and 3.11.

Table 3.9 Total Worldwide Epiwafer Market Summary: 2-in

	2000	2001	2002	2003	2004	2005
North America	47	37	42	48	54	61
Japan	55	44	49	55	61	69
Europe	31	25	28	32	37	42
Rest of world	23	19	21	24	27	31
Total	156	125	141	159	180	203

Table 3.10 Total Worldwide Epiwafer Market Summary: 3-in

	2000	2001	2002	2003	2004	2005
North America	23	18	27	39	57	84
Japan	26	21	30	44	64	93
Europe	15	12	17	25	36	52
Rest of world	10	8	12	18	26	38
Total	74	59	87	126	184	268

Table 3.11 Total Worldwide Epiwafer Market Summary: Other

	2000	2001	2002	2003	2004	2005
North America	8	6	8	11	14	19
Japan	9	7	9	11	14	18
Europe	5	4	5	6	8	10
Rest of world	4	3	4	5	6	8
Total	26	20	26	33	42	55

3.5.3 Epiwafer Industry Overview

The current epiwafer business can be analysed in a number of different ways. Unlike substrate manufacture, epiwafer manufacture is an activity still undertaken by device makers. So it can be seen as having a strong captive as well as merchant aspect. The business is still evolving towards a steadier position. Today's companies 'mix and match' their own capabilities with those of one or more merchant vendors of epiwafers. This will depend heavily on the type of device required and production maturity, for example.

Seen from another angle, R&D and production split the epiwafer business. Epiwafer supply is one of the most technologically sophisticated activities in the business. It is not unusual for a large device maker anywhere in the world to turn to an epiwafer house at certain times to assist with new product innovation and augmentation. In many respects, the epiwafer vendor is less of a product supplier than a provider of technology solutions unavailable in-house. It thus has many aspects of a design house with parallels in the application-specific IC (ASIC) business. The customer requires the prototype of a new device type and contracting this work outside to experts can save much time and perhaps also a lot of money. This would be described as the R&D stage. Subsequent to the proving of the device in the prototype product, e.g. a new type of spectroscopy instrument or CD player, the device maker may need help to set up in-house production.

The device house may also choose to second source the production to the merchant epiwafer house. This can vary in the size of the order and also in its timing. It is usual industry-wide practice to rely on one or more subcontractors to meet surges in production. This is beneficial because of the savings in plant that would be idle at other times. From here it is a short step to surrendering all production to the epiwafer house.

Other factors such as compliance with rules governing the storage and handling of the toxic materials used in epitaxy may also be a factor.

Issues as to captive or merchant epiwafer supply have other dimensions, and these are discussed below.

Another way of dividing the epiwafer business is by epitaxy type. As has already been stated, first-generation processes have a small place in the future marketplace. They have generally been superseded by MOVPE and MBE. The industry prospects for the next five years can thus also be divided into MOVPE *vs* MBE. In captive R&D and production, the past year has seen a return to pure play epiwafer business practices. This was brought into being by the merger of MOVPE-based EPI and MBE-based QED, to form IQE.

Last, it should be noted that the merchant epiwafer business is also split by device type. The optoelectronics business — of which diode lasers are a key part — has always been reliant almost exclusively on epitaxy. This shows no sign of change. Indeed, the devices with the strongest growth rate are all based on new-generation epitaxy processes. In optoelectronics the issue as to epitaxy type is less clear-cut. Older techniques such as VPE and even LPE are still popular for LEDs while leading-edge MOVPE- or MBE-type processes are required for devices such as diode lasers for DVD systems. There will be major changes in this industry over the next five years, e.g. in the perfection of efficient mass production reactors having design features common to both.

Similar principles apply to the merchant *vs* captive issues in optoelectronics. Indeed, a number of merchant epiwafer suppliers serve both opto- and microelectronic device manufacturers. Others specialize in either one or the other. There are no clear trends appearing as yet but some companies are shifting to a more diversified footing.

- EMCORE Corp is well known as a producer of epitaxy equipment. This New Jersey-based company now also mass produces not only devices such as lasers but also various types of epiwafers via subsidiary companies or divisions (see Chapter 6).

3.5.4 Merchant *vs* Captive Epitaxy Markets

In 2000 the total world market for merchant epiwafers for diode laser applications stood at US$256 million. This equates to an approximate production volume by area of over a million square inches. In other words the average price is around US$250 per square inch, i.e. a factor of ten above average prices for substrates.

The total market for diode laser epiwafer products is shown schematically in Figure 3.8. It should be noted that this graphic is a generalization rather than based on forecast data. This is because hard data on captive wafer production is proprietary. It also depends on many factors such as the technological complexity, size and delivery schedule of orders. It should also be noted that some of the world's largest single multiwafer facilities are captive. This situation looks set to continue as expansion announcements are made.

Should such companies shift their position, however, and move the larger part of their business into the merchant sector then it would cause serious worries within the merchant epiwafer marketplace given the enormous leverage it could exert from its throughput. Conversely, if for reasons such as compliance with tighter environmental regulations, or because of obsolescent equipment, captive manufacturing was closed down, then another problem materializes in the form of ramping up of additional capacity by merchant houses.

Either extreme case would seriously affect any market estimates in terms of quantitative and qualitative accuracy. As a result they are ignored for the purposes of this forecast but the authors reiterate the fragility of these data given the industry's predilection to surprise announcements of such a nature.

Overall, the epiwafer market in terms of total available business could be worth at least another 40% more than today's US$256 million, i.e. US$358 million. As an intermediate case, should captive manufacturers become capacity limited then they may offload more manufacturing to the merchant players. This may prove to be the case with Japanese device companies wishing to quickly boost epiwafer sourcing prior to installing internal capacity enhancements.

The estimates shown in Figure 3.8 are what the market was worth in late 2001. At that time it looked not unlikely that while the value of the captive fraction looked set to increase if equipment purchases are any measure — captive sales of multiwafer epitaxy machines seem about equally split between merchant and captive supply — in this forecast it is assumed that it will follow the industry trend for outsourcing manufacture and will see steady erosion of market share as the market shifts towards merchant.

The expansion of the market value will to some extent be checked by stronger competition as the market matures. This competition is intensifying as device makers pressurize epiwafer suppliers to lower prices. Prices of substrates will commensurately follow this trend. In addition, merchant epiwafers will exert commercial leverage to undercut captive supply. These trends will be felt throughout the industry including the supply of substrates, source materials and equipment. Indeed, much-improved larger wafers and multiwafers and low cost-of-ownership (COO) equipment will play more into the hands of merchant epiwafer companies.

The high profile and profitability of the sector has inevitably served to attract the attention of other players and ease the provision of start-up capital. Nevertheless, the newcomers will find it very hard to penetrate the market and secure market share. This is in large part due to the strength of the existing players. There are only so many customers — this market place is increasing, but large-scale optoelectronics corporations tend to require several years to establish themselves. Existing major players, which number only a dozen or so in the diode laser business, have already become fairly well tied-in with the early merchant epiwafer suppliers. The mutual trust and proven technical/confidential interaction that takes time to set up will work in their favour to the detriment of prospective newcomers trying to compete on price.

Overall, the merchant epiwafer market will increase via two routes:

- General expansion of epitaxy-based manufacturing of devices; and
- Shifts away from captive towards merchant epiwafer supply.

However, the potential increase in competition from newer companies getting into the epiwafer market cannot be discounted. It is common industry practice in business subcontracting to play one bidder against others so as to leverage the best deal. Well-established epiwafer players therefore ignore the threat of newcomers at their peril.

There is another key factor that is already coming into play in this arena. Environmental regulations are tightening each year and this has an impact on all epitaxy processes. Toxic hazards with respect to the handling and storage of source materials are one aspect. The other is the treatment and disposal of waste materials that can be no less toxic. The overhead costs of these operations could in time drive device companies away from future captive plant and towards subcontracting the total task. Ultimately, merchant epiwafer supply may also encompass recycling of waste materials in a closed-circuit high-purity process.

There is also an underlying trend to improve the efficiency of epitaxy processes so as to minimize wastage. This is driven by three factors:

- Increasing prices of source materials;
- Increased costs of disposal of wastes; and
- Price pressure from customers.

Because of these three factors, by the end of the forecast period covered by this report, it is likely that dramatic improvements in efficiency and yields from these operations will occur. Some companies may wish to capitalize on their experience with cost control, which can be best achieved in-house. Other companies may choose to source externally some or all of their business. Seldom is this published such is the confidential nature of this business.

However, it would take only one fairly major incident to accelerate such changes. One toxic accident would be likely to persuade most device makers to move further towards merchant outsourcing of their epiwafer supply.

Device complexity is a key factor because in the past two years the heterostructure-type devices have worked to the advantage of the merchant providers to a large extent. Captive producers have been less able to provide the optimum mix of technology and economics in the right timeframe. The availability of eager merchant providers has only exacerbated the effect.

In due course there may be a reversal of this trend as the device makers equip themselves with the means to mass produce epiwafers in-house where it will be under closer control. The challenge facing the merchant epiwafer houses is to prepare for the next generation of devices in order to be ready to ride the next wave of interest.

As befits smaller and quicker-to-react companies, merchant epiwafer companies have tended to be the first to equip with operational 'next-generation' epitaxial equipment and procedures. Thereby they have been able to secure business at a critical stage in the device market evolution. The rapidly expanding telecommunications market forced some device makers to seek outside provision of large quantities of devices or the materials to make them. In-house equipment was likely to be obsolete or not suitable and technology learning curves were often too steep to match demand.

- To raise the necessary capital several epitaxy houses have launched initial public offerings and to date most have been very successful.

3.5.5 Epiwafer Markets by Epitaxial Growth Technique

Today's merchant epiwafer marketplace is roughly equally split between MBE and MOVPE as the epitaxy technique of choice.

- For MOVPE this is shown in summary forecast in Table 3.12 and in graphical form in Figure 3.8.
- For MBE this is shown in summary forecast in Table 3.13 and in graphical form in Figure 3.9.

Figure 3.8 Worldwide Diode Laser MOVPE Epiwafer Market 2000–05 (US$ million)

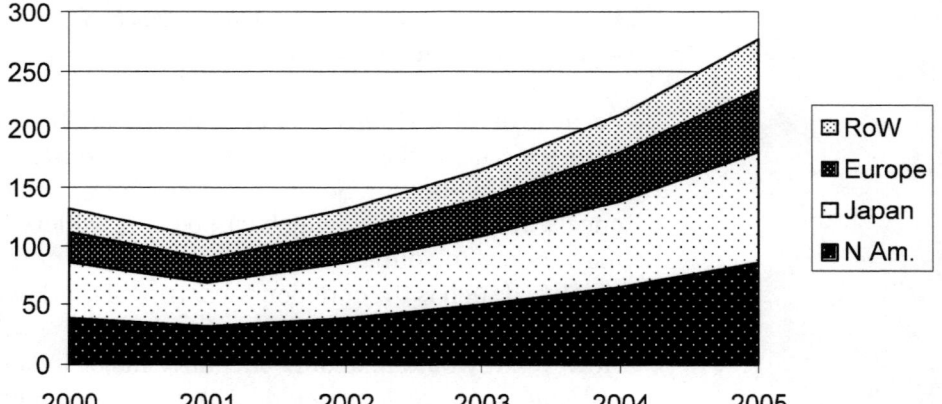

Table 3.12 Total World MOVPE Epiwafer Market 2000-05 (US$ million)

	2000	2001	2002	2003	2004	2005
North America	39	31	40	51	66	87
Japan	46	37	46	57	72	93
Europe	26	21	26	33	42	55
Rest of world	20	16	20	25	32	42
Total	132	106	131	166	213	276

Figure 3.9 Worldwide Diode Laser MBE Epiwafer Market 2000-05 (US$ million)

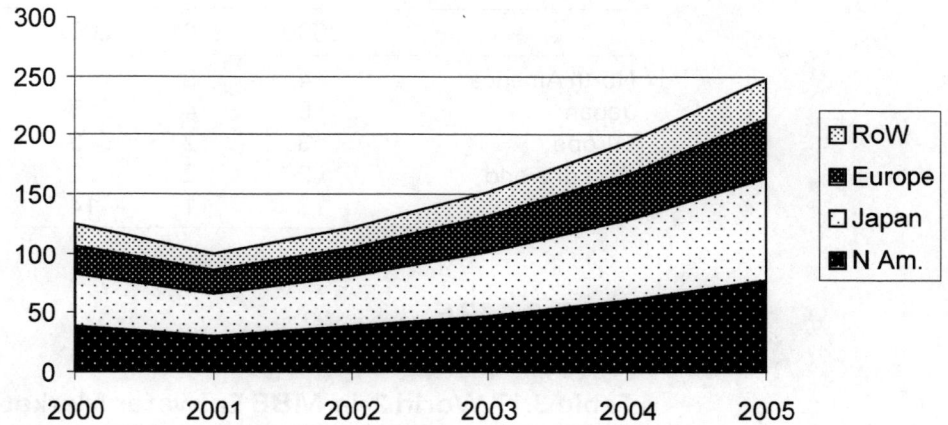

Table 3.13 Total World MBE Epiwafer Market 2000-05 (US$ million)

	2000	2001	2002	2003	2004	2005
North America	38	31	38	47	60	77
Japan	43	35	43	53	67	87
Europe	25	20	24	31	39	50
Rest of world	17	14	17	21	27	35
Total	124	99	122	152	193	248

3.5.5.1 MOVPE

Table 3.14 World 2-in MOVPE Epiwafer Market 2000-05 (US$ million)

	2000	2001	2002	2003	2004	2005
North America	23	19	21	24	27	30
Japan	28	23	25	28	32	35
Europe	16	13	15	17	19	22
Rest of world	13	10	12	13	15	17
Total	81	64	73	82	93	104

Table 3.15 World 3-in MOVPE Epiwafer Market 2000–05 (US$ million)

	2000	2001	2002	2003	2004	2005
North America	12	10	14	21	30	44
Japan	13	11	16	23	33	47
Europe	8	6	9	13	19	27
Rest of world	5	4	6	9	14	20
Total	38	31	45	65	95	138

Table 3.16 Total World Other MOVPE Epiwafer Market 2000–05 (US$ million)

	2000	2001	2002	2003	2004	2005
North America	4	3	5	6	9	12
Japan	5	4	5	6	8	10
Europe	3	2	3	4	5	6
Rest of world	2	2	2	3	4	5
Total	13	11	14	19	25	33

3.5.5.2 MBE

Table 3.17 World 2-in MBE Epiwafer Market 2000–05 (US$ million)

	2000	2001	2002	2003	2004	2005
North America	23	19	21	24	27	31
Japan	27	21	24	27	30	33
Europe	15	12	14	16	18	20
Rest of world	11	9	10	11	12	14
Total	76	61	68	77	87	98

Table 3.18 World 3-in MBE Epiwafer Market 2000–05 (US$ million)

	2000	2001	2002	2003	2004	2005
North America	11	9	13	19	27	39
Japan	13	10	15	22	32	46
Europe	7	6	8	12	18	25
Rest of world	5	4	6	9	13	18
Total	36	29	42	61	89	129

Table 3.19 World Other MBE Epiwafer Market 2000–05 (US$ million)

	2000	2001	2002	2003	2004	2005
North America	4	3	4	5	6	7
Japan	4	4	4	5	6	7
Europe	3	2	2	3	3	4
Rest of world	2	1	2	2	2	3
Total	12	10	12	14	17	21

3.5.5.3 MOVPE vs MBE Market Analysis

In this market estimation MOVPE has a slightly larger market share than MBE, as shown in Table 3.20. However, MBE is making strides as companies are equipping themselves with special phosphorus cracker accessories in order to produce multiwafer P-alloys necessary for visible lasers and so on. Another factor in favour of MOVPE is the imminent higher-volume production of violet diode lasers, which, so far, are made exclusively by this method. However, there is a likelihood that with the continuing industry emphasis on InGaP and related phosphorus-containing compounds, together with the dominance of MOVPE of GaN in the violet LED and laser sector, MOVPE will surge ahead of MBE. Against this is the huge technological knowledge base coupled with the strong desire to maintain market share that proponents of MBE will be able to deploy for technological fixes to achieve phosphorus compatibility. Such techniques are known to be in development but, with such high stakes, details remain confidential.

Table 3.20 World Epiwafer Market MBE *vs* MOVPE 2000–05 (%)

	2000	2005
MBE	48	47
MOVPE	52	53

Today there are three principal epitaxial growth techniques used to fabricate electronic devices:

- VPE.
- MOVPE.
- MBE.

There is every sign that these will continue to dominate the marketplace for the next five years with few signs of any other technique appearing to replace them. There will be further refinement and evolution rather than drastic change.

- Historically, VPE was the first growth technique to be applied to the production of electronic devices. Achieving worldwide popularity in the mid-1970s, VPE continues to be used today for diode and FET products.

MOVPE is in many respects a refinement of VPE that provides compatibility with aluminium compounds necessary for heterostructure devices such as HEMTs. The technique is now in widespread use not only for the production of electronic but also optoelectronic devices. Modern equipment is fully multiwafer capable up to wafer diameters of 6-in.

A crucial factor in the fabrication of some devices is the compatibility of the epitaxy equipment with phosphorus (so as to provide the requisite phosphorus-based alloys implicit in the diode laser structure). The first-generation laser remains a very popular device but the pointer or DVD laser that has arisen in the past two years is based on InGaAsP-type alloys. MOVPE is a mature process with respect to phosphorus-containing quaternary compound semiconductors — this is largely due to its utilization in optoelectronic device manufacture — and it is thus proving to be popular for the mass production of a whole range of devices.

MBE is also a mature process having origins in the early 1970s. Today, multi-wafer MBE is popular with several of the merchant epiwafer suppliers. Indeed, it was the chosen technique for the pioneer of the merchant epiwafer business Picogiga. This French-based company continues to focus its business exclusively on MBE with the largest single merchant fab capacity at the time of writing.

At present in the merchant epiwafer marketplace, there is a roughly equal split between MBE and MOVPE as the epitaxy technique of choice. Of the top ten merchant epitaxy suppliers worldwide, three are exclusively MOVPE-based (Kopin, EMCORE and Epitronics), three are MBE-based (Picogiga, MBE Technology and TLC) while the rest are equipped for both techniques. These latter are predominantly Japanese companies such as Sumitomo Electric Industries and Hitachi Cable. Western companies — which are truly independent merchant suppliers with no substrate or device activities — were until recently either MBE or MOVPE with virtually no VPE. The formation of IQE as the first Western merchant epiwafer supplier since Bandgap Technology in the 1980s resulted in a broader-based player in the top five companies with both techniques available. To a certain extent MOVPE rather than MBE is oriented towards optoelectronics (see Chapter 6).

The choice of MOVPE or MBE is to a large extent an unresolved question for most users of such techniques or wafers. Customers will make a choice of supplier based on many things, not the least being device performance and price. The exact technique used is of secondary importance.

The question of this choice will not be resolved for particular devices in the next five years. But it is likely that MBE and MOVPE will remain in large-scale use. VPE will continue to decline in importance and will be used only for the simplest discrete devices such as LEDs or Gunn diodes.

MOVPE will gain competitiveness from its use in the optoelectronics sector. Virtually all the new high-growth devices such as diode lasers, LEDs, solar cells and detectors rely on MOVPE. However, AlGaAs diode lasers for CD-ROM applications have been a very strong market application for MBE for some years now.

- Optoelectronics is still largely a discrete device business with very few monolithic integrated circuits available on the market. As a consequence present needs are focused on 2-in wafers. Larger wafers such as 6-in ones will only be required when the device die size becomes larger, i.e. when integration takes place.

This huge market will provide major synergistic benefit to the improvement of techniques and equipment for electronic device production. Nevertheless, optoelectronics remains largely based on 2-in diameter wafers and is some years behind electronic device manufacture that is now embracing 6-in wafers. Overall, there are signs that as a result of these trends, MOVPE may become the more competitive technology over the next five years.

The present parity in price between MOVPE- and MBE-produced epiwafers could therefore shift in favour of MOVPE. A possible counter to this is the continued large-scale investment by captive companies that prefer MBE, such as the leading Japanese companies. This commitment will ensure that MBE will continue to be a technique suitable for high-performance devices.

It should be pointed out that there is a dangerous historical precedent with respect to the choice between MBE and MOVPE. In the early 1990s the first blue laser diodes emerged. These were based on II-VI compounds and were only made possible via the MBE technique. However, by the end of the 1990s, a shift had occurred in the preferred epitaxial growth techniques. The world's first practical commercial blue diode laser was introduced in early 1999. This was based on GaN and reached commercial status via MOVPE rather than MBE.

Today, the entire blue optoelectronic device industry is MOVPE-based with little prospect of any shift to MBE or any other technique. However, the industry consensus is that GaN based on MOVPE in its present form is imperfect. It cannot be ruled out that this industry may see a further shift towards non-MOVPE processes. There are characteristics of MBE with respect to precision of layer growth that give it the potential to supplant MOVPE.

- This is especially true for diode laser manufacturing which is dominated in Japan, the homeland of such devices, by MBE.

However, with shorter-wavelength devices, particularly for DVD units, the industry has shown a preference for MOVPE fabrication. Further details of these systems are outlined in Chapter 5.

The epiwafer market will grow at an overall 24% CAAGR over the period covered by this report. However, there are significant differences within the market according to wafer diameter. The growth rates according to the forecast are summarized below.

Table 3.21 Worldwide Diode Laser Epiwafer Market by Growth Technology 2000–05 (US$ million)

	2000	2001	2002	2003	2004	2005
MBE	124	99	122	152	193	248
MOVPE	132	106	131	166	213	276
Total	256	205	254	318	406	525

3.5.6 Epitaxial Wafer Markets by Geographic Region

The geographic market for merchant epiwafers is split between the following regions:

- North America.
- Japan.
- Europe.
- Rest of the world.

The forecast for the epiwafer market split by geographic region is given in Table 3.8 and Figure 3.7. This market split is also in order of importance — Japan remains the largest overall user of epiwafers but this is predominantly captive

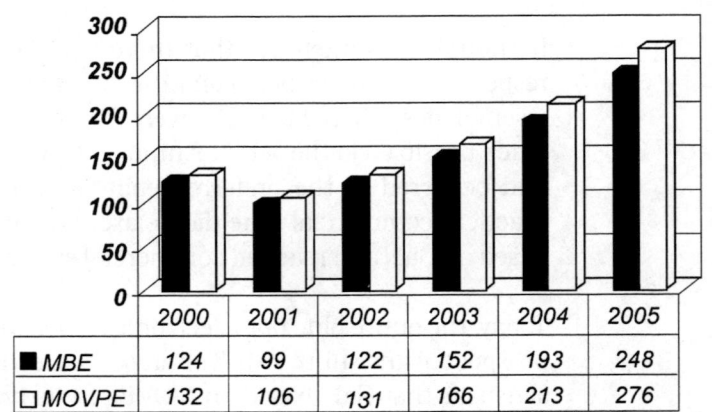

Figure 3.10 Worldwide Diode Laser Epiwafer Market by Growth Technology 2000–05 (US$ million)

	2000	2001	2002	2003	2004	2005
■ MBE	124	99	122	152	193	248
□ MOVPE	132	106	131	166	213	276

rather than the open market. However, this split differs when the market is analysed with respect to merchant epiwafer supply.

Like the substrate market, the Japanese market is to a large extent captive rather than merchant. It has become a little more open to Western suppliers but it will never likely to be fully part of the total available market. Even though there are signs of Japanese industry reorganization, the situation is unlikely to see few major changes in materials sourcing over the next five years such is the closeness of the business relationship in Japan. The implicit importance of the security of supply means manufacturers forego many of the economic advantages of the open market in favour of in-house control.

Even so the market is not completely closed to all Western suppliers. Only those offering very specialist expertise are able to exploit this business opportunity with Japanese companies. Such device makers are aware of the need to achieve maximum competitiveness by means of access at the device and systems level. Should circumstances mandate the purchase of a unique technology that is at the time unavailable in the home market, then it will be acquired by necessary means, including subcontracting to a merchant provider. However, this is usually a temporary arrangement on a per device basis. Nonetheless it may well include high non-recurring engineering charges for such a small-quantity order. Once the door is opened, a relationship may further develop and prosper over a prolonged period.

This situation prevails for many epitaxy-related products, not only epiwafers but also epitaxial deposition equipment. Several leading Western suppliers have announced purchases — and repeat orders — of equipment by Japanese clients. While some degree of anonymity is sometimes mandatory in such transactions, information is generally disseminated, though not in the press. This is in contrast to Western customers who are generally much more open to this form of product endorsement.

In contrast, the sale of epiwafer products is usually highly confidential. Seldom does this receive publicity so as to describe contracts as Western or Japanese. Such information is deemed too commercially sensitive and is infrequently published in the open literature. Nevertheless, research for this report confirms that Japanese sales of epiwafers are fairly routine for at least some of the specialist vendors.

Epiwafer sales to date have one universal characteristic. Seldom are they anything other than small but over the longer term certainly amount to a worthwhile high-value revenue. It is less usual for them to be followed by volume orders unless difficulties are experienced or second sourcing is required. There are signs, however, that the industry is shifting more towards less frequent, larger orders that are preferred by epiwafer vendors as they augment cash flow and capacity utilization.

The initial sale may be described as a prototyping service with each wafer lot likely to number less than 50. An associated one-off tooling charge is also imposed. This arrangement is not unlike that associated with ASIC manufacture. Of course, this tends to make estimation of market size that much less precise.

A device manufacturer will always seek to obtain prototype devices so as to test designs and initiate pilot production. Once proven the device structure design is then transferred to the in-house factory — usually the same equipment is utilized. This may require straightforward exchange of recipe via software thus stimulating the associated equipment market. In the main, devices are subsequently manufactured within the close control of the device maker. This is necessary so as to ensure the precision of control demanded by today's systems-level products. But it is not always practical for the complete duration of the product lifecycle and so additional orders are occasionally placed with the epiwafer vendor. The device house is also likely to revisit the supplier when further product refinement is required.

An epiwafer vendor could be said to be selling less of a physical product and more of a service. The device maker seeks the unique know-how within the collective specialists in the epiwafer company. These activities require close adherence to customer confidentiality that goes a long way to explain the low profile of this business. It also means that new entrants to the marketplace are relatively few, although this has changed in recent years. The mutual trust built up between epiwafer vendors and their customers is a mutual necessity. It is largely impervious to approaches from new players on the grounds of costs or other factors.

However, another factor has arisen that may tend to destabilize the relationship, arising from the lack of capacity to fulfil larger orders. The epiwafer vendor may struggle to match installed operational capacity with genuine customer forward planning and often opts to err on the later rather than sooner side. Until recently this has proved satisfactory. That way the epiwafer vendor is not left with costly under-utilized capacity. However, it leaves the vendor unable to fulfil surges in orders.

The currently prevailing market dynamics are causing a reappraisal of the competitive situation. New entrants to the market are able to pick up business that otherwise would have been denied to them previously.

Nevertheless, one of the best prospects for increasing market share by epiwafer suppliers is to encroach on the captive market. The aim is to persuade the device-making customer to devolve more and more of the epiwafer activity to the subcontractor. This has precedents in many other microelectronics manufacturing activities and is not limited to optoelectronics. One route encountered in today's epiwafer industry is for the merchant house to sub-let a part of its capacity, e.g. one or more of its reactors, to the device house.

- In geographic terms the pursuit of market share via increasing the share of previously captive manufacturing is more commonplace among Western vendors. Japanese vendors tend to be locked in with each other on a more formal, less changeable basis. Japanese epiwafer vendors therefore look abroad to expand market share.

The North American region is a very strong market for epiwafer products. Over the next couple of years it will have surpassed that of Japan in terms of total available market. This has come about because it represents a more open marketplace. In the USA, and to a lesser extent in Canada, the epiwafer market is a straight fight between merchant and captive supply. It is also a market open to international suppliers on a more or less equivalent basis. The only restrictions that exist relate to defence products where national sales tend to be mandatory.

North America is characterized by a small number of local merchant suppliers:

- QED (now part of IQE).
- TLC.
- Kopin Corp.
- Epitronics.
- EMCORE Corp.

New entrants include:

- Intelligent Epitaxy.
- Global Communications Semiconductor.
- Epitaxial Technologies.
- Blue Lotus Micro Devices.

These companies principally focus on epiwafers with little interest in devices. An exception to this general rule is EMCORE Corp, which began as an equipment supplier but has since broadened its portfolio to include epiwafers and devices.

The principal competition for independent epiwafer suppliers is the in-house or captive supply of the major device makers. A fair number of these have traditionally included epiwafer manufacturing capability within their own fabs. At present there remains a strong preference for what might be termed the 'belt and braces' approach where device makers rely on their own capability as well as buying externally. Sometimes this is attributable to tradition and others to practical necessity owing to device specialization, etc.

Companies with captive epiwafer capability include:

- Agilent.
- Coherent.
- JDS-Uniphase.
- Spectra Physics.
- Nortel Networks.
- Zarlink (Mitel).

These companies almost exclusively also source their epiwafer requirements from the top ten merchant epiwafer suppliers. In some cases there is a strong emphasis on one type of epitaxy while other companies rely on external suppliers for a mix of epiwafer types, MOVPE, MBE, etc.

In North America very little business is conducted via partially fabricated devices, i.e. chipped die. The business is conducted either at the substrate or epiwafer level. In these cases the wafer products are exclusively whole wafers. This is in contrast to that part of the optoelebtronics business that is conducted via LED die, for example. Such activities are found mainly with Far East-based companies. Conversely, some US companies undertake device packaging operations offshore. This is either another division of the same company — if the company is big enough — or via subcontractors. In these cases, either intact wafers or chipped die are sent offshore for packaging (and also for testing in some instances) and then returned to the maker or sent to distributors.

In Europe, the situation has many similarities with the North American III-Vs industry. This region comprises the same kind of open market contested by all merchant epiwafer suppliers. Europe is also characterized by the presence of two major epiwafer suppliers: IQE and Picogiga, plus a few smaller ones such as QinetiQ (formerly DERA) and EMF plus those that are university spin-offs, e.g. Kelvin Technology. Each has a specialization and each was a pioneer in its respective field of operations. European substrate suppliers are not active in epiwafer manufacture and most of the major device suppliers have in-house epiwafer capabilities.

- The region also has several leading manufacturers of epitaxial deposition equipment: Thermo VG Semicon in the UK and RIBER SA in France are the leaders in multiwafer MBE while AIXTRON AG in Germany is a world leader in MOVPE multiwafer machines.

All of these companies have made a point of distancing themselves from any form of epiwafer supply other than small technical samples for equipment customers during initial setting up. They are well aware of the need not to be seen to be competing with their own customers, many of whom are the world's leading merchant epiwafer vendors.

European device manufacturers rely on internal epitaxy resources but also purchase wafers on the open market. The reasons for this are varied and largely stem from tradition. They involve technology assistance or second sourcing and depend on the device type. These procedures have been in place for some years and look set to continue for the next few years.

Even those device makers that have developed their own in-house epitaxial capabilities also purchase epiwafer batches from time to time from merchant houses, depending on demand from their own customers. Incidentally, virtually all epiwafers produced by these device makers are exclusively for internal use or for sale via foundry outlets. On occasion, small quantities may be provided free of charge to university departments for the purposes of national collaborative programmes.

3.5.7 Epitaxial Wafer Markets by Application

Today the market for optoelectronic epiwafers is largely driven by the following main device types:

- Diode lasers.
- LEDs.
- Detectors.
- Solar cells.

The total available epiwafer market does of course also include electronic devices such as microwave ICs for telecommunications and consumer applications.

The earliest mass market relied on VPE production of LED structures and Gunn diodes. The present status of VPE is much diminished in favour of the more advanced processes that are capable of providing the requisite new generation of diode laser or IC devices. The newer generation 'heterostructure' devices that now dominate both opto- and microelectronic device markets would not be possible without MOVPE or MBE.

The initial success for MBE was in the mass production of AlGaAs diode lasers for compact disc players and high electron mobility transistor (HEMT) epiwafers for the direct broadcast satellite (DBS) market. This technique was able to provide the requisite precision for active layer growth. Highly uniform epiwafers over 2-in substrates were the foundation of the merchant epiwafer business.

Today, these devices are still a staple product for many suppliers albeit in more advanced forms. Such new-generation variants employing more complex epitaxial layers, finer geometries, etc., permit a relaxation in the specification (and therefore cost) of associated equipment. For example, GaInAsP diode lasers permit the higher data capacity DVD while higher gain, lower noise pseudomorphic HEMTs permit substantial reduction in the diameter of satellite dishes.

- A substantial amount of the value (though not number of wafers shipped) of the epiwafer market derives from R&D. Small runs of high-value epiwafers are still making a good contribution to the overall epiwafer market out of all proportion to their numbers. It is expected that such is the nature of this industry that this will continue for the duration of the forecast period.
- Similarly, there is a steady interest in advanced epiwafer-based devices from the military sector. For example, new types of millimetre-wave transistor are required for next-generation radars, smart weapons, etc.

For some time, the MOVPE epiwafer market was more heavily oriented towards R&D. This was for a variety of reasons including the immaturity of the performance relative to MBE. The approach has a number of advantages compared with MBE such as in the fabrication of diode laser devices and is favoured by Japanese consumer companies such as Sony and Rohm for CD-ROM. However, these generally require thicker layers and hence longer growth times; MOVPE is often the faster growth process.

The big change in emphasis in the past year has been due to a reappraisal of the use of phosphorus-based compounds required for fibre telecommunications and visible-light-emitting applications. These devices have been commercially significant since the early to mid-1990s. However, until recently they were deemed overall to be too expensive compared with older devices.

As a result of this reappraisal there has been a shift towards MOVPE-based epiwafers. MBE is less suitable for the production of InGaP. This is because of the incompatibility of high vacuum with phosphorus, as explained in Chapter 5. Many companies repositioned themselves in 1999 so as to be able to move into the HBT MMIC business. However, as described in the sections on commercial equipment, there have been large strides in the commercial deployment of phosphorus-compatible accessories, such as special crackers, so as to permit multiwafer MBE machines to compete on a more even basis with MOVPE.

3.5.8 Epiwafer Diameter Trends

In recent years the epiwafer market has experienced something of an unsettled period. This is a direct result of increased demand for these materials over the period 1999–2000 followed by a subsequent downturn in the market, a situation that prevails at the time of writing. There is every confidence that the market will resume good growth and it is more a matter of when rather than if. When that market growth resumes the leanest and fittest epiwafer suppliers will be those that reap the most benefit. Much on the minds of these companies at this time is ensuring that they are best equipped with optimal equipment capable of handling the largest wafers that are going to provide them with the optimal economics in what will be a very price-sensitive marketplace.

- Running epitaxial reactors optimally is a major challenge for the merchant epiwafer provider. This type of company is not always able to run large volumes as the captive device producers usually do.
- Modern multiwafer epitaxial machines only deliver their best process economics when fully loaded and run continually. Small runs of diode laser wafers do not always fit this criterion.
- Optoelectronic devices such as diode lasers have very small die areas. This has remained fairly constant over the past decade, in contrast to the IC industry, which sees gradual increases in die area with device functionality, and hence there the drive to large wafers is much keener.

There has been a considerable shift in emphasis within epiwafer-based materials over the years. This reflects the importance of the more modern techniques such as MOVPE and MBE over older techniques such as VPE and LPE. The former represent more advanced epiwafers and hence constitute the higher value segment, whereas the older techniques represent the more basic low-cost product. The latter is more likely to be captive rather than merchant, however.

There has therefore arisen a new set of dynamics in the epiwafer market. On the one hand, demand had been high keeping prices buoyant; on the other, yields have increased through new generations of multiwafer epitaxy systems and allowed cost reductions to be enjoyed for the first time.

Table 3.22 Epiwafer Trends by Diameter 2000–05 (US$ million)

	2000	2001	2002	2003	2004	2005
Other	26	20	26	33	42	55
3-in	74	59	87	126	184	268
2-in	156	125	141	159	180	203
Total	256	205	254	318	406	525

Pricing trends for diode laser epiwafer products are not easy to pin down. For MBE and MOVPE, representative production orders of 100-off epiwafers are typical with unit prices much higher for smaller size orders. Typically, epiwafers are not yet ordered in 1000-off quantities. This is due in large part to the relative immaturity of the market compared with that of silicon, and the continuing evolution of device structures. Such an unstable market is the direct result of a competitive market. Designs for products seldom remain frozen for long and are continually changed. These changes filter down to the epiwafer level since it is by means of this process that the most critical components in handsets are prepared.

It is not straightforward to discriminate between MOVPE- and MBE-based epiwafers on the grounds of price. Such information is proprietary and dependent on many factors, not the least being order size, device type, etc. It is also conditional on the strength of the relationship between the supplier and customer. Such long-term mutual understanding permits significant cost savings and reflects the interdependence that still exists in this business sector.

Figure 3.11 Diode Laser Epiwafer Market by Diameter 2000-05 (US$ million)

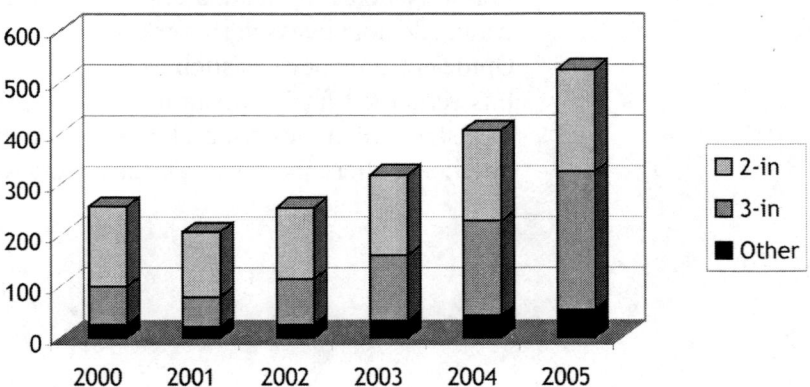

Figure 3.12 Worldwide Diode Laser Epiwafer Market by Size 2000 (%)

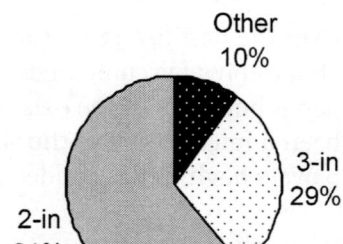

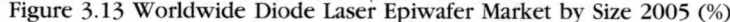

Figure 3.13 Worldwide Diode Laser Epiwafer Market by Size 2005 (%)

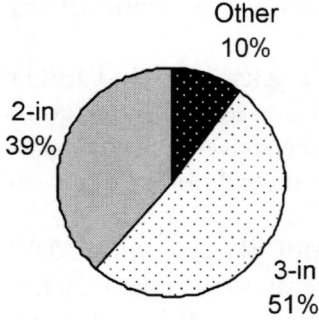

In the past year or so, the 3-in epiwafer has started to become the more important wafer size. Volumes of such wafers have now reached many thousands per annum worldwide. The 2-in wafer is still the most important commercially but will begin to see a decline over the next 3–5 years. This has the effect of pushing up prices and is exacerbated by the reduced emphasis in these products by major epiwafer providers as they wish to score the maximum economic advantage that accrues from larger wafer sizes.

The 3-in epiwafer will soon become the benchmark epiwafer size. The improved cost-effectiveness and cost-competitiveness are becoming more widely known throughout the industry. In a device marketplace that is experiencing unremitting price pressure, the move is likely to see accelerated schedules. These and other factors are forcing the wafer down a pricing curve that is likely to be steeper than previously. It should also be noted that this standardization on 3-in wafers will also be influenced by the present mixed reception to any larger sizes.

- With very little in the way of even prototypes nor any government-funded development programmes in prospect, the emergence of any 6- or 8-in (currently the semi-insulating GaAs and silicon standards, respectively) wafers is not anticipated until after the forecast period.

A point worth making here is the price pressure that is being placed on the substrate manufacturers. This is unlikely to assist the funding of further substrate development unless the manufacturers can make sufficient money to be able to plough it back into R&D.

3.5.9 MBE Epiwafer Suppliers

Over the past five years there has been little change in the complement of merchant epiwafer companies: there have been some new entrants making their debut but none of the existing players have exited the business. There has also been a major boost to the size of the installed capacity with most of the major players having either added to existing facilities or built new ones.

In the USA, the leading North American merchant MBE epiwafer supplier, QED, merged with UK MOVPE epiwafer specialist EPI to form IQE. This pure play foundry is the first non-Japanese company to offer both MOVPE and MBE products. IQE's formation could indicate a shift in emphasis in a marketplace that is less concerned with epitaxy technology and more on prices, performance and delivery. The company has subsequently acquired a substrate supplier — Wafer Technology — and added a silicon compounds epiwafer business.

Picogiga, based in Paris, France, is a global supplier of MBE-based merchant epiwafers which has acquired two companies in devices and wafers respectively (see above). Newcomers to the epiwafer supply scene include MBE-based Kelvin Nanotechnology, a spin-off from the University of Glasgow.

MBE players, unlike MOVPE players, have in several cases not restricted themselves to a single vendor. For example, Picogiga has Applied MBE, RIBER and VG Semicon MBE equipment whereas Singapore-based MBE Technology had an exclusive deal with RIBER.

In the Far East, a number of Japanese companies provide customers with MBE epiwafers but these so far have little presence outside Japan. Taiwan is populated by a relatively large number of new companies having epitaxy capabilities. However, as yet these are biased towards MOVPE reflecting their interest in optoelectronic devices.

Note also that other types of wafers are also going to increase market share. This is not only because of increased R&D but also the market presence of 4-in and larger wafers.

At present, it is not yet clear what impact, if any, the Motorola GaAs-on-silicon process will have on the epiwafer industry. One thing that is certain is that it will involve epiwafers. The prototypes relied on MBE, but MOVPE is also believed to be involved. So it is not clear which, if any, of these methods might gain a market lead as a result.

3.5.10 MOVPE Epiwafer Suppliers

The modern MOVPE epiwafer marketplace has also remained the same in terms of the dominant companies. In North America, IQE is the leader and has been seriously adding to its capacity over the past 18 months. The company has been focusing on In- and Ga-based compounds for some time and is now equipped with multiwafer MOVPE (and MBE) machines with probably the largest installation for an independent epiwafer supplier.

In Japan several companies provide MOVPE epiwafers. These include Furukawa Electric, the dominant player, as well as Japan Energy, Mitsubishi Chemical, Sumitomo Electric Industries and Hitachi Cable. These companies tend to also have MBE- and VPE-based epitaxy products. Their market focus continues to be Japan but sales are also made in North America and Europe. However, the strength of the yen and other factors have served to restrict these sales. In recent years the yen has been weaker than the corresponding foreign currencies and so this has made Japanese products more competitive. In many respects they are not offering the same kind of service as Western companies. Their epiwafers are closer to standard products rather than being customer-specific designs.

MOVPE merchant epitaxy companies, unlike those based only on MBE, tend to opt for either one or the other epitaxy equipment vendor. There are really only two multiwafer machine suppliers, AIXTRON AG and EMCORE Corp. IQE, for example, has an exclusive contract for MOVPE equipment supply with AIX-TRON.

As already mentioned, EPI, based in Cardiff, UK, has merged with QED to become IQE, offering both MOVPE and MBE. This is presently the only Western supplier to offer both services. This company also serves the microelectronics market.

EMF, based in Cambridge, UK, is an epiwafer product supplier. Its portfolio includes not only epiwafers but also epitaxial growth equipment and precursor materials. The company was formed out of one of the world's first merchant epiwafer suppliers, Epi Materials.

As far as the rest of the world is concerned, merchant MOVPE epiwafer supply is fairly small — mainly based on universities supplying small quantities of specialist products. However, in optoelectronics, Taiwan has made enormous investments in new multiwafer equipment for LEDs and diode laser products. Over 20 companies now serve this market and some of these are also entering the market for epiwafers for electronic devices.

4 Application Market Overview

This chapter examines the basic application sectors (many of which overlap) for diode lasers as well the basic commercial opportunities, changes and forces acting within each sector. The chapter also examines the market for the basic types of device according to wavelength. For each type of device, market data and forecasts are provided and future prospects described.

The highest value market in semiconductor optoelectronics is now the laser diode family of devices. Today the wavelength spectrum is now complete with the following widely available:

- Infrared — telecommunications, data and power;
- Visible I — red (DVD), green (laser pointers); and
- Visible II — violet lasers for instrumentation and analysis.

At present the most commercially important part of the spectrum remains the infrared. This is due to the matching requirements for the optical characteristics of fibre optics. It is also important because the data storage industry is still dominated by compact discs for computers, music and games.

Up to fairly recently the diode laser industry has been concerned with that device's ability to handle data streams. Whether it is sending signals down a fibre or reading a DVD, power levels of a few milliwatts are sufficient. However, the advent of diode lasers with output powers of 1 W and above have brought into being several interesting new applications arising from the use of the diode laser as a compact energy source.

Firstly, these high-power diode lasers (HPDLs) have become successful as light-pumping sources for solid-state lasers. However, their biggest opportunity is only just emerging — as compact power sources of coherent light. The wavelength range of interest is from 635 nm up to 2 μm, where they are becoming increasingly attractive for functions such as soldering and welding, as well as in microsurgery.

In some sectors of the laser industry, the importance of the HPDL has been likened to that of the transistor in the electronics industry. However, higher relative costs of HPDLs (compared with traditional laser solutions) have so far limited their use to specialist applications such as non-invasive surgical procedures and in large printing machines.

However, over the past two years further R&D and commercialization in the design and manufacture of HPDLs has taken place. Consequently, prices have fallen. This is leading to significant interest in the adoption of these devices in the marketplace.

At the time of writing, the events of 11 September 2001, and the subsequent military actions were having an effect on the worldwide marketplace. On the one hand, the terrorist attack served to further depress the total worldwide stock market, not helping its recovery. On the other hand, the military actions in Afghanistan triggered increases in defence spending within the world's largest market, North America. In that region the largest single defence contract in its history was won by Lockheed Martin Corp. The US$200 billion contract was for the Joint Strike Fighter (JSF) with anticipated international sales reaching as much as double that over the programme's life, which could be nearly half a century. Analysts were warning that such defence spending hikes could last only a couple of years. This could therefore just be a 'blip' on the overall downward trend.

In certain quarters, industry analysts were predicting an increase in spending on equipment in several key areas as a result of the recent attacks. These would not only boost offensive systems such as aircraft weapons and on-board systems but also intelligence gathering and processing. As a result, developments can be expected to include:

- Increased security across the board in the transportation industry not only at airports but also in aircraft.
- Replacement equipment for that lost in the World Trade Center with the possibility of more distributed information networks so as to minimize damage in future incidents.
- Increased interest in detection and monitoring systems for weapons of mass destruction as a result of the anthrax attacks in the USA.

However, to confirm the industry consensus that the optoelectronic components business had not been picking up in the later months of 2001, one of the world leaders in the market place, JDS Uniphase, reported a wider net loss for its first quarter and predicted weakness ahead as telecommunication capital spending and the economy continued to slump. It could still not predict the bottom of the depression and stated that it may come sometime between December 2001 and March 2002.

This chapter is concerned with diode lasers in two important applications aspects:

- Signal processing — telecommunications, optical data storage, etc.; and
- Compact energy sources — laser pumping, welding, surgery, etc.

For each application the market analysis and forecast is provided by geographic region and by component type. This data is given in tabular and graphical format.

In some cases the boundary between these sectors is less clear. For example, telecommunications and computers, where information technology is continually merging diverse disciplines as the digitization of systems continues.

Following the forecast information, a description and analysis of the factors underlying the evolution of the application sector are given. It provides an overview of the basic opportunities, changes and forces acting within these sectors for the next five years.

Overall, the optoelectronics market has recently experienced a downturn in fortunes for nearly all sectors. The exception is the military and aerospace sector, which proceeds at a slower but steadier growth than that of the other sectors. However, it is expected that once inventories have been brought into line with market demand at the systems level, the market for semiconductor optoelectronic components will resume a double-digit growth for the next four years.

4.2 Diode Lasers in the Total Laser Marketplace

This chapter includes an overview of the place of diode lasers in the total laser market. The laser market can basically be divided into two categories: diode lasers and non-diode lasers. The latter category includes a diverse range of lasers such as those based on crystals, ions, gases, and so on. While longer established than diode lasers and still a thriving business, it has now fallen to less than half of the market owing to the exceptionally strong growth of the diode laser segment.

4.3 Worldwide Market for Diode Lasers

The total market for diode laser devices in 2000 was US$5.7 billion (see Table 4.1 and Figure 4.1). This will rise to US$8.7 billion by the year 2005. This represents an overall annual growth rate of more than 17%. However, there are likely to be significant shifts in production emphasis by device type and application throughout this period.

An industry downturn was responsible for the lower 2001 market of US$5 billion. However, the market is expected to recover in the period 2001–02. This will boost the market growth from 2002 onwards; in 2002 the market will be US$5.7 billion.

Table 4.1 Worldwide Diode Laser Market by Application 2000–05 (US$ million)

	2000	2001	2002	2003	2004	2005
Automotive	126	118	138	162	192	228
Computers	645	598	672	758	857	972
Consumer	770	715	811	921	1050	1199
Industrial	441	420	494	582	688	815
Military/aerospace	240	221	253	290	335	388
Telecommunications	3343	2813	3226	3701	4248	4879
Other	130	122	142	168	198	234
Total	5696	5006	5736	6583	7568	8715

Figure 4.1 Worldwide Diode Laser Market by Application 2000–05 (US$ million)

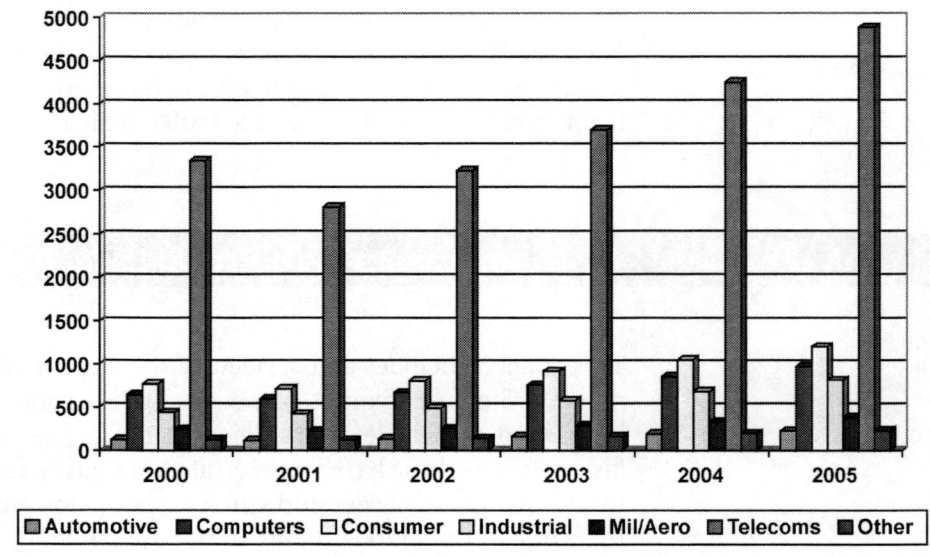

Table 4.2 Worldwide Diode Laser Market by Region 2000–05 (US$ million)

	2000	2001	2002	2003	2004	2005
North America	1811	1628	1873	2159	2492	2883
Japan	1525	1289	1472	1685	1932	2220
Europe	1448	1271	1458	1676	1928	2222
Rest of world	911	818	932	1063	1215	1390
Total	5696	5006	5736	6583	7568	8715

This report is principally concerned with diode laser materials and devices. Coverage does not include laser modules such as transceivers, optical pick-ups, etc. The following devices based on a range of III-V compound semiconductors are included:

- Gallium arsenide (GaAs).
- Indium phosphide (InP).
- Gallium nitride (GaN).

Figure 4.2 Worldwide Diode Laser Market by Region 2000–05 (US$ million)

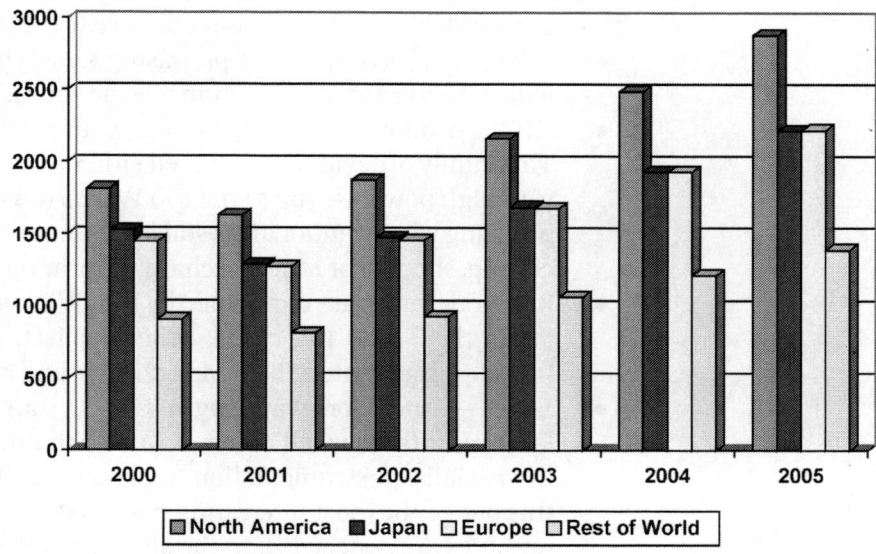

Also included are their alloys as well as other less important materials such as zinc sulphide and related II-VIs and the IV-IV compounds such as silicon carbide (SiC) where they may have an impact on the market for existing diode lasers.

4.4 The Worldwide Diode Laser Business

To summarize the overall position of the diode laser market over the forecast period the results of the market analysis are given in the following:

- Table 4.1 and Figure 4.1 show the market split by application area.
- Table 4.2 and Figure 4.2 show the market split by region.
- Table 4.3 and Figure 4.3 show the market split by device type.

4.4.1 Laser Sub-types

To summarize the overall position of the diode laser market over the forecast period the results of the market analysis are provided according to the wavelengths of the emission spectrum from the infrared (IR) through the visible range, to red and violet.

The types of diode laser covered in this report are as follows:

- Fibre optic lasers (FOL) — a generic description to cover all the transmission-type diode lasers used for fibre optic telecommunications which cover the 980–1550 nm wavelength range. There is some overlap with other diode laser types covered in this report such as HPDLs as used to pump fibre amplifiers, and also vertical cavity surface emitting lasers (VCSELs) which are finding increasing use in high-bandwidth fibre optic telecommunication applications.

- Compact disc lasers (CDLs) — which covers the long-wavelength (780 nm) read-only diode lasers and also the higher-power (30 mW) diode lasers used in recordable (CD-R) and re-recordable (CD-RW) systems.
- VCSEL — one of the most promising new types of diode laser which have wide ranging applications from telecommunications to sensors.
- HPDL — another new type of diode laser which has ›1 W output power. This family of devices can be used either singly or in arrays so as to produce very high powers — up to the 100 W plus range — for optical amplification, pumping of conventional crystal lasers or application of directed energy for cutting, shaping or other mechanical forming.
- Red diode lasers are dominated by the DVD application. These devices operate in the < 700 nm range. Other applications for this family of devices include laser pointers, barcode scanners and rangefinding.
- Violet — the short-wavelength (<400 nm) gallium nitride-based diode laser recently reached commercialization and has found growing take up in specialist instrumentation applications. Other applications will make this one of the most interesting new diode laser types as far as commercial revenues are concerned. In the short term the device is finding application in commercial printing direct to plate and over the medium to longer term, in digital VCRs.

Other types of newer diode laser are also included in the coverage but with restricted market information. These are usually sub-types of the basic types covered here. For example, the microchip laser, quantum cascade laser and Raman laser are all new sub-types of the basic IR diode laser as covered in the FOL section.

It should also be noted that while these categories are distinct there is increasing overlap for various diode laser types. For example, VCSEL development is underway for longer wavelengths for improved efficiency telecommunications, while other companies are developing short-wavelength devices. One particularly interesting development is the ultraviolet (UV) wavelength VCSEL that could rival the light-emitting diode (LED) for white light applications.

The diode laser business is an increasingly diverse and therefore complex activity rendering market analysis a necessarily non-simple task. Coverage here includes all types that have already demonstrated commercial devices and those which have the best potential for success in the forecast period.

Table 4.3 Worldwide Diode Laser Market by Device Type 2000–05 (US$ million)

	2000	2001	2002	2003	2004	2005
FOL	3588	3025	3470	3983	4571	5247
CD	689	620	671	726	785	850
VCSEL	342	308	378	465	572	704
HPDL	285	271	335	416	516	640
Red	621	590	644	703	768	838
Violet	171	193	237	290	356	436
Total	5696	5006	5736	6583	7568	8715

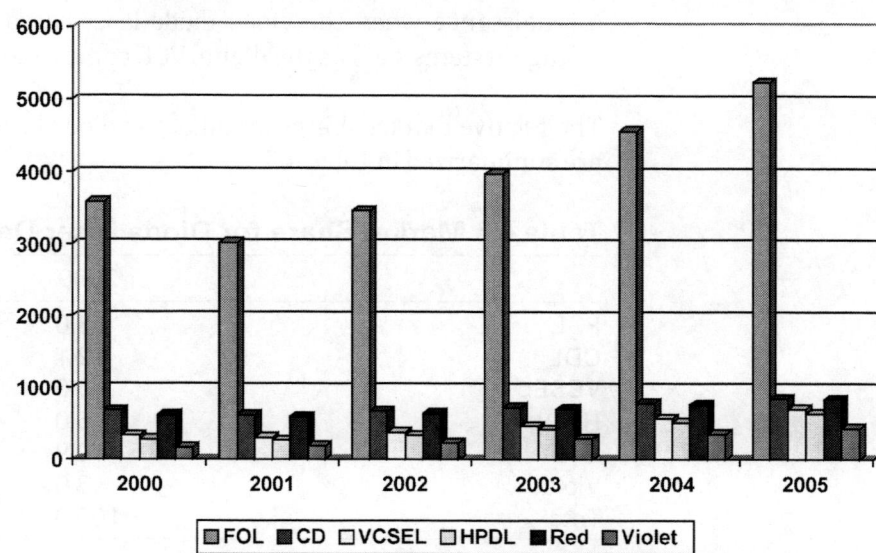

Figure 4.3 Worldwide Diode Laser Market by Device Type 2000-05 (US$ million)

While there will continue to be good growth for all types of diode laser over the longer term, some types will perform substantially better than others. This is partly due to their position in the product lifecycle, e.g. the CD diode laser family is mature and unit prices are now very low. By contrast the violet diode laser has yet to approach the middle of its product cycle and so growth rates are more robust. The relative growth rates for all of the diode laser types covered in this report are summarized in Table 4.4.

Table 4.4 Growth Rates for Diode Laser Devices by Type (%)

	2000	2002–5
FOL	–15.7	14.7
CD	–10.0	8.2
VCSEL	–10.0	23.0
HPDL	–5.0	24.0
Red	–5.0	9.2
Violet	13.0	22.5
Ave	–12.1	14.6

As a result of the different growth rates for the various types of diode laser over the forecast period, some types will improve their market share at the expense of others. Once again this is partly due to their position in the product lifecycle. As a result, it is expected that the data storage devices, i.e. the CD and DVD diode laser families, will see further pricing pressure and hence reduced market share even though unit shipments will still increase. By contrast the violet diode laser, which has yet to approach volume production, will see a near doubling of market share by 2005.

There will appear to be some erosion of the diode laser fibre optic segment. This is due in part to the rate of recovery from the industry downturn in 2000–01 which hit the telecommunication sector hardest of all. It also reflects the likely improvement of the VCSEL and HPDL share some of which will also be fibre related. Basically, the winners by the end of the five-year period will be the fibre optic telecommunication and compact light energy (i.e. HPDL) sectors.

However, this could substantially change and the data storage sector could see a renaissance when the violet diode laser-based next-generation optical data storage systems such as the digital VCR come to commercial fruition.

The relative market shares for all of the diode laser types covered in this report are summarized in Table 4.5.

Table 4.5 Market Share for Diode Laser Devices by Type (%)

	2000	2002–5
FOL	63.0	60.2
CDL	12.1	9.8
VCSEL	6.0	8.1
HPDL	5.0	7.3
Red	10.9	9.6
Violet	3.0	5.0
Total	100.0	100.0

Today's laser diode market is constantly expanding with new types being introduced on a fairly frequent basis. These types were until recently largely low-power devices for the processing of data or sending signals (or storing them). The laser industry is, however, experiencing a change in emphasis. Semiconductor lasers are becoming important as compact energy sources. This is presently less than 10% of the semiconductor laser market but has great potential for future growth.

For the purposes of this forecast, the semiconductor laser diode market is broken down according to various sub-types. IR lasers continue to dominate today's market. This is due for the most part to fibre optic telecommunications but most of the new laser types such as HPDLs also operate in the IR region.

Laser applications fall into four categories:

- Signal transmission — fibre optics.
- Optical data storage — CD- and DVD-R.
- Directed energy — materials processing, welding, etc.
- Sensing — pollution monitoring, analysis, etc.

The industry is experiencing a shift to other wavelengths in several important respects:

- To longer wavelengths for telecommunications and sensor applications;
- To shorter wavelengths for optical data storage applications; and
- To very short wavelengths for instrumentation and higher data density.

While the bulk of all these application areas involve IR wavelengths (fibre optics, dense wavelength division multiplexing (DWDM), VCSELs, HPDLs, etc.), visible (red) wavelengths are becoming very important, especially for higher data storage capacity for movies and archiving via read-only DVD and recordable DVD. These non-telecommunications applications, especially DVD data storage, are helping sustain the total market during the period of the telecommunications downturn.

The following sub-types of diode laser are included:

- IR diode lasers.
- Optical fibre systems.
- Free-space optics.
- Bragg gratings.
- Distributed feedback (DFB) lasers.
- Distributed Bragg lasers.
- External cavity lasers (ECLs).
- Fabry–Perot (FP) lasers.
- Quantum cascade (QC) lasers.
- Fibre amplifier (FA) lasers.
- Tunable lasers.
- VCSELs.
- Microchip diode lasers.
- Violet diode lasers.
- Laser pointers.
- CD diode lasers.
- DVD diode lasers.
- Erbium-doped fibre amplifiers (EDFAs).
- Raman lasers.

Diode lasers have much in common with LEDs from which they were developed. Laser diodes have become by far the most important type of laser in use today and in many respects the diode laser is replacing the traditional crystal laser in many application areas.

The evolution of the laser diode is by no means over. Not only is the generation of new market applications an on-going activity through the evolution of existing devices, but also through the creation of new types of laser diode. Owing to further refinement of its efficiency, operating wavelength and power, the laser diode continues to displace successfully traditional solid-state and gas lasers.

- The laser diode's greatest advantage is that it can be readily modulated by simply switching the drive current. Unlike for LEDs, this can be done even at very high frequencies. Hence the laser has good application for high-performance systems such as telecommunications and data storage (LEDs are limited to 100 MHz whereas diode lasers can reach better than 1 ns switching times).

As a general rule, this section follows industry practice whereby each diode laser subset is dealt with according to its operating wavelength, i.e. from IR through visible (red to green) and thence to the short-wavelength, blue-violet region — devices which have only recently become commercially significant. It should be noted that there is a degree of overlap within the categories chosen for this report. For example, the IR laser strictly encompasses a high proportion of the HPDL, EDFA and other categories because this is the wavelength where these devices currently operate.

The market for diode lasers will continue to grow strongly and be responsible for further new electronic products in data storage and communication as well as such diverse areas as pollution monitoring, rangefinding and instrumentation.

However, these applications are conditional on the further R&D of materials and devices by companies and research institutes worldwide.

Generally, today's diode lasers require low voltage and low-current drive for satisfactory operation. They exhibit good efficiency with long lifetimes and are low in cost. The drawback is that most diode lasers are temperature sensitive and, unlike gas lasers, have highly divergent beams arising from the small light-emitting area. This is not in itself a problem for fibre optics, etc., but coupling to single-mode fibre must be performed with care. To ease this, most diode lasers are packaged with a lens but this adds to the cost.

Reliability is also a key consideration for all types of diode laser. The package format is important and users must select styles with care taking into account cost over the operational life of the device. This is more critical for some applications such as space-based communications. However, for other demanding applications, such as undersea communications, this is less of a problem given that in such an environment the temperature remains fairly constant at a low value.

- Diode lasers are also able to deliver concentrated power form a compact, low-power source, hence their application to CDs and latterly for diode pumping of crystal and gas lasers and even direct diode power applications such as plastics welding, etc.

Currently, VCSELs are eliciting a lot of interest worldwide. This is because they are simpler to make than the conventional edge emitter devices that currently make up the bulk of the market.

The VCSEL market suffered due to two factors:

- Major order cutbacks leading to low demand for IR telecommunications devices, the main market for VCSELs; and
- A general downturn in manufacturing of electronic goods.

Whilst this downturn was harsh for optoelectronics in general, some types were affected worse than others. The segment has been buoyed up by several growth factors, these include:

- Rise in high-capacity optical networks;
- Wavelength shift to increase data capacity;
- Growth in other applications such as sensors; and
- Increased penetration of markets dominated by LEDs.

All sectors of the diode laser market are under strong price pressures. This serves to disguise the otherwise good progress in terms of volumes of components shipped.

VCSELs stand to become one of the key drivers of the semiconductor optoelectronics market mostly via telecommunications but as these low-cost lasers become more understood by designers they will compete with other light sources. For example, some industry observers are impressed with progress in the development of VCSEL-based white light sources.

VCSELs are a key component to the fibre optic communications sector by providing the highest performance and lowest cost solution to the technical issues facing today's high-speed optical networks. VCSEL technology will take the industry into the next generation of computing and communications, including photonic interconnects for board-to-board and chip-to-chip applications.

In the future, long-wavelength VCSELs will enable greater bandwidth and distance in multimode fibre, and lower cost light sources for single-mode fibre applications. Long-wavelength VCSELs also provide a path to overcoming eye safety problems in applications like Parallel Optical Data Links. They will provide the cost and performance benefits of today's short-wavelength VCSELs, but offer the additional benefits of transmission in the 1300 nm window.

The strong growth areas within this business are expected to be principally in diode lasers especially for telecommunications, data communications and optical data storage applications. These sectors have the potential to almost double in size by the end of the forecast period.

In two decades the diode laser industry has been transformed from small-scale runs to generating production runs of many hundreds of millions of low-cost units per year. This has arisen from the major worldwide effort to make diode lasers much more cost-effective and reliable for telecommunication, consumer and computer markets. This has taken time and a great deal of effort by many companies who have become major players in the industry worldwide.

The growth cycle for mainstream semiconductors has stabilized to some degree whereas that for diode lasers has been growing strongly year on year for the past five years. This reflects the fact that within this area there are product types that are seeing exceptional growth in both the short and longer term. Overall, they will return to sustained positive growth above the average economic growth of the worldwide market, making diode lasers one of the most impressive performers across the microelectronics industry.

However, this continuing success has been adversely affected by a mismatch in supply and demand. Earlier there was the situation of the market being incompletely fulfilled by virtue of the lack of sufficient supply of key components. This also served to strengthen the average selling price (ASP) for most diode laser types.

The dynamics of the market over the past year have, however, seen an abrupt downturn. But this is expected to return to positive growth at some point in the next five years. A strong upward trend will then be likely to continue to the end of the forecast period. The timing of this resurgence had yet to become clear at the time of writing but the consensus — which is used here — is that it will resume in 2002 but not return to the 1999–2000 level until at least one year later.

Examples of applications that will have the strongest influence on the growth of the diode laser sector include telecommunications, and data communications and optical data storage. These are in turn fuelled by the continued enormous growth of the Internet which in itself boosts sales of infrastructure and office/domestic/mobile products, DVD movie players and games consoles, archival data storage, high-resolution displays, multimedia systems, instrumentation and new military/aerospace systems.

The emphasis of the industry has clearly shifted away from signal processing as the principal application area for diode lasers, and moved to applications such as fibre optics and DVD players, and to a two-tier business owing to the advent of HPDLs.

Over 2000–01 the laser diode market was affected worst of all in the semiconductor optoelectronic components marketplace. This was due to two factors: major order cutbacks leading to low demand for IR telecommunication devices, and a general downturn across all electronic goods sectors. The laser diode segment lost one-fifth of its value over the period 2000–01. It will not return to this value for at least another two years.

Whilst this downturn was severe it could have been worse. Several growth markets have buoyed up the segment. These include:

- A rise in optical data storage via DVD success — DVD-ROM and DVD-players which include IR lasers for retro-compatibility; and
- Growth in high-power diode lasers which are mainly IR devices.

Contrary to these positive factors all aspects of the diode laser market are under strong price pressures. This leads to amelioration of otherwise good growths in terms of volumes of components shipped.

Table 4.1 shows the regions having greatest market share are Japan and North America. Here the large-scale use of lasers is driven by volume manufacture of modules, sub-systems and systems for telecommunications and data storage. These are split between original equipment manufacturers (OEMs) and contract electronic manufacturers (CEMs). However, there is a general trend for shifting manufacture from the OEMs to CEMs. This is accompanied by a geographical shift, i.e. the CEM being located in a lower-labour-cost region, e.g. Mexico or South-East Asia.

North America remains the largest merchant market for diode lasers, but it is Japan's market that is dominated by captive production.

Demand from North America and Europe is expected to grow over the next five years. This is due in no small part to the fact that the USA and Europe represent the largest markets for diode laser products. Demand for consumer electronics applications is currently dominated by Europe and Japan, but it has emerged strongly in North America. Emphasis in the rest of the world region is growing strongly too. Captive manufacturing continues to be important in regions such as China and in Taiwan.

Applications arising from the communications industry continue to be strong in all geographical regions. Japanese companies remain the largest *producers* of diode lasers for optical data storage for consumers by some margin, and dominate the *merchant* market for these types of discrete devices. In other regions no new merchant suppliers dealing exclusively with discrete devices have emerged.

- In North America, the leading merchant suppliers include Agilent.

In Europe diode laser manufacturing in volume is restricted to only a handful of companies; for example, Infineon, Thales Optics and Alcatel. Philips handed over its diode laser discrete activities to JDS-Uniphase. However, Europe is home to leaders in merchant epiwafer manufacture such as IQE and a few smaller diode laser makers such as Dilas and Modulight.

In addition, diode lasers have benefited from an on-going penetration of other sectors. But it is in the innovation of new products that their unique combination of properties provides a particular advantage in this regard. Nevertheless, in this and other roles, these devices are constantly under price and technological pressure.

The order of ranking of regions is unlikely to change over the period of this report. The strength of prospective industry growth derives from not only a mature consumer/computer manufacturing base but also from the telecommunications manufacturing sector. In Japan there will be a continued emphasis on vertically integrated manufacture of equipment by OEMs, who also sell components on the merchant market. Most of these major players in recent years reported adverse financial results but have now been helped by stronger component sales including optoelectronic devices.

The leading region in terms of both manufacture and market for devices will remain North America. This region includes several world-class players at each tier of the market from substrates to finished devices. These will continue to be the engines of growth, even though these are undergoing restructuring, the effects of which have yet to be worked through. The region also has a high number of very capable specialist suppliers and start-ups with their own dynamics for growth. It is expected that some consolidation will occur within these companies as further merger and acquisition activity is worked through. At present there is a renewed enthusiasm for start-ups that in turn are likely to be acquired by corporations seeking to rapidly acquire the requisite technology and market share.

Another engine for growth is found in the rest of the world category. Previously, this was taken to represent subsidiary countries where manufacturing proceeds at a lower level of activity. Today, however, this region includes countries with a high representation of OEMs such as Taiwan and South Korea. At present, the fraction of manufacturing devoted to diode lasers is small but growing fast. In Taiwan, for example, the number of companies supplying optoelectronic components has grown enormously in the past five to ten years.

Taiwan is already a world leader in several sectors such as consumer and computer products, so individuals and companies there collectively target highly attractive sectors using diode lasers. They will strive to become self-sufficient in the requisite key components rather than rely on sources in Japan where cost-competitiveness and leading edge designs do not always favour manufacture.

The clear leaders in the applications analysis for diode lasers remain tele-communications and optical data storage. These areas were the underlying factor in the dramatic escalation of the industry. These and other products will continue to be engines for growth for the forecast period of the report.

These factors will play a strong role in ensuring that the vital new markets and new subsets of existing markets will be able to further drive up demand.

The following sections proceed in order of importance of total market value according to Table 4.1.

IR diode lasers have been one of the most successful optoelectronic devices. Two important basic types are:

- Fabry–Perot cavity which has a cleaved facet at either end; and
- DFB laser, suited to lower-cost applications, having an internal grating to select wavelength.

DFB types have found application in long-haul fibre optic communications. They suit high bit-rate, 1.55 μm telecommunications. Other applications include wavelength division multiplex (WDM).

Environmental sensing is an important application sector for diode lasers (and LEDs). For example, Epichem has fabricated mid-IR LEDs and diodes lasers with its Admiral Consortium partners Aixtron AG, RWTH Aachen, Université Montpellier II and the IP-Czech Academy of Science.

The antimonide-based devices emit in the 3.3–4.5 μm range, which is suitable for applications in environmental sensing and the non-invasive biological monitoring of toxins and contaminants such as CO_2 and CH_4.

4.4.2 Market Applications

In the following, each application sector is dealt with in turn, a market forecast and overview of the prevailing business conditions are given together with references to pertinent commercial devices and examples of their manufacturers and applications. The sector analysis is covered in order of their importance in terms of market value and numbers of devices required:

- Telecommunications.
- Consumer.
- Computer.
- Automotive.
- Military/aerospace.
- Industrial.
- Other.

4.5 Telecommunications Markets for Diode Lasers

The market value of devices in the telecommunications sector was estimated to be over US$3.4 billion in 2000. An industry downturn — which affected the telecommunication sector and hence diode laser sector particularly badly to the tune of 16% — was responsible for the lower 2001 market of US$2.8 billion. However, the market is expected to see positive growth restored in the period 2001–02. This will boost the market growth from 2002 onwards and increase by an annual average rate of 15% up to a 2005 value of US$4.9 billion. The market estimation and five-year forecast are summarized in Figure 4.4 and Tables 4.6 and 4.7.

Table 4.6 Diode Laser Telecommunications Market by Region 2000–05 (US$ million)

	2000	2001	2002	2003	2004	2005
North America	1039	909	1048	1210	1396	1612
Japan	843	652	744	849	969	1107
Europe	934	790	908	1043	1199	1379
Rest of world	528	461	526	599	683	780
Total	3343	2813	3226	3701	4248	4879

Table 4.7 Diode Laser Telecommunications Market by Device Type 2000–05 (US$ million)

	2000	2001	2002	2003	2004	2005
FOL	3107	2592	2962	3385	3869	4422
CD	13	12	13	14	15	16
VCSEL	79	71	87	107	132	162
HPDL	80	76	94	116	144	179
Red	56	53	58	64	71	78
Violet	9	10	12	14	18	22
Total	3343	2813	3226	3701	4248	4879

Of course, IR diode lasers presently have the strongest market share of the telecommunications segment. This sector has achieved enormous success due to two basic areas of application: telecommunications and data communications. While telecommunications will be likely to remain the province of IR wavelengths coupled with optical fibres, data storage has begun to move to shorter wavelengths.

- The bulk of telecommunications device applications will remain the province of IR diode lasers for the long term. The technology is never stationary with manufacturers continually developing better devices for lower prices.

Cost-effective manufacturing of laser diodes has been one of the most challenging tasks in the past decade and continues to preoccupy the top ten suppliers. This investment is worthwhile because the markets are huge and capital

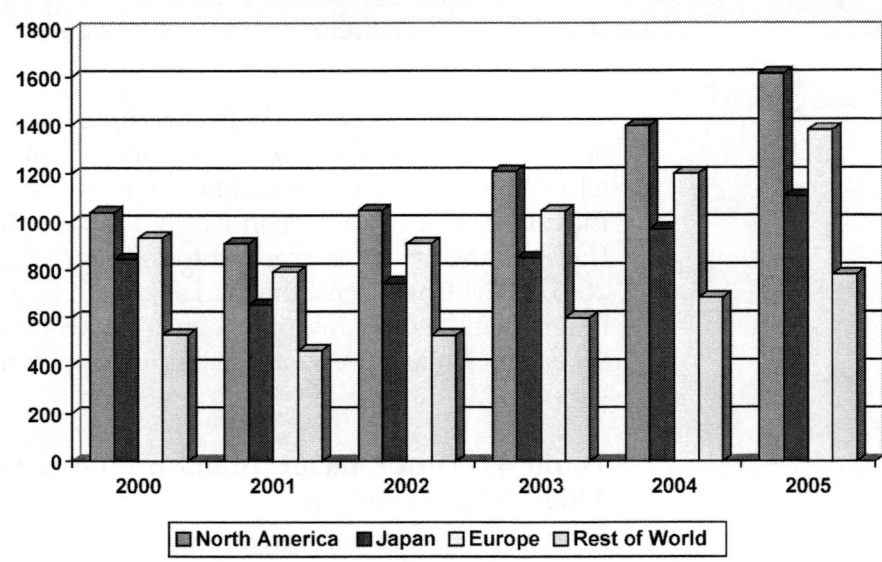

Figure 4.4 Diode Laser Telecommunications Market by Region (US$ million)

expenditure for the very specialized, highly engineered automated diode laser lines is fairly quickly amortized.

The telecommunications market for diode lasers is generally agreed to comprise two main categories. Basically, component requirements fall into either signal transmission or amplification, i.e. erbium-doped fibre amplifiers (EDFAs). The former is by far the dominant application area amounting to around 80% of the market at the present time.

- A growth area for telecommunications diode lasers is in DWDM whereby DFB laser diodes in the 1530–1565 nm range are required. This family of devices is much in demand as a solution for the insatiable demand for higher bandwidth for Internet and multimedia data communications and looks set to see spectacular growth over the next five years.

Diode lasers are set to enjoy further strong growth in the telecommunications sector but there are a few question marks about the actual magnitude of this growth. For example, there is much discussion as to the prospective success of the next stage in fibre optic telecommunications, that of fibre-to-the kerb or fibre-to-the-home (FTTH).

However, sooner or later the present infrastructure will have to reach the consumer directly and this will be likely to become a staged process as older lines are replaced in turn. Without FTTH the full spectrum of two-way services and other bandwidth-hungry applications may not be able to reach their full potential.

- There is thus every expectation of the practical realization of FTTH at some point within the next five to ten years.

As mentioned previously, IR lasers have the biggest market share for the telecommunications business. Detectors are, of course, very important but these simpler devices carry lower unit price. While the detector market does to some extent track that of lasers it is much lower in value.

Fibre optic telecommunications and data communications are the foundation of the laser diode success. While data storage applications now demand increasing numbers of shorter-wavelength lasers, communications require more IR devices.

- The trend is moving towards longer wavelengths as required by the needs of new forms of fibre optic transmission.

Manufacture of these and related devices is a technology-intensive task. It remains an important preoccupation for most of the major optoelectronic component supplier companies. The laser market for this sector is, however, divided into only two general areas:

- Signal transmission.
- Signal amplification (pump lasers).

At present signal transmission is the more important area commanding over three-quarters of the market. Pump lasers for EDFAs are growing in importance.

- Driven by greater needs for the Internet and multimedia, this DWM area is expected to see further strong growth over the next five years.

There is still a question concerning the size of directions of this growth. So, too, the type of laser that will take greatest market share. The edge emitting laser (EEL) — mainly the DFB type — is presently the dominant type. The VCSEL is gaining ground strongly, effectively taking market share from the DFB. This is due to the large unit cost differential between the types.

However, all future growth is predicated upon the growth of the fibre optics market. Much expectation is made of the penetration of fibre networks. Right up to the home, fibre has been mooted as the ideal information highway of the future, not only for office, but also for domestic usage.

- For data and interactive TV, for example, fibre is a very good provider of bandwidth. The problem that always stood against such developments is that associated with installation costs.

With novel new approaches to high-bandwidth data transmission in prospect, such as wireless millimetre wave or free-air lasers, the future success of fibre is not as guaranteed as it may once have looked. Equally well, the future double-digit growth of the required optoelectronic components may well not appear as industry observers have forecast.

Overall, the next five years will see great market activity in this application sector. However, this growth will see average prices remaining under severe pressure in a competitive environment.

- Communications is ranked as the second highest application sector in terms of value, ahead of computers but behind consumer.

Communications systems provide one of the strongest driving forces for the growth of the diode lasers industry.

The following sections discuss applications in each of the main sectors of the communications industry.

4.5.1 Data Communications

Bandwidth capabilities in response to the growing use of multimedia and the Internet have been constrained by the 'first-mile' problem. In other words, the inability to send data from the business to the local switching office or Internet access point at broadband frequencies. With today's technologies utilizing wire- or cable-based infrastructure, communications are limited to just a few hundred kb/s up to a few Mb/s.

While fibre optic cabling provides virtually unlimited bandwidth, it is expensive. Fibre installation costs can be up to US$600 000 per kilometre and take months or years to completely install. Alternatively, broadband wireless technology is relatively inexpensive, quick to install and provides data rates up to 144 Mb/s (OC-3).

Wireless LANs are therefore potentially a great business opportunity for equipment providers. The demand is clearly there and present options are limited. Wireless LANs require high-frequency microelectronic components in just the same way as cellular telephones and hence this sector provides potentially a market for devices approaching that of handsets for mobile communications.

Wireless data communications is made up of three principal sectors:

- LANs (local area networks).
- WANs (wide area networks).
- MANs (metropolitan area networks).

WANs provide two-way communication, independent of location, and are thus similar to cellular and paging systems, except that they have data and voice capability. In due course they will also have video capability. WAN terminals consist of wireless modems built into mobile telephones and portable computers.

- Since it is unlikely that data-only networks will be built (besides the narrow-band paging systems), WANs will use the existing and developing cellular networks.

4.5.2 Fibre Optic Communications

In this market forecast the most important device sector in terms of value is that for telecommunication applications. Herein, the family of components having specific application to fibre optics communications is evaluated and forecast. The research and computation shows that the value of the FOL devices sector was US$3.6 billion in 2000.

The industry downturn was responsible for the 2001 market being worth US$0.5 billion less than the previous record-breaking year. In fact, the 2001 market was worth US$3.0 billion. However, the market is expected to recover thereafter. This will boost the market growth from 2002 onwards which will increase by an annual average rate of 15% up to 2005.

The value of this important device family will increase to US$5.2 billion by 2005. This makes the sector the highest value in the industry representing 63% of the total market in 2001 but this will have fallen to just over 60% by 2005 as a result of the poorer recent industry performance.

The market estimation and five-year forecast are summarized in Figure 4.5 and Tables 4.8 and 4.9.

Table 4.8 Worldwide Market for Diode Lasers for the Fibre Optic Telecommunication Market by Region 2000–05 (US$ million)

	2000	2001	2002	2003	2004	2005
North America	1127	986	1137	1312	1514	1747
Japan	911	711	812	927	1059	1210
Europe	984	833	957	1100	1264	1452
Rest of world	566	495	564	644	735	838
Total	3588	3025	3470	3983	4571	5247

Figure 4.5 Worldwide Market for Diode Lasers for the Fibre Optic Telecommunication Market by Region 2000-05 (US$ million)

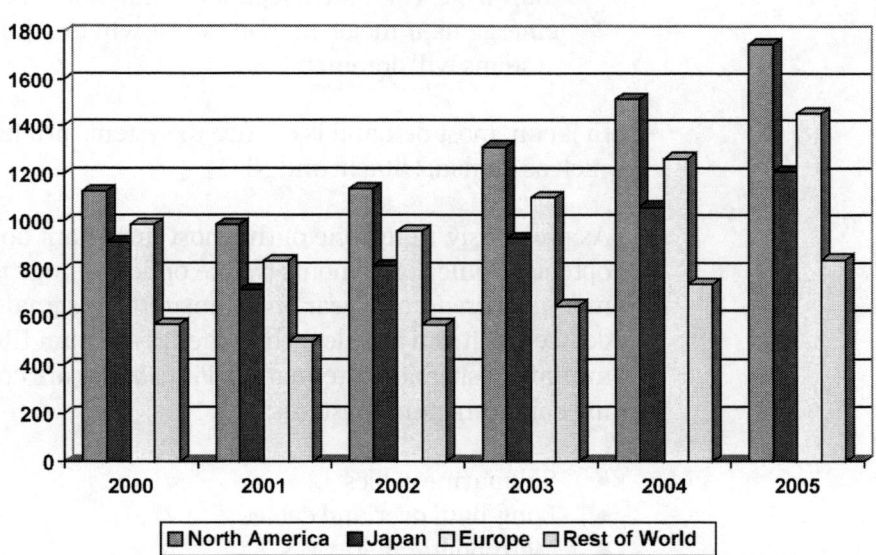

Table 4.9 Worldwide Market for Diode Lasers for the Fibre Optic Telecommunication Market by Application 2000–05 (US$ million)

	2000	2001	2002	2003	2004	2005
Automotive	18	16	19	22	26	31
Computers	72	65	76	89	105	123
Consumer	179	161	190	223	262	308
Industrial	108	97	114	134	157	185
Military/aerospace	90	81	95	111	131	154
Telecommunications	3107	2592	2962	3385	3869	4422
Other	14	13	15	18	21	25
Total	3588	3025	3470	3983	4571	5247

Fibre optic communications systems include long-haul, local loop and fibre optic LANs or WANs (LANs/WANs) (FO-LANs). Component demand is characterized by low volume and high prices. The key benefit for these applications is its ability to handle wide-bandwidth data communications.

The majority of demand in the future will arise from increasing deployment of fibre in the local loop by CATV and telephone companies. FO-LANs are not expected to emerge as a major market, since wireless (microwave and IR) and wired systems will dominate.

- Future demand will come from increasing deployment of fibre-in-the-local-loop by CATV and telephone companies. FO-LANs are not expected to emerge as a major market, since wireless (microwave and IR) and wired systems will dominate.

In Japan, most demand is captive to systems producers with internal capability such as Fujitsu, Hitachi and NEC.

As previously, noted one of the most important applications for semiconductor optoelectronic components is fibre optics. A huge number of IR laser diodes are manufactured each year for transmitting signals down optical fibres. These devices emit at a wavelength in the 1.3–1.55 μm IR region. This is at the point of minimal dispersion/attenuation. Fibre optics are commonly found in the following communication systems:

- Submarine cables.
- Long-haul overland cables.
- Metropolitan networks.
- LANs.

The diode lasers and fibre optics markets have therefore been inextricably linked since their early days.

- Lucent Technologies recently made a breakthrough in the manufacturing of optical fibre, harnessing a previously untapped region in the fibre spectrum. The new process has led to the introduction of a revolutionary new optical fibre, called AllWave Fibre, which provides 50% more usable wavelengths than today's conventional fibre.

However, fibre optics have yet to fully displace conventional copper-based transmission media. One of the great hopes for further market expansion opportunity for diode lasers and other optoelectronics components, the so-called 'fibre-to-the-curb (kerb)' (FTTC) programmes, are having to be postponed as telephone companies reappraise copper technologies such as xDSL.

- This has happened largely because the installation costs for FTTC continue to be relatively high especially for retrofitting of fibre. However, in many places worldwide, new buildings are being constructed that are fibre ready, so as to minimize the costs in anticipation of customer demand for high-bandwidth communications such as two-way interactive TV and video-telephony.

Over 1 million US homes are being connected by FTTC, with cable systems growing fastest of all. However, the price pressure forced by the need to reduce the price of these systems has had a severe impact on the producers of optoelectronic components and this shows little sign of letting up for the next two years. Whilst very attractive from the volume point of view, the fibre telecommunications market has within it some of the most competitive sectors in the optoelectronics market.

For the moment, enhancements of the POTS copper-based system has forced a reappraisal of the fibre-in-loop approach and reduced the penetration of FTTC. However, this is likely to only delay rather than prevent the arrival of FTTC given the tremendous increase in demand for multimedia communications. While broadband services are now possible based on copper wiring with such systems as ASDL offering two different asymmetrical transmission data rates (2 Mbit/s to the home but only 384 kbit/s on return) these are likely to decline in due course before the end of the period covered by this report.

4.5.3 Plastic Optical Fibre

There is renewed interest worldwide in developing mature cabling technologies based on multimode plastic optical fibre (POF). Multiple vendors are marketing products and a POF standard now exists at the asynchronous transfer mode (ATM) forum. The thrust of development stems from the desire to increase operating distances, to lower system-wide installed costs and to improve overall performance. The use of light to send data or the optical transmission of data is a great advance. Fibre optics is the most eloquent and simple medium for such transmission. Copper wire is unsuitable for high-speed data transmission because of its limitations and susceptibility to interference. Glass fibre's very small diameter and its fragility carries with it a high cost of installation.

Currently, almost all LANs are based on copper, a material that cannot support the bandwidth requirements of multimedia and Internet technologies. In addition, copper is vulnerable to electromagnetic interference and can be easily tapped, making it a poor choice for secure environments. Despite some improvements, antiquated wire-based technology is accepted because there has been no practical alternative.

At data transmission rates of 300 Mbit/s to 3 Gbit/s, POF can operate faster than copper wire. But unlike copper, POF provides for a secure environment that can support the bandwidth and performance requirements of ATM communications, Gigabit Ethernet, as well as of multimedia and Internet technologies.

The material cost of POF falls directly between the cost of high-speed copper and glass. POF is versatile and rugged making it suitable for anyone looking for an easy-to-install, high-bandwidth, low-cost fibre optic replacement for outdated copper cabling. The performance of this new fibre will cover the critical last mile of the telecommunications network.

For more than 30 years plastic optical fibre has proved successful and reliable. The early 1980s saw the first use of POF in industrial controls and simple signalling applications. Signal and illumination were also common uses of POF. However, by the late 1980s, with the need for bandwidth in data communications becoming apparent, a new application for POF was found.

The advantages of POF can be summarized as follows:

- Uses visible light so exploits low-cost LEDs and detectors;
- POF diameter is 740–1000 μm so lenses are not needed to couple light;
- Unterconnection needs only simple tools;
- Cheap plastic connectors are acceptable;
- POF is bio-compatible and disposable (cf. glass).

Reversing the comparison, glass fibre is fragile and brittle, meaning it is vulnerable to breakage at the connection. It is also less pliable. Once cost and availability considerations are added to the mix, for short-haul applications, glass fibre offers no practical advantages over POF. In contrast, POF is ideally suited to ATM operation, conveying information more rapidly than copper and offering almost unlimited capacity to meet the high-speed transport requirements.

Moreover, glass fibre installation and management require extensive and costly technical expertise and equipment. For example, a routine glass fibre termination could range from 4 to 20 minutes depending on the application. Plastic fibre cable termination can be done in about a minute. Based on a network installation of 100 terminations, the labour time saved can be substantial.

From a market standpoint, POF has been positioned as the technology that fills the price/performance void between glass and copper systems. Presently, POF represents approximately 1% of the data communications wiring market, which is estimated at US$2 billion. Growth trends in this marketplace are expected to continue at an accelerated rate. The cost of POF can be as much as 40% less than the typical cost of approximately US$0.30 per foot for multimode glass fibre. When high cost and product availability considerations are added to the mix for short-haul applications (glass fibre is obtained in set amounts and gives rise to waste), plastic fibre is on a par with glass fibre.

With more and more home systems — particularly home entertainment systems — being digital, and with more and more households having their own PC, all of these will be needed to be linked together in a high-fidelity all-digital network. POF is one form of media being used to link these devices together.

However, attention is focused on POF 'Fibre-to-the-Home' and within the home as a seamless, all-digital 'Home LAN' for leisure, learning and security. Allied to this are communications applications in the small offices and home offices (SOHO) arena. Corporations and businesses are also seen as using POF to implement affordably secure, high-bandwidth LANs for both internal and external communications.

Mobile workers will be able to purchase POF-equipped automobiles capable of supporting two-way information transfer, Internet access, and a variety of mobile computing tasks. Air travellers will sit in seats integrated to a POF network, offering each passenger a menu of business and entertainment choices, such as Internet access and connectivity to home-based computer systems, movies, video games and shopping catalogues. The military may also be provided with 'wearable computing', lightweight, water-resistant POF 'smart suits' that can send, retrieve and store mission-critical information.

4.5.4 Free-space Lasers

Free-space optics (FSO), or telecommunications systems based on free-space lasers (FSLs), have attractions as an alternative to existing signal conduits such as fibre or wireless. FSO offers high data rates, about 1 Gbit/s, and can be as much as one-fifth the cost of underground fibre. It has the potential of becoming a strong niche market in telecommunications over the next five years. The Strategis Group estimates that global FSO equipment revenues could grow from US$100 million to US$2 billion by 2005. Despite its attractions, FSO has drawbacks that may mean it remains a niche alternative.

New-build offices and conurbations are enjoying fibre from the outset — this is still costly but less so than retrofitting. But retrofit is where the biggest market opportunity lies. For high-bandwidth data communications for interactive digital services, FSLs offer one of the few cost-competitive solutions. As with satellite dishes, aesthetic considerations always have to be borne in mind. Fibre is effectively out of sight but that benefit comes with a high price tag.

Today's telecommunications systems are dominated by two principal applications: fibre optics and wireless, i.e. RF. FSO fits into the second category and yet has features of the first. Many of the aspects of FSO are related to fibre optics. The important difference being that air, or free space, is the transmission medium rather than the glass of the fibre optic.

The technology uses laser light in the IR to transmit optical signals between two points via free space. This requires devices not dissimilar to those used for the transmission through fibre optic cable except that the signal is transmitted through free space and not via optical cable capable of transmitting data, voice or video.

The advantages of the system are said to be:

- No need for trenching;
- Time and labour saving — quick and efficient installation;
- Does not require radio permits and licences as required for microwave systems;

- Bandwidth equal or superior to fibre systems, much better than RF;
- Unlike cable, FSO is a recoverable and non-fixed asset; and
- FSO does not cause electromagnetic interference with other equipment.

However, the wavelength of such transmissions are affected in different ways by the environment, e.g. weather. The three most significant conditions that affect laser transmission are:

- Absorption;
- Scattering; and
- Shimmer.

All three conditions attenuate the transmitted energy affecting reliability and the bit error levels.

Fibre optic telecommunications have become big business but fall short of maximizing their full potential. This is less of a performance issue as being a cost-related one. Currently there is much debate on the amount of fibre in use — so-called 'lit' fibre, against that which is available but pending use, so-called 'dark' fibre. This situation is itself a result of the high cost of installation of the fibre optic cables. Too much is installed so as to be ready for increased demand promptly and without having to lay new cables. This substantial cost has an impact on the overall competitiveness of fibre.

It has also prevented the deployment of the so-called 'fibre-to-the-home'. Revenues from services such as interactive TV have yet to be seen as outweighing the initial high costs of fibre installation. Upgrading existing analogue, copper wire-based networks for the UK's NTL is proving a major problem. A good deal of investment is required before any revenue will come from the new digital services.

No such problems exist for the dominant satellite TV service provided by Sky TV. If effective competition is to exist, NTL and the like have to find a way around the copper-replacement problem. FSLs could be the answer. As a result they would trigger a demand for many optoelectronic components. These devices would not, however, be the same as those required for fibre. The latter are a big success for III-V diode lasers and detectors because of the perfection of the requisite materials and process technologies.

Such devices precisely and efficiently match the characteristics of fibre optics transmission, namely emission at the minimum attenuation windows at 1.3 and 1.55 µm wavelengths. The attenuation characteristics of 'free space', i.e. air but also the vacuum of space as a promising application for FSO, are inter-satellite communications. FSO systems have already begun to be deployed as communications media in a number of key applications around the world. This is important because it demonstrates the effectiveness of FSO as an option for achieving high-bandwidth, reliable communications.

There had been doubts as to its use. These arose from such concerns as:

- Weather disruption.
- Signal blockage, e.g. by birds.

- Eye safety.
- Stability and reliability.
- Security.
- Line-of-sight.

All of these issues, say the proponents of FSO, have been dealt with and prove to be no obstacle to effective communications over short ranges, e.g. for building-to-building.

Installation is little different from fitting a microwave dish. The pair of units is attached to convenient line-of-sight poles, for example, and signal and power leads attached. Following alignment, transmission and reception is locked in and users report performance as least as good as that from fibre or radio systems.

At the moment there are only a few companies offering FSO systems. The large corporations that provide fibre or radio systems seem to be happy to continue to work on making the existing systems ever more competitive. This is all well and good but however small the cost of the fibre or associated components, the problem will always be the cost of laying the cable. Digging up the roads, etc., is never going to be cheap. But FSO systems on the roofs of offices and houses are an altogether more cost-effective solution.

In addition, one of the advantages of FSO is that it does not require specialist new components. Those systems deployed today are constructed from off--the-shelf lasers and detectors which are little different from those used in fibre optics. Some companies are, of course, making component offerings targeting the FSO market. For example, Princeton Lightwave Inc. (PLI) recently launched a device having an output power of 340 mW from a 1550 nm DFB laser coupled into a single-mode fibre. This power level is more than twice the previous record of 165 mW, the company says.

PLI's WaveHarp product line is a family of advanced DFB lasers that emit high power in a narrow spectral band. As a wavelength-stabilized pump, the Wave-Harp takes the bold approach of incorporating the grating on the laser chip, eliminating the need for an external fibre Bragg grating and reducing customers' costs. As a source laser, it can reduce the need for amplifiers in metropolitan DWDM networks, leading to significantly reduced system cost and complexity. However, its most interesting application may be in FSO networks.

While transmitters for FSO applications have been shifting into the 1500 nm eye-safe wavelength range, thus far they have been limited to one wavelength channel for transmission of the signal. PLI's WaveHarp products offer exceptional wavelength stability, enabling coarse wavelength division multiplexing (CWDM) configurations with four or eight channels for transmission and a corresponding number of channels for reception. In addition, the high power output of the WaveHarp transmitters enables generous power budgets for network designers, potentially leading to less expensive networks on a cost-per-bit basis.

- The high wavelength stability, high power, and the compact footprint of WaveHarp enables cost-efficient WDM in FSO access networks, PLI claims.

There has been research into the FSO market. A report by the Strategis Group, *Free-Space Optics (FSO), Light by Air*, predicts that worldwide FSO equipment revenues will reach US$3.8 billion by 2006, up from approximately US$200 million in 2001. It says that there are numerous upcoming opportunities available to FSO equipment manufacturers and service providers. FSO is not a new technology but one whose 'time has come'. FSO provides transmission speeds of up to 2.5 Gbit/s, and is secure, safe to the eye and reliable. It can be used by CLECs and ISPs to 'conquer the last mile', by network operators to close a SONET ring or for network-to-network interconnection.

Furthermore, relying on just one access technology to reach the customer is going to have to change, the Strategis Group says, with access technology diversification being the key. Last mile operators need to reach the customer by any means necessary, whether it is by FSO, copper cable, broadband wireless or dial-up. FSO is a viable alternative available to service providers who understand that there is more than one way to provide service to the customer, it concludes.

FSLs are also being seriously considered for space communications. Both NASA and ESA have systems that exploit the special attractions of FSL communications.

- The ESA advanced relay and technology mission (ARTEMIS) data relay satellite is the first to exploit FSL technology. It will demonstrate FSL technology in near-Earth orbit for the first time via the optical payload experiment inter-satellite link connecting with the Earth observation satellite Spot 4. The ESA semiconductor laser inter-satellite link experiment (SILEX) uses a 60 mW laser at 800–850 nm to link the pair with a 50 Mbit/s data link.

The most challenging aspect of FSLs in such applications is pointing and locking-on with the beam. With an angular divergence of just a few microradians this translates to a foot square at a distance of 100 miles.

NASA has an FSL experiment planned for the International Space Station (ISS). The optical communication demonstration and high-rate link facility comprises a satellite uplink from a ground station having a 2.5 Gbit/s data link. This is based on a 200 mW EDFA at 1550 nm and illustrates one of the advantages of FSLs in the adoption of commonly used terrestrial data communications equipment.

4.5.5 Fibre Amplifiers

An important new application for IR diode lasers is the area of fibre amplifiers. Based around double-clad rare-earth doped fibre, they include a special core that is externally optically pumped by a diode laser.

This is a key component in the exciting new market for high-power telecommunications laser-systems based on double-clad rare earth doped fibres. These new fibre amplifiers and lasers feature a single-mode fibre core which is pumped through a high numerical aperture cladding layer. This allows the high powers generated with multimode semiconductor lasers to be efficiently converted to single-mode 1550 nm power for use in communication systems.

- High-power fibre amplifiers and lasers are an enabling technology for multi-channel DWDM systems, remote pumping of undersea networks and optical satellite communications.

The continuous pressure to expand the bandwidth of communication networks is pushing network equipment vendors to increase the number of wavelength channels transmitted through a single fibre. The optical power required to pump EDFAs used in these networks increases in proportion to the number of wavelength channels transmitted.

4.5.6 DWDM

Dense wavelength division multiplexing (DWDM) systems have at their core a DFB diode laser operating at 1530–1565 nm. These devices have seen considerable increase in market share as DWDM networks spread. However, they may at some future date be at least in part replaced by arrays of VCSELs.

Following on from fibre optics and the need to reduce the interconnections and associated amplifiers, optically amplified WDM systems have emerged. These exploit the technology of fibre Bragg gratings.

Driven by the rapid increase in Internet use and the introduction of applications such as digital television and videoconferencing, the demand for a significant increase in bandwidth is growing at an unparalleled rate. Only two years ago, the 16-channel DWDM system was introduced commercially. Today, it is already insufficient to meet high-capacity requirements for the transfer of information. Increasingly, products with up to 80 channels are starting to appear on the market. As they do, it is clear that a race has begun among systems manufacturers to provide ultra-high-capacity DWDM.

- DWDM increases optical fibre carrying capacity by transmitting data through different wavelengths. Today, DWDM can squeeze 80–160 data streams along one fibre. While it is already used to transmit data over long distances, it has yet to be deployed on a large scale in metropolitan areas, where the bottleneck exists.

There is a growing demand for optically amplified DWDM transmission networks for new fibre optic components and many are based on the unique combination of benefits offered by fibre Bragg grating technology. These deliver cost-effective means of performing many functions required in DWDM systems, and because they are fabricated directly in the fibre, alignment and assembly cost can be significantly less than competing devices that use bulk optics and fibre.

- One of the most promising recent developments in optical technology, the widely tunable laser, is offering both an architectural paradigm shift and a cost reduction in WDM transport.

A key growth area for telecommunications diode lasers is in DWDM whereby DFB laser diodes in the 1530 to 1565 nm range are required. This family of devices is much in demand as a solution for the insatiable demand for higher bandwidth for Internet and multimedia data communications and looks set to see spectacular growth over the next five years.

- The US ATP (Advanced Technology Program) on Advanced Processes For Photonic Manufacturing is geared towards developing new technologies to enable batch processing of integrated optoelectronic devices, leading to substantial reductions in packaging, testing costs and time-to-market. Part of this project includes a team looking at DWDM applications. This includes the development of an automated laser diode test system used in fibre optic communication applications.

SONET-based networks are unable to deliver the bandwidth, scalability and provisioning times required by customers in metropolitan markets. DWDM is quickly becoming the desired architecture for new networks. Metropolitan DWDM networks can deliver a wide variety of services — SONET/SDH, ATM, Internet Protocol (IP) and Gigabit Ethernet (GbE).

Initially service providers will deploy metropolitan DWDM solutions with ring topologies and then steadily move to meshed architectures. So DWDM equipment vendors will have to make this migration as smooth as possible.

WDM systems are presently dominated by edge-emitting diode lasers, but VCSELs promise to fulfil the need for multiple laser sources which are closely spaced for well-defined wavelengths.

WDM has begun to make an impact on the fibre optics industry. In turn it has had an impact on the market for semiconductor optoelectronics. The effect it has had is not a simple one. It has created new markets, e.g., diode lasers for optical pumping, plus corresponding detectors. Such sub-systems may be fairly small in relation to, for example, optical pick-ups, but they are high value. Thus WDM will make a very positive contribution to the market. However, it is likely to be less than appears at first glance. While some of this market will be new, a good part will be a substitute for existing parts. The net effect will, however, be a boost to the market in this case.

- It has been estimated that a relatively small amount of today's fibre optics cabling is actually carrying data/voice traffic. As demand picks up more of the dormant — so-called 'dark' fibre — will be 'lit'. This will significantly boost demand for components.

Another method of increasing bandwidth per fibre is by opening additional bands. L-band (long wavelength) could be considered as the method using EDFAs and can be extended from the C-band.

Essential to making any increase in capacity a reality is the optical fibre amplifier, which is the key element and the enabler of DWDM. The foundation for DWDM is the EDFA. Like the 16-channel DWDM, EDFAs may no longer provide enough bandwidth to meet escalating requirements. New amplifier technology promises to increase the amount of usable bandwidth and enable the development of a new generation of high-capacity DWDM.

- Because WDM greatly boosts the capacity of existing fibre it replaces the equivalent of many non-WDM fibres. The market may thus see some tempering of growth even when demand returns.

There is also uncertainty as to the applicability of WDM to all fibre networks. Installation is relatively costly but has significantly more implicit data capacity than older networks. The industry continues to try to match supply and demand — upon which profitability keenly depends — and in some respects WDM has further unbalanced the equation.

4.5.7 Tunable Lasers

One of the more recent innovations in diode lasers is the advent of devices that can be operated at one wavelength and then 're-tuned' to operate at another. These tuneable lasers have become very important because of the advent of DWDM. This system has massively increased data transmission rates. Today, networks can handle many close-packed data streams up to and beyond 10 Gbit/s. Channel counts are now up to the 200 per fibre mark.

DWDM is particularly appropriate for the long-haul fibre optic sector. But it is also of some interest to metropolitan networks. For a variety of reasons, such systems may exhibit wavelength incompatibilities. There is thus a need for changing the operating wavelength of the diode laser. Various methods are available to achieve this but none are ideal. To assist further market growth in this important sector, several companies are developing tuneable laser sources.

Basically there are three ways to achieve laser tunability:

- Temperature tuning of edge emitters;
- Current tuning of distributed Bragg reflector lasers; and
- Mechanical tuning of a VCSEL in combination with a micro-electro-mechanical system (MEMS).

The third option is the newest and most innovative. It also offers the best mix of characteristics.

The US company Bandwidth 9 has developed a 2.5 Gb/s tuneable laser module for metropolitan networks. Metroflex is the VCSEL's mirror and is adjusted using a MEMS cantilever. Because the VCSEL is inherently a single-wavelength device, even small changes to its cavity spacing via one of its mirrors gives a good spread of controllability over emitted wavelengths.

- The Metroflex is based around a 1539–1610 nm VCSEL targeting EDFA applications. The InGaAs VCSEL is based on an indium phosphide substrate and has a proprietary metamorphic top DBR mirror. This whole structure is produced in a single growth run. The system is completed with a simple GaAs/AlGaAs cantilever. This is moved by electrostatic force. The deflection changes the cavity length and hence the wavelength.

This system is not only rapid and stable but also easily locked onto the required wavelength, one of the main considerations for this application. The total unit of the Metroflex is compact and priced so as to be competitive with other approaches. The company also plans to deploy other metropolitan network products based on this technology. These include optical cross-connects (OXC) to handle high-density terminations where long haul meets the metropolitan network.

Another area of interest for VCSELs is bump-bonding arrays of devices onto a sapphire substrate. This is not only optically transparent but also compatible to high integration density for electronics and detectors. These arrays are being developed by Peregrine Semiconductor for short-reach multi-mode fibre applications where parasitics are a problem.

4.5.8 EDFAs

EDFAs are commonly used in optical network systems because of their simplicity and low cost and their optimum performance and efficiency. The notable feature of an EDFA is its ability to amplify an optical signal by the incorporation of a short length of single-mode erbium-doped fibre. Signal amplification is a passive process with the exception of the pump laser and auxiliary monitor and control electronics.

EDFAs have boosted the wider deployment of WDM, but it was highly reliable pump lasers that enabled development of the optical amplifier. The 1480 nm pump lasers were the first to be designed. Higher power 980 nm pump lasers then began to be used when JDS Uniphase/SDL introduced lasers with long lifetime units.

Today's high-power EDFAs can have three or more stages compared with the single stage common when EDFAs were first introduced. The first stage is pumped using 980 nm lasers, which allow the most signal gain without introducing noise. Later stages typically use 1480 nm lasers with more efficient pumping (lower cost per milliwatt of output), significantly boost signal power relative to noise and offer better distributed (more uniform) gain.

The multistage, high-power EDFAs require large amounts of erbium-doped fibre so as to provide greater pumping power to excite the erbium and amplify the signal. A 1480 nm laser can pump these long spools much harder than a 980 nm laser without introducing non-linearities and other deleterious effects.

EDFAs coherently amplify 1550 nm signals through the conversion of either 980 nm or 1480 nm pump laser light:

- 980 nm laser diodes are made almost exclusively in North America or Europe.
- Japanese companies are predominant in the production of 1480 nm devices.

DWDM is a major factor influencing EDFA design and hence the performance of the laser diodes that pump them. More channel count mandates correspondingly higher total pump laser power. A trend of increasing WDM channel counts is likely to boost the average number of pump lasers per high-power optical amplifier.

4.5.9 Raman Lasers

Raman amplification offers an alternative to EDFAs for applications needing amplification over broad wavelength regions. Unlike EDFAs, Raman gain is not limited to specific wavelength regions. In due course, applications may extend over the entire transparent band — from 300 nm up to 2000 nm — and even into the visible band.

In the short term, for DWDM applications Raman gain is appropriate for the 1300–1700 nm low-loss region, whereas an EDFA's useful range covers 1525–1620 nm, a narrow, approximately 100 nm window.

Raman laser-based amplification is important in very long-haul networks so as to permit greater distances without electrical regeneration. Based on distributed Raman amplifiers (DRAs), signals go into the fibre with power levels below the level at which non-linearity penalties become significant.

- Without DRAs, a DWDM signal can traverse approximately 6 × 80 km spans without regeneration, roughly 500 km. With pumps powering DRAs lower-power signals can be transmitted giving better signal to noise so a full DWDM signal can reach as many as 12 spans, i.e. 1000 km, without the extra cost of electrical regeneration.

Pump power efficiencies have improved 100-fold, with new power laser diodes supplying greater than 1 W of output power to the fibre. New diode array cladding-pumped lasers are reaching output power levels of 10 W and above. Manufacturers can also now produce lasers with a wider range of wavelengths.

The marked improvements in fibre and packaging techniques have also helped propel Raman laser applications. Fibre optics used as the gain medium can now be made with smaller diameters to reduce spool sizes. These fibres have lower loss, reducing the amount of gain medium needed for amplification. Overall, the economics of using Raman amplification is now reaching parity with EDFAs. However, the advantages of Raman amplification have yet to be fully explored and exploited. So Raman is a technology destined for innovation and further market success.

4.5.10 Fibre Bragg Gratings

Fibre Bragg gratings (FBGs) are a relatively new family of products in fibre optics. They were first reported in the late 1970s but have only recently begun to be commercially important. Their appeal stems from two basic application areas. Firstly, they have the potential to overcome some of the limitations of fibre optics for high-bandwidth networks. Secondly, their unique set of characteristics make them attractive for a wide range of sensing applications from monitoring the health of the human body to the structural behaviour of buildings.

There are three principal uses for FBGs:

- Controlling.
- Combining.
- Routing light.

FBGs also allow multiple wavelengths to be combined in a single fibre for WDM systems. Specific wavelengths can also be routed from a fibre so as to add or omit specific signal wavelengths from a WDM system. Alternatively, the processing of such signals has also been developed so as to permit the sensing of a growing range of parameters such as pressure and temperature to name but two.

- The highest growth rate will occur in the generic grouping of networks. This application group includes all aspects of fibre-based telecommunications and data communications for high-bandwidth networks.
- The other main application sector for FBGs is in sensors.

A key step in the evolution of networks from digital to all-optical is inevitable as the number of bandwidth-hungry Internet applications continues to grow. These new, dynamically reconfigurable networks require compact, versatile and cost-effective photonic devices. Within five years, photonics will be to the communications industry what electronics is to information technology.

- FBGs will be playing a key role in the all-optical networks which will be providing telecommunications service, having faster provisioning of customer services and operational simplicity, plus dramatic increases in the usable capacity of existing fibre and which will be better suited to satisfying the ever-growing demand for higher bandwidth and dynamic scalability.

The rate of growth will be uneven over the various chosen application sectors. The highest growth rate will occur in the generic grouping of networks. This application group includes all aspects of fibre-based telecommunications and data communications for high-bandwidth networks. This sector will not only have the highest growth rate but also reach the highest market value.

The other main application sector for FBGs is in sensors. Their application is gaining momentum and becoming more diverse with time. In this area the coverage will include various fields from monitoring of buildings, bridges, nuclear power stations, etc. to the use of FBGs in personal health monitoring. FBGs are finding wider use as permanent remote sensors in so-called smart structures, i.e. FBGs embedded in concrete structures such as bridges and buildings.

Applications include marine structures, aircraft and road vehicles. Optical fibres are about ten times the diameter of structural carbon fibres and they are laid as an integral part of the laminated structure. Tests show that optical fibres have no significant effect on the structural integrity of a mast. Also, FBG sensors withstand composite fabrication and cure as well as prolonged fatigue while still functioning satisfactorily.

- The technology is expected to transform design and lay-up of composites but it has far-reaching applications across the board in civil engineering, in offshore structures and in the aerospace and automotive industries.

British Aerospace is developing distributed strain sensors based on optical fibres to monitor panels and structures in aircraft. The sensors may also have applications in the intelligent monitoring of the manufacture of composite materials. Aerospatiale is developing high-pressure fuel tanks made from composites containing FBGs.

Worldwide, a growing number of companies and laboratories are researching applications of distributed FBG-based strain gauges. Most European research into FBG sensors is addressing the issues involved in the monitoring of structures.

- CEA-LETI, together with EDA and Framatome in France, are experimenting with FBG extensiometers fitted to the containment shield of a nuclear reactor. They claim that the sensor is capable of detecting a change of 2 microstrain, which is going to be very important in the long-term monitoring of the structural integrity of so safety-critical a system.

The FBG has become established over the past five years. To feed this growing market, a number of specialist companies have sprung up supplying a steadily expanding range of components based on FBGs. In Canada, where FBGs first originated, there are unusually high numbers of important start-up suppliers. For example, KROMAFIBRE was founded in February 1997, with the aim of developing and manufacturing the best in FBG-based devices for the new generation of optical fibre communication systems using WDM. Saint Laurent-based KROMAFIBRE founder, Francois Ouellette, is one of the pioneers in FBG technology.

Recently, US instrumentation specialist company PerkinElmer bought a 13% stake in FBG specialist Bragg Photonics of Canada in a multi-million-dollar deal. Bragg Photonics' fibre grating technology will 'significantly reduce the time-to-market of our high-powered transmitters and 980 nm pump lasers,' says PerkinElmer. Bragg Photonics is developing products for DWDM, optical add-drop multiplexing, laser wavelength stabilization, dispersion compensation and optical cross-connect switches. The company says that all-optical network architectures using its all-fibre components allow more efficient bandwidth expansion for the Internet, intranets and enterprise networks.

Quebec-based O/E LAND has launched a new wide-range tunable FBG. The wavelength can be tuned over 20 nm with just a fingertip. The new tunable FBG has a totally new and simple tuning mechanism compared with existing tunable FBG products. It is easy to fabricate, is low in cost, is easy to control electrically and offers good reliability and long lifetime. Tunable FBGs have many applications in dynamic add/drop DWDM filter, wavelength router/switch, tunable fibre laser, dynamic gain flattening, dynamic dispersion compensation and other fibre optics applications.

The USA has its fair share of players, and not surprisingly several of these companies are based in California. For instance, Diglions of Sunnyvale, California, recently acquired FBG technology and patents from Foster-Miller of Waltham, Massachusetts. Digilens has also joined forces with Foster-Miller's R&D team to launch a family of OEM optical switches for telecommunications.

In March 2001, 3M Telecom Systems Division announced that it was going to triple its output of polarization-maintaining fibre and FBGs. Expansion of the group's FBG plant in Austin, Texas, has just been completed. 3M's engineers have also unveiled a proprietary controlled-flow process for volume production of chirped FBGs more than 10 m long. The breakthrough is reinforced by a fibre encapsulation process that is claimed to provide a tenfold increase in the grating's mechanical strength (from 50 kpsi to > 550 kpsi).

JDS Uniphase, one of the world leaders in optoelectronic components, is also now active in the FBG sector. It was the first company to introduce passively stabilized in-fibre Bragg gratings (in-FBGs) in 1995 and has continued to build in this capability and can now offer gratings in a range of packaging options for research, development and OEM use. It has been able to draw upon FBG experience at the Australian Photonics CRC since 1990. Team members have themselves made key contributions to the technology including the basic mechanisms of photo-refractive index change and lifetime stability of in-FBGs.

In the UK, Winchester-based Southampton Photonics offers a range of innovative, high-performance optoelectronics products including FBGs. Southampton Photonics was formed by senior members of the world famous Optoelectronics Research Centre (ORC) at the University of Southampton, together with experienced engineering and business managers with a track record of bringing new technology to market.

4.5.11 VCSELs

The VCSEL diode family is among one of the most interesting in today's marketplace. The value of this market sector was US$342 million in 2000. Because of the aforementioned industry downturn this fell by 10% to US$308 million in 2001. The VCSEL market is expected to return to positive growth from 2002 onwards increasing by an annual average rate of 23% up to 2005. This is one of the highest growth rates of the diode laser market. The value of this important device family will therefore increase to US$704 billion by 2005. This means the sector's overall 6% share of the total market in 2001 will increase to just over 8% by 2005 as a result of broader industry application, especially with a resurgent telecommunications market.

VCSEL diode lasers are gaining more and more interest worldwide. This arises from being simpler to make than the conventional edge emitter devices that currently make up the bulk of the market. The market estimation and five-year forecast are summarized in Figure 4.6 and Tables 4.10 and 4.11.

Figure 4.6 Worldwide VCSEL Market by Region 2000–05 (US$ million)

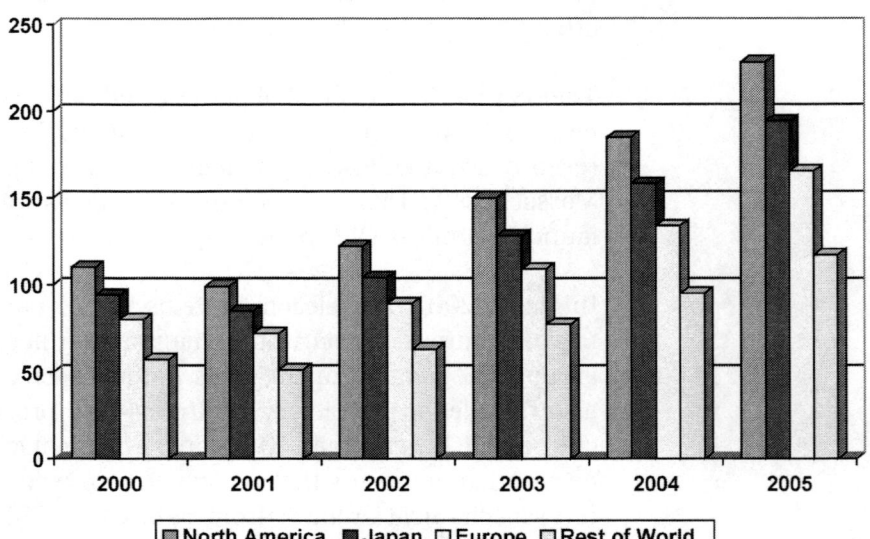

Table 4.10 World VCSEL Market by Region 2000–05 (US$ million)

	2000	2001	2002	2003	2004	2005
North America	110	99	122	150	185	228
Japan	94	85	104	128	158	194
Europe	80	72	89	109	134	165
Rest of world	57	51	63	77	95	117
Total	342	308	378	465	572	704

Table 4.11 World VCSEL Market by Application 2000–05 (US$ million)

	2000	2001	2002	2003	2004	2005
Automotive	31	28	34	42	52	63
Computer	63	57	70	86	106	130
Consumer	51	46	57	70	86	106
Industrial	65	58	72	88	109	134
Military/aerospace	31	28	34	42	52	63
Telecommunications	79	71	87	107	132	162
Other	22	20	25	30	37	46
Total	342	308	378	465	572	704

VCSEL production is critically dependent on technology investment. While the device has come to be seen as a good competitor to edge-emitting diode lasers, practitioners must have good control over epitaxial growth. The VCSEL offers a high degree of process simplicity akin to that offered by mainstream silicon processes such as on-chip test.

In fact, the VCSEL has stolen market share over the edge emitter and looks set to continue to do so over the next five years. The next stage in the market development of this device family is creation of new markets via its unique characteristics. While presently dominated by telecommunications, the VCSEL market is moving into new directions which, though continuing to be dwarfed by telecommunications, are none the less interesting to several of the new start-ups. Such applications include sensors, optical microsystems and optical computers.

The advantages include:

- Single longitudinal-mode optical output;
- Very small size;
- Very low power consumption;
- Mode-hop free wavelength tenability; and
- Two-dimensional array capabilities.

Thus the VCSEL presents a real challenge to the well-established edge-emitter laser. They also present a real threat to LEDs for some applications. This is because VCSELs have been shown to be more efficient than LEDs; they are also faster and have lower beam divergence. Such advantages have drawn the attention of manufacturers of printing equipment, displays and novel sensors.

In summary, prospective applications for VCSELs include:

- Fibre optic data links.
- Sensors, e.g. proximity or gas sensing.
- Encoders.
- Laser rangefinders.
- Laser printing.
- Bar code scanners.
- Optical storage.
- Frequency sources, e.g. for atomic clocks or GPS.
- Spectroscopy.
- Fibre coupling.
- Gas sensing.
- Medical diagnostics.
- Optical microsystems.
- Short-range communications via plastic fibre.
- All-optical terabit per second data networks.

The main interest is driven by telecommunications and in particular the wavelengths coupled to fibre optic attenuation windows. This was previously 1.3–1.5 µm, but today this is moving to longer wavelengths. Conversely, R&D is steadily growing for very short wavelength devices such as violet light emitters.

The VCSEL, which emits light in a cylindrical beam vertically from its surface, offers significant advantages when compared to the edge-emitting lasers currently used in the majority of fibre optic communications devices.

As yet not successfully commercialized, long-wavelength VCSELs will enable greater bandwidth and distance in multimode fibre, and lower cost light sources for single-mode fibre applications. Long-wavelength VCSELs also provide a path to overcoming eye-safety problems in applications like Parallel Optical Data Links. They will provide the cost and performance benefits of today's short-wavelength VCSELs, but offer the additional benefits of transmission in the 1300 nm window. Short-wavelength VCSELs, emitting at 850 nm, have enabled a cost-effective transition to gigabit networks.

Emerging applications, such as Gigabit Ethernet and Fibre Channel, require transmission distances of a few hundred metres over multimode fibre at data rates greater than a gigabit per second. In these applications, VCSELs are the emitter of choice because LEDs cannot meet the speed requirements, and because VCSELs offer fundamental cost advantages over edge-emitting lasers.

- Despite success in these applications, today's 850 nm VCSELs have limitations. VCSELs operating at 850 nm are not optimally matched with standard multimode or single-mode optical fibres that perform best with sources emitting at 1300 nm. These VCSELs are facing significant distance limitations in multimode fibre at high speeds, and are not compatible with standard single-mode optical fibre.

Long-wavelength VCSELs offer the promise of improved bandwidth and distance in multimode fibre, and are well suited for single-mode fibre applications. The spectral purity of single-mode long-wavelength VCSELs is similar to that of

DFB lasers, so relatively long-distance transmission is achievable. Future VCSELs and VCSEL arrays operating at 1550 nm may make excellent sources for WDM applications.

VCSELs operating at 1300 nm will be the most cost-effective single-mode source for applications like WANs, interoffice communications, FTTH and FTTC. Long-wavelength VCSELs also offer an economical path to overcoming some significant challenges around eye safety in the emerging parallel optics market. Most manufacturers are struggling to develop parallel modules that meet the IEC Class 1 international eye safety standard, and are opting instead to develop products that meet IEC Class 3A. Because 1300 nm light is much less damaging to the eye than sources at 850 nm, more power can be introduced into the fibre, and IEC Class 1 eye-safe operation can be achieved without compromising link performance.

As the data-carrying capacity of copper coaxial cable approaches its limit, the demand for a widespread and high-speed optical distribution of data becomes ever more acute. A particularly attractive solution to this problem is based on the parallel transmission of data over multimode plastic fibre ribbon using arrays of VCSELs as the optical source.

The combination of the VCSEL's intrinsic high speed, planar geometry and almost ideal beam profile, coupled with the exceedingly low cost of plastic fibre, make this a potentially very low-cost and high-data-rate (>1 Gbit/s) distribution system. Although such systems are being developed using a range of different wavelengths, there are no 650 nm VCSELs that are currently capable of meeting realistic system specifications.

It is interesting that the economics of the VCSEL threaten new and established markets for the LED. But it does not stop there: the first near-UV (380 nm) solid-state microcavity laser has been demonstrated in prototype by scientists at the US Department of Energy's Sandia National Laboratories working with colleagues at Brown University. Currently, the invention at in the laboratory stage, powered by bigger, more conventional lasers — a method called optical pumping. The next step is electrical pumping, a more commercially useful way to power such devices. This will require developing connectors that will transmit electricity to the tiny lasers. The group expects to be able to demonstrate this capability in one to two years.

- Based on AlGaN, Sandia chose to add indium which brought the VCSEL efficiency to a tolerable starting point of 20% even though it pushed the wavelength emitted further into the near-UV range.

VCSELs are a leading contender for several reasons. Blue LEDs, already commercially available, also give off white light through phosphors, but the light so produced is considered a cold white because it is overbalanced toward the blue end of the spectrum. Combinations of red, blue and green LEDs can produce white light but may not be cost-effective. VCSELs create a more monochromatic and directional beam.

- Like white LEDs, UV VCSELs coated with phosphors offer the prospect of generation of white light for indoor lighting as a supplement or replacement for fluorescent tubes and incandescent bulbs.

Sandia says that such solid-state emitters could last five to ten times longer than fluorescent tubes and will be far hardier. Also, grouped several hundred to a postage stamp-sized chip, they will have an aesthetic value: instead of a single, bulky tube, the chips will be arranged in any configuration one might wish on ceiling, wall or furniture.

- Easily portable UV laser light also interests Sandia researchers because it causes weapons-grade fissionable materials and dangerous *E. coli* bacteria to fluoresce very efficiently in the visible spectrum, aiding in the detection of attempted thefts and preventing the spread of natural or human-caused epidemics.

4.6 Consumer Markets for Diode Lasers

The consumer sector is amongst the most important in today's marketplace. The value of this market sector was US$770 million in 2000 but because of the industry downturn this fell by 7% to US$715 million in 2001. The market is expected to return to positive growth from 2002 onwards increasing by an annual average rate of nearly 14% up to 2005. The value of this important device family will therefore increase to over US$1.1 billion by 2005. These market data are summarized in Figure 4.7 and Tables 4.12 and 4.13.

Figure 4.7 Worldwide Diode Laser Consumer Market 2000-05 (US$ million)

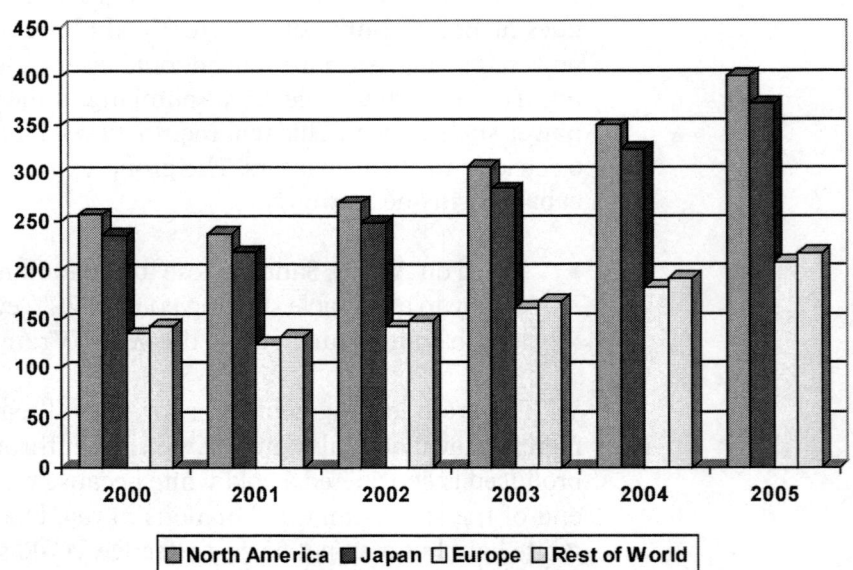

North America ■ Japan □ Europe □ Rest of World

Table 4.12 Worldwide Diode Laser Consumer Market by Region 2000–05 (US$ million)

	2000	2001	2002	2003	2004	2005
North America	257	238	270	307	350	401
Japan	236	219	249	284	325	372
Europe	135	125	142	161	183	209
Rest of world	143	132	149	169	192	218
Total	770	715	811	921	1050	1199

Table 4.13 World Diode Laser Consumer Market by Application 2000–05 (US$ million)

	2000	2001	2002	2003	2004	2005
FOL	179	161	190	223	262	308
CD	249	224	244	265	289	314
VCSEL	51	46	57	70	86	106
HPDL	34	32	40	50	62	77
Red	217	206	226	246	269	294
Violet	39	44	54	67	82	100
Total	770	715	811	921	1050	1199

For the purposes of this report, the consumer market is taken to include:

- Home entertainment (TVs, VCRs, games consoles, etc.).
- Domestic appliances (refrigerators, ovens, cleaning equipment, etc.).
- Domestic lighting (general illumination as well as new forms of specialist lighting).
- Displays and signs (information displays in shops, etc.).
- Toys and novelties (laser pointers, electronic pets, etc.).
- Others, e.g. home security.

In the future, this market will also include domestic data storage systems for entertainment, i.e. next-generation recordable DVD for domestic use, the so-called solid-state VCR.

- The first recordable DVD system for domestic use was demonstrated by Samsung in late 1999.

The coverage of the consumer application area for diode lasers continues with the visible diode laser market which is basically made up of the strong CD and DVD markets for music, movies and games, and other markets such as laser pointers.

4.6.1 Introduction to Visible Laser Diodes

In the market for diode lasers one of the most important areas is that of lasers that emit in the visible part of the spectrum. However, this market has seen major changes in the past two years. Laser pointers have fallen out of favour to be replaced by DVD as the dominant market application for visible diode lasers.

In its fairly short history — half a decade — two notable characteristics of the visible diode laser market are evident:

- Domination by laser pointers boosted the 1999 market value to well over US$0.5 billion. This spectacular market growth was quickly followed by a correspondingly rapid fall as the popularity of the novelty devices faded.
- The market for red lasers really took off as sales of DVD players, DVD-ROMs for PCs and games consoles achieved commercial success. These were the markets for which the visible laser technology was developed but it had been delayed by failure of the equipment companies to agree on a common industry standard.

In this analysis and forecast two factors are conspicuous:

- The dominant market for visible diode lasers is now DVD equipment; and
- The laser pointer has shrunk to only a few millions of dollars per annum.

Other key factors in the visible diode laser market are as follows:

- Continuous severe price pressure to reduce unit prices.
- Retro-compatibility for DVD equipment requires the incorporation of IR diode lasers in the optical pick-up which serves to boost the total market.
- Violet diode lasers (which are treated separately below) are the newest key growth area for visible lasers.
- Markets for other wavelengths (orange, green, etc.) have so far not taken off. Demand is presently met by pumped lasers.
- Existing markets such as barcode readers are likely to remain unspectacular and fairly stable.

This marketplace is becoming one of the most competitive in optoelectronics. At the time of writing, average selling prices are seeing a decline and manufacturers are hoping to be able to shift the emphasis of their product base to more stable lines such as diode laser-based barcode readers, levellers, gunsights, laser markers, etc.

For the purposes of this report the visible diode laser marketplace is conveniently divided by the principal wavelength regions, i.e. into red and violet types. Because of the importance of DVD to the consumer sector via movie players and games consoles, the red diode laser family is covered in this section. The violet diode laser will surely become important for the home entertainment (and other) markets in due course but at present its main application is in the industrial sector so it is covered under that application area.

Similarly, the market for diode lasers for CD equipment, although still very important for the consumer market, is covered in the computer application section of the report. The coverage presented in this section regarding DVD also alludes to the use of these devices for data storage although brief mention is made, where appropriate, in the computer application section.

4.6.2 Red Visible Lasers

The principal application for diode lasers operating in the red part of the spectrum is in DVD players and recorders. It also includes other applications such as barcode scanners, etc. The red diode laser family was worth US$621 million in 2000. Because of the aforementioned industry downturn, this fell by 5% to US$590 million in 2001. The red diode laser market is expected to return to positive growth from 2002 onwards increasing by an annual average rate of 9% to 2005. This is not one of the highest growth rates of the diode laser market and is the result of the steady maturing of the sector and the increased pricing pressure typical of this application. The value of this device family will increase to US$838 million by 2005. This means the sector will decrease in overall share from almost 11% of the total market in 2001 to just under 10% by 2005 as a result of broader industry application, especially with a resurgent telecommunications market. The market estimation and five-year forecast are summarized in Figure 4.8 and Tables 4.14 and 4.15.

Table 4.14 Worldwide Red Diode Laser Consumer Market by Region 2000–05 (US$ million)

	2000	2001	2002	2003	2004	2005
North America	204	194	211	230	251	274
Japan	188	179	196	215	236	259
Europe	137	130	141	154	167	182
Rest of world	92	87	95	104	113	123
Total	621	590	644	703	768	838

Figure 4.8 Worldwide Red Diode Laser Consumer Market by Region 2000–05 (US$ million)

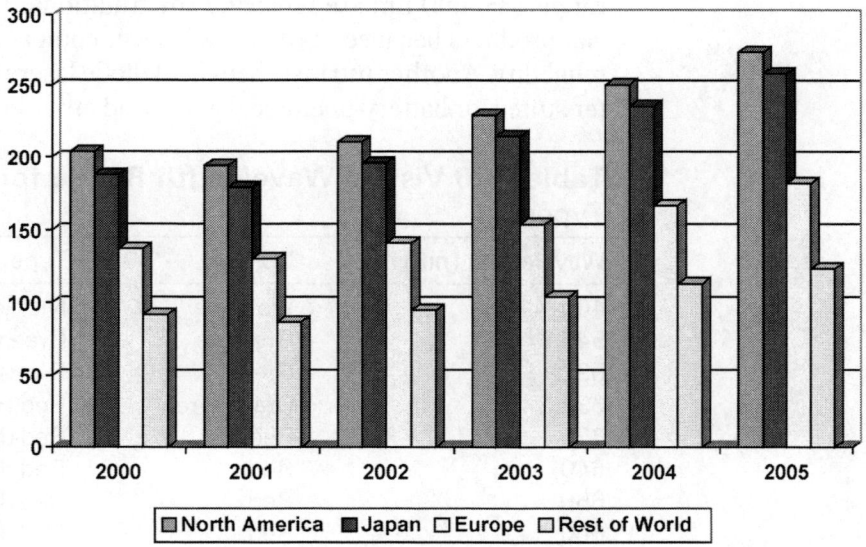

Table 4.15 Worldwide Red Diode Laser Consumer Market by Application 2000–05 (US$ million)

	2000	2001	2002	2003	2004	2005
Automotive	25	24	24	25	26	27
Computer	186	177	193	210	229	250
Consumer	217	206	226	246	269	294
Industrial	68	65	71	79	86	95
Military/aerospace	43	41	45	50	55	60
Telecommunications	56	53	58	64	71	78
Other	25	24	26	29	31	35
Total	621	590	644	703	768	838

With a typical 3–5 mW power, visible laser diodes emit at a wavelength in the 635–670 nm range. They are found in all modern red laser pointers, newer barcode scanners, laser light positioning devices and in DVD players/drives.

The first direct injection laser diodes (i.e. electrically pumped monolithic semiconductors), developed in the 1960s, were pulsed devices emitting at near-IR wavelengths (and possibly only with cryogenic cooling), around 750–800 nm. As the technology has matured, room-temperature CW laser diodes have become readily available and the range of wavelengths has expanded to include visible red (670 nm), orange-red (635 to 650 nm), and pushed further into the IR (up to about 2 μm).

Diode lasers continue to find new product applications as the wavelength is pushed further into the visible spectrum. The latest generation of visible diode lasers operate at or near 635 nm. This wavelength, being equivalent to a helium–neon (HeNe) gas laser, is highly visible to the human eye. VLDs in the range 635–690 nm are replacing the traditional HeNe laser in many commercial products because of their lower cost, compact size and superior long-term reliability. Another intrinsic benefit of diode lasers is that they are generally better suited for battery-operated devices and other low-voltage applications.

Table 4.16 Visible Wavelength Range for Diode and Other Laser Types

Wavelength (nm)	Colour	Type
400–415	Violet	Nichia violet diode laser
543.5	Green	Green HeNe laser
532	Green	Diode-pumped solid-state laser
632.8	Orange-red	Red HeNe laser
635	Red	Red diode laser
640	Red	Red diode laser
650	Red	Red diode laser
660	Red	Red diode laser
670	Red	Red diode laser

4.6.2.1 CD/DVD

Visible diode lasers are used extensively in commercial products. Common visible diode lasers have a maximum optical output power of 3–5 mW. Due to the sensitivity curve of the human eye, a wavelength of 635 nm (see Table 5.4) appears at least four times brighter than an equivalent power level at 670 nm. Thus, shorter wavelength laser diodes will be best where maximum visibility is important. However, these are currently much more expensive — but this will change as DVD technology continues to take off.

For red diode lasers, at 680 nm, there are problems with internal power loss in the laser facets as the power level increases. This becomes a positive feedback situation with the energy causing more damage, which causes more absorption and so on. Researchers have proposed various solutions to overcoming this problem, for example:

- Toshiba has been mass producing visible laser diodes for some time. The Japanese company recorded a world first with the TOLD9421 LD, which was the first 5 mW LD emitting at 650 nm.
- Targeted at the replacement market for HeNe gas lasers, the TOLD9521 LD emits 3 mW of optical power at 635 nm for portable barcode scanners, pointers and alignment equipment.
- In Rohm Electronics' portfolio of optoelectronic devices is a range of visible red laser diodes suitable for DVD-ROM and laser pointers. These devices are based on InGaAlP double heterojunction structures and combine high-accuracy operation with minimal light dispersion.

It is worth noting that such devices usually incorporate a photodiode for monitoring and control of output power.

DVD is one of the most important new applications for visible (red, i.e. 630–650 nm) diode lasers. This is a competitive market sector dominated by Japanese suppliers and one that promises to become at least as big as the CD-ROM market in the course of the next five to ten years.

The present structure of the laser diode data storage market by wavelength and power can be broken down as follows:

- CD-ROM: 780 nm at 5 mW.
- CDR-R/-RW/MD: 780 nm at 30 mW.
- DVD/DVD-ROM: 650 nm at 5 mW.

The CD-ROM represents over three-quarters of the market for data storage market applications. This is, however, a fairly mature market but one that could continue. Newer variants such as the re-writable CD-R and CD-RW formats have reinvigorated a market that had been saturated with approaching full penetration of the computer data storage market and music reproduction markets.

It is not unlikely that the basic 780 nm, 5 mW CD-ROM diode laser which retails at around one dollar per optical pick-up will endure for the remainder of the period covered by this report. Certainly, it is likely that the CD-ROM will constitute at least half the market by 2003 owing to its very competitive cost structure (for hardware and discs) for software and music applications. In addition there is the matter of a huge user base which will be reluctant to discard the many CDs and CD-ROMs it has accumulated.

- Whilst these markets have recently been adversely affected by the general downturn in the computer and consumer electronics industry, growth is irresistible. The introduction of the DVD will be one of the factors that will continue this upward growth for diode lasers.

With the exception of visible laser pointer products, most laser diodes will continue to be exported from Japan, where, owing to highly automated processes, manufacturing costs are at the minimum. There will continue to be a sizeable market while OEMs leverage the laser diode for new products or new applications of established products. The laser diode has few competitors due to its unique combination of characteristics.

DVD will grow to dominate the market by mid-decade. This is due to factors such as the switch from VHS videotape playback to DVD movies and the switch from CD-ROM to DVD for the PlayStation2. A similar transition is also underway for office and home PCs.

Increasingly, standard specification PCs include DVD-ROM rather than CD-ROM even though software title availability is presently limited. But even by this time the total market will still be dominated by the various CD formats. This derives from the CD-R's ease of use for archiving, data back-up and music recording.

- It will be some years before the shift to DVD is complete and this will be conditional on the commercial success of recordable DVD formats. Of course, DVD will be reverse-compatible so as to handle the majority of older CD-ROMs.
- Sony does not expect that US sales of its PlayStation2 console will affect sales of CD players, as has happened in Japan. The consoles play game discs and DVD movies, a major selling point in Sony's home country, where DVD players still carry premium prices. More than 50% of Japanese consumers over the age of 30 who bought a PlayStation2 did so primarily because of its DVD capability.

Data storage in the form of the CD-ROM and the CD player have been amongst the biggest successes for optoelectronics components in the past decade. These sectors represent a very large market and look set to continue to do so for the next five years. Akin to the telecommunications market, this market is driven by an insatiable demand for data handling. The advent of truly multimedia data processing demands many gigabytes of data storage in as small a space as possible which can be accessed as quickly as possible. CDs fulfil this role admirably but would not have been developed were it not for the semiconductor laser diode.

- Panasonic Disc Services Corporation (PDSC), a subsidiary of Matsushita Electric Industrial, and Universal Music & Video Distribution, a division of Universal Music Group, formed a joint venture company to manufacture DVD optical discs and music CDs in North America.

The next phase of the expansion of the DVD market will be the recordable format which is already beginning to become commercially significant in key areas worldwide: Such is this success that companies like Panasonic have had to double US DVD production capacity to 5 million units per month.

Hewlett-Packard has announced the first DVD+ReWritable drive (DVD+RW), the HP DVD Writer 3100i drive, which reads and writes to DVD+RW discs that have 3 Gb of storage capacity, equivalent to 100 minutes of high-quality digital video. In addition to DVD+RW media, the HP DVD Writer 3100i can read DVD-ROM, DVD movie, CD-RW, CD-R, CD-ROM and CD audio.

Unlike other rewritable DVD formats, such as DVD-RAM, DVD+RW drives use disc media similar to CDs. But DVD-RAMs require a media cartridge and so have limited compatibility. For example, they cannot be read in a DVD-ROM drive unless the drive is altered to accept a cartridge. Nevertheless, based on capacity per disc, DVD+RW media provide users with a huge cost benefit over other storage alternatives.

- At approximately US$30 per DVD+RW disc, the cost per megabyte is less than a few cents. DVD+RW offers high capacity with fast access time.

Rewritable DVD technology could become the universal rewritable storage technology for the desktop. Industry observers expect the number of DVD-ROM drives sold to have tripled since 1998 to nearly 20 million per annum. With the addition of its DVD+RW product category and as the DVD-ROM category grows, HP expects to see large amounts of growth in the DVD+RW market.

Unlike the CD-ROM, DVD has from the outset been available in dual format, i.e. read and write. Much use is anticipated for DVD-RAM and in due course for very high-capacity data back-up and archiving (e.g. multimedia) write once and rewritable DVD. So far, rewritable DVD take-up has been fairly limited largely because of the higher prices of the writer as well as the scarcity of blank DVD disks.

The DVD Forum — which ratifies the various DVD specifications — plans to create a logo called DVD Multi, which will identify products compatible with virtually all the official DVD formats. The plan will cover both consumer electronics and computer equipment, and aims to diminish consumer confusion and uncertainty.

In recordable DVDs, Pioneer's DVD-RW format has the support of most of the Japanese players. DVD Multi Recorders will write on DVD-RAM, DVD-RW and DVD-R discs. Hitachi, Toshiba, Panasonic and Samsung plan to launch decks using DVD-RAM discs for video recording. The 4.7 Gb single-sided disc medium is comparable to that for PCs, but store two hours of high-quality video.

The latest, second-generation rewritable discs will be playable in most existing DVD-ROM, DVD players and the new RAM-capable decks. Despite the advance of DVD in computers and home entertainment, Sony envisages a use for higher-capacity CDs, especially since computers can now handle bigger audio, video and still image files. Because of the projected lower running cost, the discs may work out slightly cheaper than other rewritable optical disc formats such as DVD-RAM.

But the data storage market is seeing more diversity than at any time in its history. Existing optical formats have a huge user base and are likely to see off all contenders such as solid-state flash memories and hard disk drives. It is not unreasonable to anticipate an all-optical home entertainment system by the end of the decade. Similar developments are just around the corner for business applications of data storage.

In the data storage market, optical (i.e. CD/DVD) represents the best value in terms of cost per megabyte, e.g. MP3, the popular compressed audio format that enables music to be downloaded from the Internet in a reasonable time. These tracks can then be played on portable pocket-sized units. The audio is recorded on solid-state, CMOS-based flash memories as used for digital cameras, PDAs, etc. These are convenient for re-recording but are expensive.

- However, MP3-only players may prove to be a short-lived fashion accessory. The CD-ROM is already moving into this territory with enhanced portable CD players that can playback MP3 as well as normal music files. Each CD can hold 650 Mb compared with 64 Mb of a flash card and is one-tenth of the price.

4.6.2.2 Laser Pointers

As previously mentioned, the laser pointer is a product that has seen considerable market success which has subsequently diminished in importance in favour of the DVD.

Historically, the laser pointer diode laser came about largely as a result of a breakthrough in epitaxial growth technology.

- These devices would not be possible to produce at such a low price without quaternary compound semiconductor alloys and complex device structure technology produced by new-generation processes.

This technology came at a time when the first truly multiwafer mass production epitaxial reactors were up and running. The owners of these machines were anticipating a market boom for other products, the DVD in particular. However, the commercial introduction of this product was delayed by reasons other than technical — basically it was about the inability of various interested parties to reach a unanimous decision on international standards. The equipment and processes were looking for a market and found it through judicious marketing of a new novelty, the laser pointer. Within a short time these machines meant for mass producing DVD laser diodes were switched to manufacturing laser pointer diode lasers.

With the present resurgence of interest in the DVD market, laser pointers have served their usefulness and the machines have been switched back to their original purpose. So perhaps it is of less interest to these companies as to the fate of the laser pointer. They served their purpose at the time and no doubt saved many companies into the bargain. The learning curve for the mass production of the critical component in any DVD system has been experienced and much learned which has made DVD visible red laser diodes an even better commercial product and ensured the further success of this entertainment and data storage system.

What was previously one of the most spectacular market successes for visible laser diodes, the laser pointer (635–670 nm) has seen its market value peak and fall. Growing out of a fairly limited market for study presentation aids to become a truly mass-market product in the space of just a few years, they became a fashion accessory and then a victim of safety issues and then disfavour in the novelty marketplace.

While the commercial success of the laser pointer was due to dictates of consumerism it also had much to do with technology, i.e. epitaxial growth technology based on MOVPE whereby highly complex multi-layer device structures based on quaternary semiconductor alloys can be very uniformly and reproducibly prepared over large-area wafers.

- These techniques are in large-scale use with manufacturers in Taiwan and South Korea (rather than Japan, the origin of most other diode lasers) where most of the 20 million plus per annum laser pointers were manufactured.

However, more versatile versions have been coming onto the serious market, which could sustain the market, albeit at a lower level, for a few more years. These have more colours and special effects and retail at even lower prices. As a result this marketplace has become one of the most competitive in optoelectronics. At the time of writing, ASPs were seeing a steep decline and many manufacturers were hoping to be able to shift the emphasis of their product base to more stable lines such as laser diode-based barcode readers, rangefinders, etc.

The laser diode has a 3–5 mW maximum output at 670 nm (deep red) while later ones operate in the 650–635 nm range (red to orange-red, which appears many times brighter to the human eye).

- Early laser pointers had the laser diode in its own 5 or 9 mm can package mounted (probably press-fit) in a metal casting which in addition to holding the optics acted as a heat sink. In an effort to reduce costs, the newest ones have the bare laser diode chip mounted directly to a heat sink.

Red laser pointers are by far the most common and are now quite inexpensive. Prices of less than US$5 are common and are falling further. However, except for various shades of red (depending on wavelength), all other colours are expensive. In fact, there is really only one other colour of any practical consequence — green, which is produced by a different type of laser from the simple diode lasers used in red laser pointers.

- Currently, nearly all green (532 nm) laser pointers are based on diode-pumped solid-state frequency-doubled (DPSSFD) laser technology.
- The DPSSFD laser is based on the intracavity frequency doubling a Nd:YVO4 (vanadate) chip using a KTP crystal inside the laser cavity.

Presently the prices for green pointers are much higher (averaging about US$300). As for other colours there is little need for anything beyond green since its wavelength is near the peak (555 nm) of the human eye response curve. A blue pointer might cost US$2000 for one using DPSSFD technology but it is not as bright as a US$5 red pointer. There are no yellow or orange laser diodes or practical DPSSFD alternatives. Violet — which would be really hard to see — could use the Nichia violet (400–415 nm) laser diodes but they cost around US$1000. Presently, there is no technology capable of producing a variable colour laser pointer.

4.6.3 Violet Visible Lasers

Violet visible lasers are one of the few diode laser device types to show positive growth in the period 2001–02, of 13%. Thereafter when the industry picks up again the growth rate will be at least 22.6%. However, it is likely that this growth will turn out to have been conservative. Much will depend on the successful commercialization of the solid-state digital VCR.

When economical, the shortest wavelength laser diodes — those emitting violet light — will represent the enabling technology for yet another revolution in the storage capacity of optical drives (at least a factor of two better than DVD). Compared with the 4.5 Gbyte capacity of one surface, one side of a DVD, a DUD (Digital Ultra Disk) drive would hold about 13 Gbytes based on the wavelength difference alone.

At present violet lasers are proving popular in low-volume/high-price market sectors such as instrumentation and printing. The first high-volume/low-cost application is likely to be the solid-state optical-based digital video recorder to replace the tape-based VHS VCR.

Violet laser diodes operate at an emission wavelength of 400 nm with an output power of 5 mW. This is achieved with an operating current of 40 mA at a voltage of 5 V. The lifetime of the laser diodes is more than 10 000 hours at room temperature.

Violet laser diodes are good news for next-generation optical storage (beyond DVD) and high-resolution laser printers, but those wanting highly visible wavelengths (e.g. 555 nm green and full-colour displays) may have to wait a little longer.

Shorter-wavelength laser diodes should also find applications in:

- Higher-resolution laser printers and similar devices;
- Underwater communications where blue wavelengths would be ideal; and
- Compact full-colour displays with the addition of green laser diodes.

These laser diodes are being sampled by equipment companies to design next-generation super-high-density data storage products such as CD-ROM and DVD. Other prime applications for these devices are a non-tape VCRs for domestic use as well as office laser printers and information display units.

At present, semiconductor devices, e.g. LEDs, find their principal application in the area of illumination — e.g. for large outdoor displays for sporting events. The use of a laser to provide general illumination is a novel and intriguing consideration. The principle of conversion differs little from that used for white LEDs, i.e. a down-conversion phosphor is required. UV light emitted from the VCSEL is 'stepped down' to longer wavelengths. The net result is a broad-spectrum emission that appears as white light to the human eye.

- These UV VCSELs are also of great interest for other applications, as they are a compact solid-state alternative to the conventional gas-discharge tube.

The next five years will see substantial R&D activity in this application sector. However, this growth will see average prices remaining under severe pressure in a competitive environment.

4.7 Computer Markets for Diode Lasers

The computer sector is a diverse application marketplace for diode lasers but it is dominated by the need for diode lasers for optical data storage. The value of this market sector was US$646 million in 2000 but because of the industry downturn this fell by over 7% to US$598 million in 2001. This market is expected to return to positive growth from 2002 onwards increasing by an annual average rate of nearly 13% up to 2005. The value of this important device family will therefore increase to over US$972 billion by 2005. This market data are summarized in Figure 4.9 and Tables 4.17 and 4.18.

Figure 4.9 Worldwide Diode Laser Computer Market by Region 2000–05 (US$ million)

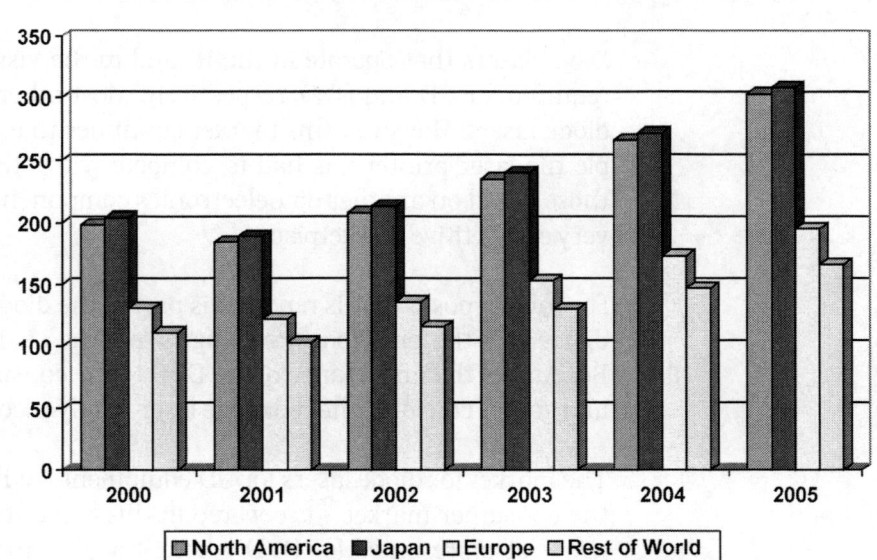

Table 4.17 Worldwide Diode Laser Computer Market by Region 2000–05 (US$ million)

	2000	2001	2002	2003	2004	2005
North America	198	184	208	235	266	303
Japan	204	189	213	240	272	308
Europe	131	122	136	153	173	195
Rest of world	111	103	116	130	147	166
Total	645	598	672	758	857	972

Table 4.18 Worldwide Diode Laser Computer Market by Device Type 2000–05 (US$ million)

	2000	2001	2002	2003	2004	2005
FOL	72	65	76	89	105	123
CD	272	245	267	290	316	344
VCSEL	63	57	70	86	106	130
HPDL	23	22	27	33	41	51
Red	186	177	193	210	229	250
Violet	29	33	40	49	61	74
Total	645	598	672	758	857	972

4.7.1 Introduction to Diode Lasers for the Computer Market

For the computer market, diode lasers are important for two principal application sub-segments:

- Optical data storage, i.e. CD and DVD for reading and writing of data; and
- Other applications which include laser printers, networks, wireless data links, some of which overlap with the telecommunication sector.

Diode lasers that operate in the IR and in the visible part of the spectrum are required for CD and DVD, respectively. Most other applications also require IR diode lasers. However, this market continues to experience changes, for example the laser printer has had to compete with other types such as ink-jet and those based on another optoelectronics component, the LED, which makes for a very competitive marketplace.

For the purposes of this report this part of the diode laser marketplace has been divided by the principal wavelength regions, i.e. into IR, red and violet types. Because of the importance of the DVD to the consumer sector via movie players and games consoles, the red diode laser family is covered in that section.

The market for diode lasers for CD equipment – which is still very important for the consumer market – is covered in this computer application section. Coverage is also presented for DVD where it applies to use of these devices for data storage although other mention is made, where appropriate, in the consumer application section.

4.7.2 CD Diode Lasers

The compact disc laser (CDL) family is amongst one of the more mature but nonetheless interesting in today's laser marketplace. The value of this market sector was US$689 million in 2000. This is will decrease because of the afore-mentioned industry downturn and other factors such as pricing and reduced demand as the computer industry switches over to DVD. Its market value decreased by 10% to US$620 million in 2001. The CDL market is expected to return to positive growth from 2002 onwards, increasing by an annual average rate of 8% up to 2005. This is one of the lower growth rates of the diode laser market and would have been lower if not for the recordable CD (CD-R) segment. But across the board, the sector is under considerable pricing pressure even though volume shipments have been increasing at more than triple this rate. The value of this important device family will therefore increase to US$850 billion by 2005. However, the sector will decrease its overall share from over 12% of the total market in 2001 to just under 10% by 2005 as a result of price pressure. This market data are summarized in Figure 4.10 and Tables 4.19 and 4.20.

Table 4.19 Worldwide CD Diode Laser Market by Region 2000–05 (US$ million)

	2000	2001	2002	2003	2004	2005
North America	220	198	215	232	251	272
Japan	207	187	202	219	238	258
Europe	132	118	127	136	146	157
Rest of world	130	117	127	138	150	163
Total	689	620	671	726	785	850

Figure 4.10 Worldwide Diode Laser Computer Market by Region 2000–05 (US$ million)

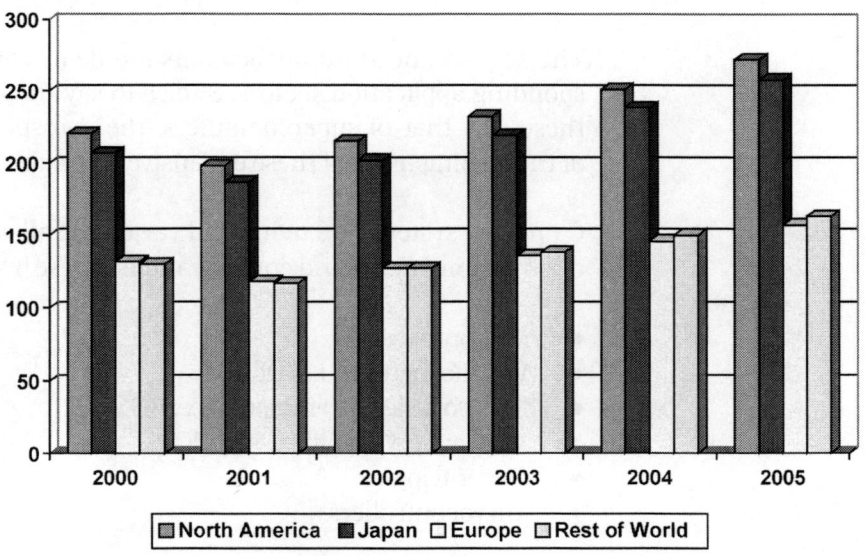

Table 4.20 World CD Diode Laser Market by Application 2000–05 (US$ million)

	2000	2001	2002	2003	2004	2005
Automotive	10	9	10	10	11	11
Computer	272	245	267	290	316	344
Consumer	249	224	244	265	289	314
Industrial	62	56	59	62	66	69
Military/aerospace	55	50	52	55	57	60
Telecommunications	13	12	13	14	15	16
Other	28	25	27	29	32	35
Total	689	620	671	726	785	850

The borderlines between the computer sector and that of others such as consumer and telecommunications are becoming less clear-cut. As noted elsewhere, optical data storage is popular for music and video for the consumer sector as well as for professional and leisure software in the computer sector. Also, the ever-increasing demands for multimedia via the Internet and the ability to interact with other computer users is blurring the lines between the computer sector and that of telecommunications. All three are moving towards a truly seamless system where the computer could form the centrepiece of a home/business information processing system By information it is meant everything from on-line banking to music and movie entertainment and appliance control.

Computer equipment that incorporates one or more devices tends to be restricted to a few basic areas:

- Supercomputers.
- Specialist applications such as defence/aerospace.
- Data communications.

The second and third applications are dealt with in more detail in the corresponding application sectors. Suffice to say that there is some overlap between these, e.g. that of supercomputers, the aerospace/defence industry being one of the leading users of these expensive systems.

Computer systems are many and varied but all comprise a basic combination of circuit components and configurations. These building blocks are as follows:

- Microprocessor.
- Volatile memory, i.e. DRAM.
- Non-volatile (NV) memory, i.e. ROM.
- Application-specific ICs (ASICs).
- General logic.
- Microcontrollers.

Collectively, these device families make up more than 90% of the worldwide semiconductor device market and are predominantly based on silicon circuits or variants thereof. Given lack of market demand and paucity of suppliers, this situation is likely to persist for the five-year forecast period covered by this report.

In today's military circles, very high-speed data processing is becoming ever more important. For example, modern warfare has become more reliant on reconnaissance in real time. Processing of signals at very high speed is also required for the processing of radar signals from phased array radars. Interception of short- to medium-range ballistic missiles such as Scuds is also high on the agenda of many armed forces around the world.

One of the most important segments for laser diodes is that for data storage. Taken in total with the closely affiliated consumer applications, i.e. music CDs, DVDs and games consoles, this represents the largest collective market for laser diodes.

The computer data storage market has recently expanded from CD-ROM to include DVD-ROM and writable CDs, i.e. CD-R (write-once) and CD-RW (read/write many times). It is thus a dynamic and expanding market that will remain so for the duration of this forecast and beyond. Recordable DVD has yet to make a comparably widespread commercial impact. Once formatting and other matters have been fully sorted out, this will provide yet another step function in the demand for diode laser optoelectronic components.

The CD-ROM is for now the dominant data storage medium. This applies to the provision of software as well as the backing up and archiving of data.

- This market sector has also been boosted by the leisure side of computing. A PC now enables the small-scale manufacturer of music (and other CDs) for private or public entertainment.

A key factor in the success and likely continued dominance of the CD format has been the availability of very low-cost blank discs. This represents by far the cheapest archiving medium in terms of dollars per megabyte. It has enabled the CD-ROM to more or less see off contenders for archival formats such as the Zip Drive.

- Moreover, the CD-ROM is not standing still, with numerous new variants having been proposed. This coupled with appropriate compression software has greatly expanded the capability of this robust medium.

However, one clear trend has eroded the CD-ROM market as regards PC retail. It is now more likely that a DVD-ROM drive will be included in the US$1000 home/office PC or laptop. Because this reads CD-ROMs or DVD-ROMs with equal ease, a CD-ROM is no longer required. The market has grown and brought about a dramatic reduction of the unit cost of the DVD drives.

This has further served to enhance the DVD's penetration and displacement of the CD-ROM market. Therefore, today's US$1000 PC is likely to include a CD drive only as a writer. The PC thus requires two optical pick-ups and provides market demand for a pair of diode lasers and their accompanying detectors.

- The move from CD-ROM to DVD-ROM has necessitated new diode laser technology. This is basically a shift in wavelength and also a higher power so as to provide the read/write function.

There are a number of challenges facing the business of laser diode supply. In the past few years, diode lasers for telecommunications applications have eclipsed those for data storage by market value (though not by volume of devices shipped).

The present structure of the laser diode data storage market by wavelength and power can be broken down as follows:

- CD-ROM: 780 nm at 5 mW.
- CDR-R/-RW/MD: 780 nm at 30 mW.
- DVD/DVD-ROM: 650 nm at 5 mW.

Philips says that the number of audio CD recorders sold around the world since October 1997 by Philips, Marantz and Pioneer had reached 1 million units by June 2000. It is expected that new companies who are backing audio CD will accelerate market growth even further, with predictions suggesting that hardware sales exceeded 3 million by the end of 2000.

The audio CD continues to grow in popularity, with around 800 million CD players in the world, and more people every day extending their access to the format via portable players and in-car systems according to companies like Philips Business Group Audio. Philips Electronics recently unveiled plans to increase manufacturing capacity of CD-RW products to exceed 1 million units per month by the middle of 2002. The increased output will be accomplished through further output growth in its production facilities in Hungary and through additional expansion in Asia Pacific.

- The market for CD-RW drives was projected to grow to 35 million drives by the end of 2001.

The CD-RW platform was developed and pioneered by a group of six companies, Hewlett Packard, Mitsubishi Chemical, Philips, Ricoh, Sony and Yamaha.

Industry observers predict that CD-RW, or its follow-on combination CD-RW/DVD-ROM drive, will be the dominant high-capacity rewritable removable storage device in PCs within a few years.

- The industry has an installed base of more than 1 billion CD players (275 million CD-ROM drives and 800 million audio CD players). Nearly 3 billion CD-R and CD-RW media were expected to reach consumers in 2000.

However, just to confuse things further, music buyers will have another format choice by the end of 2001. Sony will introduce in America its SA (Super Audio) CD player, already available in Japan. The format promises better audio quality with nearly twice as much data on a disc. Players will be in the US$2000 price range.

Sony also has plans for a double-density format for CD-ROM/CD-R and CD-RW discs, boosting capacity from 650 Mb to 1.3 Gb on a single 120 mm disc. Sony claims manufacturers can employ their current CD technologies and production facilities to manufacture double-density CDs (DD CDs) at a low running cost.

Networks are on the verge of another computer revolution with the anticipated introduction of low-cost interconnections such as Bluetooth. At present virtually all networks are based on copper wire or fibre optic links. There is already a small but useful market for wireless links, e.g. for exchange of data with laptops, etc. These use optoelectronic devices and are therefore line-of-sight. Also short range, Bluetooth is less directional and one component (if so equipped) automatically communicates with all others within range. This development is poised to become important throughout the computer industry. It is, however, not likely to provide any major market boost for optoelectronic components. It relies on microwave frequencies and will supplant all other approaches for low- to medium-data-rate applications. Bluetooth will thus have a negative effect on optoelectronics-based wireless links such as those between a computer and printer.

- There are a number of question marks concerning the widespread market penetration of Bluetooth. Industry observers are confident of it happening; however at the time of writing, it is still not known when it will happen.

Over the longer term, probably not until the end of the forecast period and beyond, the diode laser will see some commercial application in all-optical computers. There are distinct performance advantages, not the least being very high processing without conversion from optical to electronic and back again many times within the computer unit.

Another key longer-term area of commercial impact for diode lasers is in optical data storage using holographic techniques. These have been around for several decades but have yet to be applied in the general archival data industry sector. Potentially, holographic techniques have many orders of magnitude better data storage and retrieval over the CD- or DVD-based approaches but as yet, amongst other disadvantages, they are too bulky.

4.7.3 Higher Density Optical Data Storage

There is at present no requirement for the higher data storage capacity obtainable from shorter-wavelength, i.e. blue-green, laser diodes. It is anticipated that such devices may start to become available from the middle of the forecast period and onwards. However, this will be only for high-end computer workstations and similar high-specification 'price is no object' computer systems such as supercomputers for defence projects. It is likely that this period could be longer for two principal reasons: lack of demand and unavailability of the requisite disc technology.

At present, the marketplace for DVD based on red lasers has yet to saturate and is likely to satisfy demand for at least five more years. As yet, there is some commercially available software based on DVD although it is foreseen that this will change once the customer base of DVD-based PCs has become commercially important. Providers of software and hardware to this sector are unlikely to be willing or able to launch new products based on blue laser diodes for some time to come. They have made large investments in the present generation of DVD products and seek to gain a return on this before embarking on the next generation of products.

There are various technical reasons as to why the blue laser diode DVD will be some way off. These include development of the associated disc: present discs are optimized for longer-wavelength light.

- Blue laser DVD will require a different technology that will take several years to develop. Since the availability of the first blue semiconductor diode laser in 1999, manufacturers of discs have been able to start the necessary R&D which will lead to the optimized DVD disc product. This will take at least three years but is likely to take longer given the present low level of demand.

The basic CD-ROM market has focused on ever-faster read rates with units now available at better than 20 times read rate. Step by step these units become the industry standard, only to be displaced by faster models. But this will soon approach a limit and does not, of course, do anything about the present storage limit. Key to the on-going evolution of this market is improved performance optical pick-ups based on laser diodes coupled with comparable development in electronic and mechanical components.

Comparable technical achievements have led to the market introduction and on-going success of recordable and re-writable CD-based data storage for computers. These require different types of laser-based optical pick-ups. Basically, two lasers have to be integrated into the pick-up, one to write and one to read. The successful development and commercial mass production of these units are the keys to the availability of competitively priced CD writer products.

The basic requirement that had to be met was integration of a basic read laser diode with a higher-power one to write (and re-write) the data onto the special type of blank CD-ROM. This has been achieved and factories set up to mass produce optical pick-ups have been built based on special highly automated machines. This technology gave a boost to the laser diode industry; not only for the newer higher-power IR diode lasers but also for the basic unit for the read function. This sector is now set to follow the CD-ROM market with faster speeds and lower prices. However, it is also constrained by the ultimate data storage capacity of the CD.

- While acceptable for data archiving and software dissemination — and to a growing extent for copying of music CDs for personal use and home recording — it falls well short of full-motion video presentations of any duration.

It is likely that the CD-R and CD-RW market will remain strong for the period covered by this report. Although towards the end of the five-year forecast period it is expected that DVD will have significantly displaced CD-ROM (but not CD-R) in higher-end computer applications. This is not as a result of technical factors — such as optoelectronic component development — but rather due to the large user base and other factors. The momentum of the CD format is sufficient to sustain it against inroads from competing data storage formats.

- For example, by 2000 CD-R had already begun to replace magnetic media such as the Zip Drive or variants of the 3.5-in floppy. The very cost-competitive pricing of the blank CD-R (and to a lesser extent CD-RW) discs ensured this shift.

CD-ROM-based products are also reverse compatible which made it possible for computer OEMs to offer a single data storage unit in home and office PCs beginning in 1999. The DVD drive is thereby able to play back (though not record) either DVD (e.g. movies) or CD-ROM (e.g. software or music) as required.

At the time of writing, DVD is achieving good market penetration in the home entertainment sector. It is also beginning to achieve some presence as a data storage medium in PCs but is suffering from public unfamiliarity and lack of commitment from software vendors. Most computer utility or leisure software is only available on CD-ROM. Companies are reluctant to move up to DVD until a more substantial user base has become established. Industry observers also noted that in the competitive software publishing arena an additional format was undesirable. This situation looks set to remain for another two years.

While DVD has some commonality with CD-ROM, it is based on a shorter-wavelength laser diode (in conjunction with special data compression software) and so is be able to provide many gigabytes of storage capacity. This has required design and tooling up for an entirely new type of optical pick-up. The laser diode (and to a lesser extent also the photodiode) has moved from the IR (650 nm) to the visible red (610 nm) wavelength region. Such devices were already in commercial use for such applications as barcode readers but extensive engineering was required before a competitively priced optical pick-up could be mass produced. It is not unreasonable to expect that a similar route will be taken by the violet laser diode as it moves towards commercial realization in the computer market. As DVD and other high-density storage media become more widespread, violet diode lasers have gained increasing interest for their minimized focusing spot that is necessary to realize the data densities required of new media.

Like the CD-ROM, DVD is also available in dual format, i.e. read and write. Much use is anticipated for DVD-RAM and in due course of very high-capacity data back-up and archiving (e.g. multimedia) write once and re-writable DVD.

- Rewritable DVD take-up has been fairly limited not only due to the lack of agreement on industry standards — commercial equipment has been launched but there are multiple approaches such as DVD-RAM, DVD-R, etc., each with its own advantages and drawbacks.

The recordable DVD systems will in due course be sorted out and a standard emerge. Once these are commercially available they will follow the traditional downward price trend of forebears such as CD-R and DVD-R.

In the next five years it will be a combination of factors that will have to progress to a certain point before a data storage format can supplant its predecessor. It is vitally important to have the requisite laser diode component. The first samples of such a device have only just become available but are not yet optimized for an application such as PC-DVD. This will take several years to reach fruition. However, even this does not guarantee the success of a new storage medium.

The development of the violet diode laser came at a time when the data storage industry was finding it difficult to agree on an industry DVD standard. As a result, it was regarded by many as a fine technical success but with little to contribute for a few more years. Indeed, with DVD only just beginning to find favour, it is unlikely that these lasers will see significant market success in the immediate future as far as data storage is concerned.

There is one factor that could accelerate this process: demand from consumers for a digital VCR based on disc rather than tape (i.e. VHS). Sales of DVD have been strong but in many cases the large user base of conventional video recording has set a precedent for a similar system of record and playback from a disc-based system. If the producers of such equipment develop such units then they will also be adopted for use in computers. The aim would be to develop a manufacturing process optimized so as to make the recordable DVD units as cheaply as possible.

- Merging the consumer and computer market would effectively double the available market or halve the time to acceptable critical mass. Availability of such a large-volume market would accelerate the development of competitively priced DVD units based on violet lasers.

The method of development and manufacture of CD-ROM laser optical pick-ups is also important. These are almost exclusively done in-house by the large Japanese companies such as Sony and Hitachi. This is necessary given the high level of engineering required in the task to make them so cheap. It is not unlikely to expect that this procedure will also be adopted for any DVD system based on blue diode lasers. Herein lies a problem insofar as the only maker of such diode lasers at the moment, Nichia, does not have any intention of either making this technology more generally available or of entering the DVD market itself.

The makers of these systems will therefore have to develop their own route to the blue diode laser. This is likely to be already underway but has to run the gauntlet of avoiding the infringement of patents with Nichia. The company has shown it can be done but has cornered this market. It is at present unknown how wide the window is for the practical realization of such devices by a non-Nichia route. Upon this will depend many things, not least the timing and size of the market of the violet diode laser-based DVD for computers and other applications. An example of this commercial line of development comes from a Japanese company well known for its CD- and DVD-based products using diode lasers, Matsushita Electric Industrial

In September 2001 Matsushita Electric Industrial announced it had developed the world's first high-power, high-efficiency, second-harmonic generation (SHG) blue laser ready for use in next-generation optical disc recording systems. The laser achieves a high power level of 30 mW, 15 times greater than previous SHG lasers, whose 2 mW has not been enough for disc recording. Moreover, the new laser produces a wavelength of 410 nm, shorter than the previous 425 nm, considered too long for next-generation optical disc systems. The higher power and shorter wavelength were made possible by using a high-powered wavelength-tunable IR laser diode and then transforming the IR light into blue laser light.

4.7.4 Laser Printers

Another key application area for the diode laser is in the area of printing. The office and home market is very competitive, with laser printers having a proportionately higher unit price than other types such as the ink-jet and LED-based printers.

- Modern laser printers use laser diodes producing anywhere from 5 to 50 mW and beyond depending on their resolution and speed (pages per minute). High-resolution laser imagers, typesetters and plotters may use laser diodes producing 150 mW or more.

Available in two basic formats, colour and monochrome, the laser printer is favoured for the office environment but manufacturers with an eye on the larger low-end market are beginning to offer suitably configured competing machines. For example:

- In the summer of 2001 Hewlett-Packard, one of the pioneers of the laser printer, introduced the HP LaserJet 1000 printer. Since it was on sale for a retail price of US$249, the HP LaserJet 1000 printer was claimed to be the lowest-priced monochrome HP LaserJet printer ever introduced.

According to Lyra Research, sales of personal laser printers are increasing. Analysts forecast that the estimated revenue potential for the personal laser market was likely to exceed US$3.4 billion worldwide in 2001. Moreover, during the next three to five years, analysts predict continued robust growth for this market segment.

Observers of this market previously predicted that inkjet technology would make personal laser printers obsolete. However, with the growing number of small and medium businesses, not only have the sales of inkjet printers continued to rise, there is continued demand for high-quality, monochrome printing as black output continues to be the staple for office printing.

4.8 Industrial Markets for Diode Lasers

The market value of diode laser devices in the industrial sector was estimated to be US$441 million in 2000. This will drop to US$420 million by 2002. The growth in the period 2001–02 is likely to fall by 4.6% as a result of the industry downturn. For the period 2002–05 this will transform into positive growth of 17.8% per annum, reaching US$815 million by 2005. As a result the industrial market sector will have reached a value of US$815 million by 2005. The market estimate and forecast are summarized in Figure 4.11 and Tables 4.21 and 4.22.

Figure 4.11 Worldwide Diode Laser Industrial Market by Region 2000–05 (US$ million)

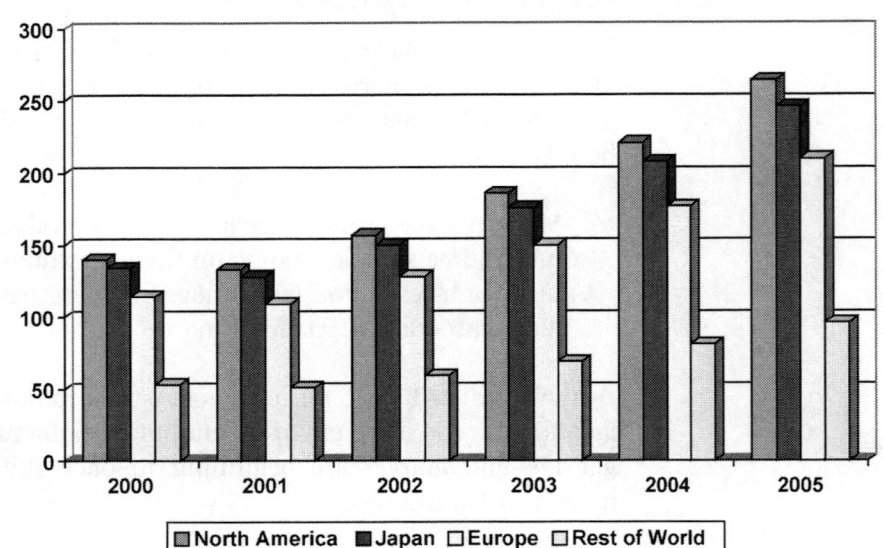

■North America ■Japan □Europe □Rest of World

Table 4.21 Worldwide Diode Laser Industrial Market by Region 2000–05 (US$ million)

	2000	2001	2002	2003	2004	2005
North America	140	133	157	186	221	264
Japan	134	128	150	176	208	246
Europe	114	109	128	150	177	210
Rest of world	53	51	59	69	81	96
Total	441	420	494	582	688	815

Table 4.22 Worldwide Diode Laser Industrial Market by Device Type 2000–05 (US$ million)

	2000	2001	2002	2003	2004	2005
FOL	108	97	114	134	157	185
CD	62	56	59	62	66	69
VCSEL	65	58	72	88	109	134
HPDL	63	60	74	92	113	141
Red	68	65	71	79	86	95
Violet	75	85	104	128	157	192
Total	441	420	494	582	688	815

In industrial sector, diode laser devices are required for a diverse range of applications. However, individually each of these applications is smaller than, for example, in the telecommunication or consumer sector, but amount to solid collective market share.

- The industrial sector is characterized by low volumes of devices shipped but with prices that are typically fairly high compared with, for example, consumer or computer market applications.

4.8.1 Introduction to Diode Lasers for the Industrial Market

Industrial applications include all forms of instrumentation, robotic systems, factory controls and automated assembly, inspection, test and measurement, diagnostics, condition monitoring, and other applications covering a wide range of subset market sectors. The laser provides a set of functions that cannot be met by any other light source; it is thus responsible for a growing range of interesting commercial products.

These applications make use of the unique mix of characteristics of the diode laser and are likely to continue to expand by the following routes:

● Innovative new equipment to exploit new variants of the laser such as wavelength or output power; and
● Taking market share away from the traditional laser, e.g. solid-state or gas lasers, by virtue of lower power, convenience, portability, etc.

In fact the industrial arena is a good example of where diode lasers have taken away significant market share from traditional lasers. Two examples of this are the violet laser for instrumentation previously dominated by helium–cadmium lasers and the use of direct diode higher-power diode lasers instead of crystal or gas lasers.

The market will steadily grow as the industrial world continues to demand improved capabilities from instrumentation, control, automated facilities, etc., as well as larger volumes of standard equipment. The manufacturing industry is continually stressing the need for improved production efficiency and quality control, which will result in the further penetration of calibrated monitoring equipment. Similarly, the increasingly strict regulation of waste management and environmental emissions is also requiring the installation of further fail-safe, continuous monitoring and control systems to manufacturing processes.

Overall, the next five years will see good market activity in the industrial application sector. Growth will see a better stability for average selling prices but for a lower-volume environment.

Makers of equipment for the industrial sector represent a relatively smaller demand for optoelectronic components. The business is characterized by low-to-medium volumes of units but ones with a premium in terms of quality and performance and hence cost.

● Diode lasers and detectors for instrumentation plus optical data archiving via CD-R, etc.
● Violet diode lasers for analytical instruments for chemistry and nuclear physics.
● HPDLs are largely confined to this sector as regards machining and processing of materials.

The industry sector comprises specialized activities with factories usually operating within the parent company's home country or close to customer. Some sub-systems and modules may be sourced from manufacturers in lower-cost labour regions.

This sector is diverse in terms of applications from the exotic, such as satellite monitoring, to bench-top test equipment. Much of this equipment is also portable requiring low-power components that are lightweight and high performance. In terms of value this segment represents a more dependable market for optoelectronic components.

4.8.2 Application of Violet Diode Lasers

As discussed above, the bulk of the laser diode market is clearly within the longer-wavelength part of the spectrum. The successes of the laser diode have chiefly been in the IR wavelength range and have only recently begun to move into the visible light range. This is in contrast to the non-diode laser industry which has a complete spectrum of wavelengths to satisfy every requirement for concentrated, coherent light sources: from IR through the visible and UV and even into the X-ray region, there is now a laser for every application.

Key trends in the violet laser diode segment are as follows:

- The market is presently small but growing fairly strongly. However, this is mainly for instrumentation and printing equipment.
- Nichia recently launched a new device having high enough power to be used in optical data recording. This could be a stage in the development of the solid-state digital video recorder (DVR) and related products.
- Existing red laser-based DVD data storage has yet to reach its full commercial potential and will not need the violet laser diode for a few more years.
- Philips and Sony have demonstrated a third-generation optical disc-based archiving system, to replace the VCR. With 2.2 Gbyte capacity and with a data rate of 35 Mbit/s it uses the same phase-changing recording principle as rewritable CDs.

The January 1999 introduction of commercial samples of a violet laser diode from Nichia Chemical Industries — the originator of the blue-green GaN LED — has set in motion the plans for future generations of data storage products.

- However, with DVD yet to hit its stride it is unlikely that the violet laser diode will be required for high-density data storage for some time to come.

It is more likely that this interesting device will find applications in other areas such as instrumentation and over a later timeframe in an all-solid-state video recorder. In the longer term, the interest in data storage products may switch to other formats such as three-dimensional holographic systems. These have the potential for a tenfold data density increase compared with the fourfold increase from switching from the red to the blue wavelength. However, these systems are still under development with no commercial product having been demonstrated at the time of writing.

In addition, III-nitride diode lasers can expect further competition from the latest generation of frequency-doubled lasers. An important newcomer to the optoelectronics market is the microchip laser. This device has already been used to prototype laser-based displays and related products that were scheduled for the GaN diode laser. Microchip lasers are a compact source of coherent light available at competitive prices over the green-blue to UV wavelength range.

The introduction of commercial samples of blue LED lasers in early 1999 was followed by announcements of new instrumentation capable of exploiting their unique characteristics. There are several blue laser-based instruments known to be in development. Unexpectedly, it was two German companies, PicoQuant of Berlin and TuiOptics of Martinsried/Munich, that announced the first products based on the Nichia laser. Such companies had an outstanding need for compact sources of lower-cost, coherent blue light excitation sources to complement similar longer-wavelength diode laser sources. The applications of interest are not data storage or displays but rather analytical instrumentation for biochemical spectroscopy and plasma physics.

Nichia's violet diode lasers are rated at 5 mW CW at 400 nm. However, these devices require somewhat different operating power than their red or amber counterparts. The operating voltage is 5–6 V at a current of 40 mA; this is not a problem but does require some adaptation of associated driver circuitry. This is more than made up for by their time resolution, which is at least as good as mainstream detectors but comes at a tenth of the cost.

TuiOptics had a need for a tunable blue laser for applications in plasma physics such as monitoring fusion experiments. With the help of the Nichia laser it has built an instrument that is lower in cost and more compact than earlier models. The DL-100 shows short pulse duration at a high repetition rate with high pulse power or true single frequency, and tunable or frequency-stabilized performance with an unmatched amplitude stability. There are many other less exotic applications for such an instrument, e.g. in trace analysis and atomic absorption spectroscopy.

Time-resolved fluorescence spectroscopy is based on nanosecond timeframe fluorescent decay processes. An instrument to study these requires a light source capable of providing 100 ps pulses. These were previously reliant on traditional lasers such as frequency-doubled dye lasers which are not only inefficient and bulky but also very costly. At first PicoQuant adapted blue-green Nichia LEDs and these worked fairly well but had limited capability as regards pulse generation. Switching to the Nichia laser enabled PicoQuant to create an instrument that is both economic and compact with a pulse peak power of up to 400 mW.

It is likely that once these instruments reach the market other companies will be looking to utilize the particular advantages of diode lasers and there will be a steady replacement of existing sources such as HeNe gas lasers. The diode lasers also include highly stable amplitude unmatched by conventional light sources. Such a capability will make violet diode lasers attractive for use in imaging, microscopy and printing.

The violet laser diode is the world's first compact source of coherent light in the 210 nm range. This has particular application in instrumentation, e.g. in chemical analysis and in the drug industries. These applications have previously utilized He–Cd gas lasers for short-wavelength light emission but are less convenient and, in the longer term, should prove much more costly compared with cheaper violet diode lasers.

The availability of compact diode lasers over the entire visible spectrum is nearly complete. As a result, the capability of analytical instrumentation has been significantly enhanced. The next step in this evolution can proceed in two directions: it will mean either arrays of individual diode lasers, each tuned to a specific wavelength, or one laser that is tunable over the whole spectrum. So far, the former approach is the more practicable. A number of instruments have been launched with multi-spectral capability that formerly required multiple instruments.

In less than five years, violet diode lasers have become an important new tool in the analytical instrumentation marketplace. This is not the big business for which many were hoping, but it is providing useful revenues for several companies in Europe.

The violet laser is carving out a niche in two ways:

- Filling the gaps in the spectrum for spectroscopy and related applications; and
- Ousting traditional lasers, especially the HeCd, from existing markets.

The violet laser diode market is now underway. However, it is worth one-tenth of what some said it would be by now — a few tens of millions of dollars compared with the hundreds of millions for today's red laser diodes for DVD, which is the standard for high-density data storage.

Nichia has recently launched a new series of violet laser diodes having the requisite 405 nm peak wavelength and a 30 mW maximum optical power output required for next-generation DVD optical data storage for PCs, movie players, games consoles, laser printers, scanners, etc. The increased optical power output of the NLHV 3000 series enables the violet laser to not only read data but also to write data.

Blue-violet laser diodes with wavelengths of 410 nm had been long awaited for use in next-generation optical disc drives. Practical blue-violet laser diodes are not expected to be available for a few more years.

A CD drive, for example, uses a near-IR laser diode with a wavelength of 780 nm and an object lens with an NA of 0.45. Modern DVD drives use red diodes from 635 to 650 nm, with an NA of 0.6. The new blue-violet laser diodes offer the even shorter wavelength of about 410 nm. This shorter wavelength means that surface recording density should be, theoretically, 2.5 times higher than existing DVD. The major target for the new drives is the VCR market.

- Rewritable discs will be more important than play-only discs, because it will mean inexpensive recording of high-definition broadcasts in the home. These drives are likely to replace the analogue VCR and a home optical disc recorder could appear in about 2002, capable of recording at least three hours of imagery on a single disc.

The caveat for the prospective introduction of solid-state video recorders is that they are already here in the form of home computers fitted with the requisite TV reception card plus a hard drive with, for example, a 20 Gb storage capacity. Moreover, the present generation of DVD players based on much longer-wavelength laser diodes has taken much longer to take off commercially than the industry had anticipated. This alone has had the effect of postponing the timetable for the demand let alone the commercial acceptance of blue laser-based data storage equipment for consumer or computer applications.

- With its spectral and time characteristics the blue laser diode is particularly suitable for biomedical applications. For example, the detection of fluorescein- or rhodamin-labelled substances can be realized very efficiently. The blue laser light source is aimed at satisfying an outstanding need for compact and affordable excitation sources. Previously, flash lamp-driven frequency-doubled Ti:sapphire solid-state lasers were required and these are unwieldy and expensive.

This discussion would not be complete without mention of the prospects for UV light-emitting lasers. Very short-wavelength UV light-emitting lasers have been available for some time albeit in the form of bulky argon gas lasers or similar systems. These are applied to the needs of semiconductor mask inspection and confocal microscopy.

Other applications for blue-green lasers include data storage, instrumentation and solid-state video recording, none of which will ever become commonplace commercially with solid-state lasers. These systems like their longer-wavelength laser diode-based antecedents mandate a compact source of short-wavelength light.

- In the field of solid-state lasers, work continues on developing so-called SHG lasers. In this respect companies have used crystal rod lasers having a frequency-doubling capability pumped by diode lasers so as to give UV light output.

Light Age in New Jersey, USA, has created a laser which operates in the 650–700 nm range with a minimum power of 300 mW. This is pumped by 680 nm diode lasers whose light is beamed into a beta-barium borate in the form of a V-shaped folded cavity to produce 380 nm continuous output. This approach has many advantages for such applications as CD mastering, laser-induced fluorescence, hologram generation and blood-flow cytometry. Higher power versions are being developed which would suit applications such as laser radar (LIDAR) remote sensing and pollution monitoring, circuit board processing and semiconductor mask inspection.

The latter applications will over a five- to ten-year period become important applications for semiconductor diode lasers in the very short-wavelength region. Even though companies have begun to approach the concept of commercial violet light-emitting lasers, at the time of writing no comparable announcements had been forthcoming as to the development of even prototype diode lasers in the UV region. This is despite being theoretically possible given the bandgap of gallium nitride-based compound semiconductor alloys.

UV laser diodes based on nanocrystalline zinc oxide thin films have been described by Japanese researchers at the Tokyo Institute of Technology. The laser action results from the propensity of zinc oxide to form nanocrystalline domains that act as laser cavities. The ZnO films are pumped with laser light and radiated in the UV region. This is said to open the door to laser diode technology beyond blue-emitting GaN systems. Better performance is expected when the technique is applied to multilayered heterostructures, which could lead to a UV diode laser.

At present, the first commercial samples of the violet diode laser are available from Nichia Chemical Industries. However, the number of actual products which are commercially available is very small — just a few specialized analytical instruments (see industrial section). Despite the great anticipation that preceded this development there has been little in the way of public announcements of products in the consumer arena. This is not surprising for various factors:

- Commercial secrecy — any blue laser diode product development is proprietary and unlikely to be announced until the product is ready for commercial launch.
- Understanding what the new diode laser is capable of and building a design around it is a lengthy process.
- Present-generation products — DVD in particular — are based on longer wavelength (i.e. red) visible laser diodes. This market is still immature and so there will is a reluctance within the industry to move to another system until it has completed its market cycle.

In October 2001 Matsushita Electric (Panasonic) introduced the world's first 50 Gb blue laser re-writeable dual-layer optical disc technology. The system is capable of recording four hours of high-definition moving pictures to a DVD-size disc. However, the system does not use the Nichia type of GaN-based violet diode laser but rather it achieves its high power and shorter wavelength via a high-powered wavelength-tunable IR laser diode. The IR light is then transformed into blue laser light. By using a high-efficiency SHG device (made in collaboration with another Japanese company, NGK Insulators), the size of the blue laser is also compact (just 0.3 cm^3), a quarter of the size of previous packages. The SHG device splits the wavelength of semiconductor light into two distinct wavelengths. For example, 425 nm blue laser light is produced using an SHG device to split the 850 nm wavelength light of an IR semiconductor laser diode. By employing a high-precision assembly technique for miniaturized SHG lasers, this new laser also achieves one-third the noise level and a 90% reduction in wavelength variation compared with GaN laser diodes.

Demand from the public for a digital VCR based on discs rather than tape (i.e. VHS) is likely to follow after the DVD market has gained critical mass. Sales of DVD have been strong but in many cases the large user base of conventional video recording has set a precedent for a similar system of record and playback from a disc-based system. If the producers of such equipment develop such units then they will also be adopted for use in computers. The aim would be to develop a manufacturing process optimized so as to make the recordable DVD units as cheaply as possible. Merging the consumer and computer market would effectively double the available market or halve the time to acceptable critical mass. Availability of such a large-volume market would accelerate the development of competitively priced DVD units based on blue lasers.

The development of this market will depend on the captive development and manufacture of CD-ROM laser optical pick-ups. Today, these important components are almost exclusively made in-house by large Japanese companies such as Sony and Hitachi. It is not unlikely to expect that this procedure will also be adopted for any DVD system based on blue diode lasers. The only maker of such diode lasers at the moment, Nichia, does not have any intention of either making this technology more generally available or of entering the DVD market itself. The makers of such equipment will therefore have to develop their own route to the blue diode laser. This is likely to be already underway but has to run the gauntlet of avoiding the infringement of patents with Nichia. The company has shown it can be done but has cornered this market. It is at present unknown how wide the window is for the practical realization of such devices by a non-Nichia route. Upon this will depend many things, not the least the timing and size of the market of the blue diode laser-based DVD for computers and other applications.

At the moment, with the availability of lower-cost devices, the range of applications for laser diodes that emit in the visible range is increasing. Blue diode lasers are also expected to follow this trend.

The aesthetic attraction of a *blue* laser pointer is self-evident. Nevertheless, it is unlikely at present for the new Nichia device to find its way into this market segment. There are several reasons for this:

- The aforementioned prospective international banning of the laser pointer on grounds of safety; and
- The unavailability of the key MOVPE growth technology to mass produce these devices at costs comparable to the very low prices of red laser pointers.

4.8.3 High-Power Diode Lasers

Another type of diode laser in the industrial market is the HPDL. The HPDL is considered here because of the future importance of these devices in industrial processing in particular, even though they have broader applications. At the moment, the most important application of HPDLs is the 'pumping' (i.e. acting as a coherent light source replacing flash lamps, etc.) for traditional lasers as used in two main areas: fibre telecommunications and industrial processing such as welding.

In the IR laser diodes market, potentially one of the most important devices is the HPDL. In 2000 this market was worth US$285 million. This fell in 2001 to US$271 million but will resume good growth in 2002, increasing to US$640 million by 2005. The HPDL market estimate and forecast are summarized in Figure 4.12 and Tables 4.23 and 4.24.

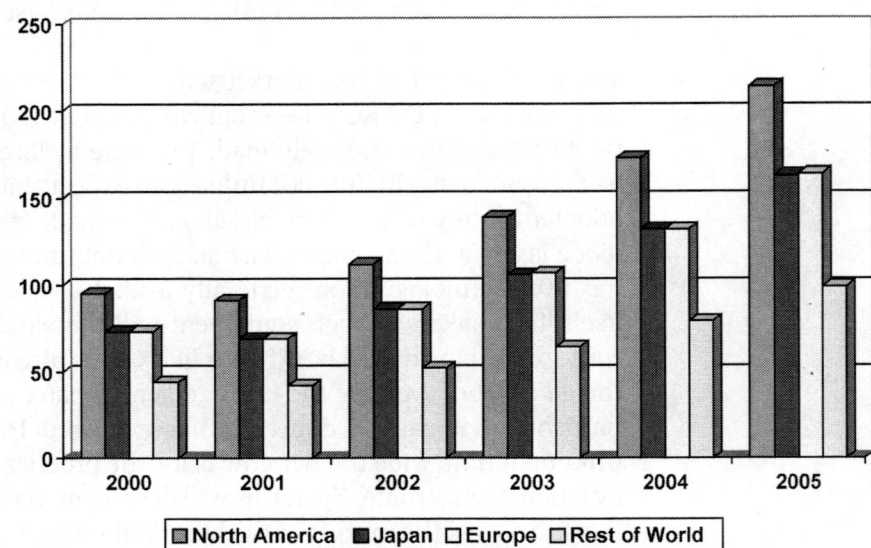

Figure 4.12 Worldwide HPDL Market by Region 2000-05 (US$ million)

Table 4.23 Worldwide HPDL Market by Region 2000–05 (US$ million)

	2000	2001	2002	2003	2004	2005
North America	95	91	112	139	173	214
Japan	73	69	86	106	132	163
Europe	73	69	86	107	132	164
Rest of world	44	42	52	64	79	99
Total	285	271	335	416	516	640

Table 4.24 Worldwide HPDL Market by Application 2000–05 (US$ million)

	2000	2001	2002	2003	2004	2005
Automotive	37	35	44	54	67	83
Computer	23	22	27	33	41	51
Consumer	34	32	40	50	62	77
Industrial	63	60	74	92	113	141
Military/aerospace	14	14	17	21	26	32
Telecommunications	80	76	94	116	144	179
Other	34	32	40	50	62	77
Total	285	271	335	416	516	640

The solid-state laser (SSL) such as the Nd:YAG for lower-power applications has come under threat from HPDLs. Clearly, the HPDL is one of the most strongly growing areas for laser diodes and has already begun to outpace that of conventional laser types. Moreover, the HPDL is mostly used for diode pumping of SSLs but is beginning to also see many new markets develop as a result of the availability of compact power sources.

Market value fell in 2001, to US$271 million. This was due to two main factors:

- The general market downturn in materials processing and manufacturing, the principal users of HPDLs; and
- Price pressure on this new variant of the diode laser.

HPDLs will, however, see a quicker return to expected growth. The market will recover more quickly than some other laser diode types and reach US$335 million as soon as 2002. Good growth should then continue until 2005. Nevertheless, while applications for this device family will continue to expand, price pressure will increase. Volumes of devices shipped will increase at a proportionately higher rate than that for the market value.

At present, HPDLs have found solid, albeit fairly small niche applications in printing, surgery, soldering, etc. Market expansion is awaiting the reduction of a previously adversely high price. However, the dollar per watt pricing benchmark for HPDLs is following a fairly steep downward curve, which will ensure the further extension and growth of this market.

Thus it seems that the laser diode has a very promising future with developments of existing types and novel new diode lasers with all the consequences these promise for electronic products.

This report has been concerned with the diode laser's ability to handle data streams. Whether it is sending signals along a fibre or reading a DVD, power levels of a few milliwatts are sufficient. HPDLs are diode lasers with output powers of 1 W and above and interesting new applications arise from the direct use of these compact energy sources.

Although a relatively new sector, HPDLs have nevertheless become very successful as light-pumping sources for solid-state lasers. Yet their biggest opportunity is only just emerging — as compact power sources of coherent light. The wavelength range of interest is from 635 nm up to 2 μm, where they are becoming increasingly attractive for functions such as soldering and welding, as well as in microsurgery.

In some sectors of the laser industry, the importance of the HPDL has been likened to that of the transistor in the electronics industry. However, higher relative costs of HPDLs (compared to traditional laser solutions) have so far limited their use to specialist applications such as non-invasive surgical procedures and in large printing machines. However, over the past two years further R&D and commercialization in the design and manufacture of HPDLs has taken place. Consequently, prices have fallen. This has led to a significant increase in the adoption of these devices in the marketplace.

It is clear that the HPDL is one of the faster-growing market sectors for laser diodes and has already begun to outpace that of some conventional lasers. In fact, the HPDL market has been less affected by the downturn than many other types of diode laser. However, even this sector has suffered as the indirect application of diode pumping has been hit by the downturn in demand for telecommunications components.

The key factor is that, due to the reduction in price per watt of light output, the HPDL market is now able to begin a period of serious growth. This is proceeding along two lines: by displacing existing lasers and also by developing new markets stemming from the availability of these compact energy sources. By 2003 technical development of the HPDL is expected to have reached a point where the wall-plug efficiency has been boosted, allowing the applications market to really expand.

The French company Thomson-CSF (now Thales) was one of the pioneers, but in the past five years the number of companies focusing on HPDLs has increased significantly. These now include:

- Alfalight, USA.
- Coherent, USA.
- DILAS Diodenlaser, Germany.
- Diomed, UK.
- Fisba Optik, Switzerland.
- HPD, USA.
- IQE, UK.
- Lasertel, USA.
- Nuvonyx, USA.
- Opto Power Corp, USA.
- Polaroid, USA.
- SDLI, USA.
- Siemens, Germany.
- Sony, Japan.
- Thomson, France.
- Unique Mode, Germany.

It is noticeable that, in contrast to the optical storage laser diode market, it is North American and European companies that dominate this sector. Surprisingly, laser manufacturers in Japan and South-East Asia have so far made only limited progress in HPDLs. Interest is on-going in these countries, but much less so than in the West.

However, there is interest in the use of HPDLs for diode pumping of solid-state lasers (e.g. Sony Semiconductor's SLD402S 790–840 nm laser array family for diode YAG-pumping applications, which has a maximum power of 22 W).

In the West, two basic types of company have arisen. Firstly, the vertically integrated companies such as Coherent and SDL (now part of JDS Uniphase) in the USA, which make everything from the laser materials through to complete systems. Secondly, there are the equipment companies that buy components and specialize in certain areas (such as Diomed in medical applications or the Swiss company Fisba Optik in very sophisticated lower-power systems with precision optics for soldering, plastics welding, etc.).

The overall leaders in the HPDL business are the US players. For example, Opto Power Corp specializes in fibre-coupled laser modules and open arrays and has demonstrated surface hardening and drilling of stainless-steel ribbon, for example, as well as solid free-form fabrications via sintering with metal and ceramic powders using fibre-delivered diode power. These demonstrations indicate that many additional applications — including cutting, soldering, marking, printing, welding and coating — can be performed by currently available commercial diode lasers when the energy is coupled to achieve high-brightness spots.

In Europe, the leading manufacturer of laser dies is Infineon Technologies. These products are purchased by many systems companies worldwide. For example, DILAS is taking Infineon's kilowatt-class laser diode stacks to make much higher-power units for machine systems companies such as Trumpf (who use them for welding and hardening equipment in vehicle production lines, printers, etc.).

There is a determined collaborative effort in Germany. In a programme headed by the Fraunhofer Institute, companies and research institutes have just completed the NOVALAS programme and are about to initiate a follow-up programme on HPDLs. The Fraunhofer Institute has also just spun-off a company — Unique Mode — to develop and market commercially miniature 1–10 W, 80–1480 nm HPDLs.

DILAS claims to have HPDLs with the lowest cost per watt of any kind of laser. These units are very compact, with an array of air-cooled HPDLs capable of 25 kW of power within a space the size of a shoebox. These higher-power (›1 kW) units are competing strongly in manufacturing and materials processing operations such as welding and tool hardening.

Direct diode lasers (rather than diode-pumping lasers) are by far the more exciting area for future growth and — as some observers have already suggested — they may come to supplant entirely conventional lasers in a number of applications. The availability of well-defined high-power optical output in a compact form is one of the main reasons for the interest in HPDLs. Product development has been rapid and higher-power (i.e. kilowatt) units are now in widespread use in many machining and materials processing tasks worldwide. These include:

- Laser welding.
- Heat treating.
- Brazing.
- Laser sintering.
- Printing/reprographics.
- Cladding.
- Plastic welding.
- Soldering.
- Paint stripping.
- Continuous seam welding.
- Laser forming.
- Spot welding.
- Photodynamic therapy.
- Cosmetic surgery.
- Diode pumping.
- Illumination.
- Surface melting.
- Composite forming.
- Epoxy curing.
- Free-form fabrication.

There are two directions for the HPDL market to take: the replacement of traditional processes such as those based on SSLs and gas welding and in the many new applications opening up where an SSL would be less convenient and too costly (e.g. PCB soldering, dentistry, etc.). Another key area of success has been in the medical field, not only for keyhole surgery but also for cancer therapies and cosmetic treatments (e.g. Diomed of Cambridge, UK, specializes in medical applications of HPDLs).

Advantages of diode laser systems for medical applications include their small size, portability, higher efficiency and low cost-of-ownership. These not only ensure a thriving market but they are also responsible for the broadening of the medical applications market. In keeping with the majority of HPDL applications, the most popular is the 810 nm wavelength, but other wavelengths are now becoming important. Applications began with surgical lasers but the shift is now to ophthalmic and cosmetic/aesthetic applications, together with photo-dynamic therapy (PDT). PDT is a relatively new area that was created by SSLs and is now moving over to the convenience of HPDLs. This application uses the diodes not for their power but rather for their precision of wavelength control. Other related applications include cosmetic surgery procedures such as removal of unwanted hair, birthmarks and tattoos.

Price depends on a number of factors, but a large contribution comes not from the laser but from the micro-optics (more so for finer beams and shorter working distances). Price is also very dependent on the output power, since a 5 kW system will be orders of magnitude more expensive than a 5 W system.

HPDLs have demonstrated electrical-to-optical conversion efficiencies that range from 30 to 60%. The result is a system (including power supplies and water cooling systems) that can have a wall-plug electrical-to-optical power conversion efficiency of > 25%. Consequently, a 4 kW CW laser diode system can consume less than 16 kW of electrical power. This efficiency translates into a lower cost of operation for the user. When comparing the amount of output laser energy for a given electrical input, the efficiency of the direct diode laser is much higher than that of conventional laser systems.

Of great significance is the fact that HPDL direct diode units have compact air cooling. This means that they often only need a single mains plug for operation. Not only does this confer great portability and versatility but also very low maintenance (of the order of tens of thousands of hours between failures).

While delivered optical power is increasing (now exceeding 100 W), heat-sinking of the device is even more important. Not only does this mandate suitable design considerations, but also the appropriate engineering for competitive manufacturing.

Another relevant point is that the HPDL operates with a centre wavelength of 810 nm. This is in the near-IR region and is therefore invisible to the eye. HPDLs operate at a wavelength shorter than that of Nd:YAG (1.06 μm) and CO_2 (10.6 μm) lasers, commonly found in industrial applications. The lower operating wavelength results in a higher absorption rate with many metals (especially aluminium, which has an absorption peak at the HPDL's wavelength of 810 nm).

One of the few drawbacks of the HPDL at present is its unsatisfactory beam quality. The associated problems of optical focusing result in lower than desired power levels impinging on the chosen site. Many researchers are addressing this problem, and progress includes the introduction of tapered resonator/ amplifier devices.

Another important development relating to the HPDL is the fibre laser. This variant involves coupling the HPDL to an optical fibre to direct precisely the light remotely from its source. This promises to make the HPDLs even more versatile and attractive to a wide range of today's applications and those yet to come.

Overall, the well-established SSL market is growing only steadily, while the smaller and less mature direct diode market is at least double that of SSLs. An important point is that the direct diode HPDL has yet to accumulate any history —there will be an industry reluctance to adopt these devices until the necessary experience has been gathered. However, there is no doubt that the direct diode HPDL is proving to be a very viable tool with many applications through its compactness and low cost of ownership.

A relatively new sector, HPDLs have become very successful as light pumping sources for SSLs. However, they have begun to see strong take-up as compact power sources of coherent light in their own right. This is as a result of their usefulness for such operations as soldering and welding as well as in microsurgery.

Further R&D and commercialization in the design and manufacture of HPDLs has taken place in the past two years, and as a result prices have fallen. This has had the effect of causing a significant increase in the take-up of these devices within the marketplace.

Finally, another important development relating to the HPDL should be mentioned — the fibre laser. This variant involves coupling the HPDL to a fibre optic so as to direct precisely the light remote from its source. Overall, the well-established solid-state laser market is growing by around 10% and the smaller and less mature direct diode market is at least double that.

4.8.4 Quantum Cascade Lasers

Originally developed by scientists at Bell Labs, the quantum cascade (QC) laser is mooted to be an important new laser diode type which will be important in sensing applications requiring mid-IR wavelengths such as pollution monitoring and drug detection.

QC lasers are one of the newer members of the laser diode family. They provide emission at specific wavelengths in the mid-IR region (3.5–20 μm).

- QC lasers have good direct-current modulation properties with a theoretical upper limit in the several hundred gigahertz range making them potentially very attractive for very high-speed data communications.

In particular, the QC diode laser has application as a free-space optical communications medium. This is because of its wavelength range that coincides with an atmospheric window of minimum attenuation.

Bell Labs described the QC laser as a revolutionary light source because it is the first laser in which the wavelength is determined by the thickness of the active materials rather than by their chemical composition. The wavelength can be pre-selected anywhere in the mid- and long-wavelength IR regions. This is the broad, invisible range in which heat sensors work and where most gases and vapours leave telltale light-absorption fingerprints.

The QC laser has potential commercial applications in such areas as pollution monitoring, industrial process control, auto emission diagnostics and medical testing. In pulsed mode, it works at room temperature and above and is hundreds of times more powerful than conventional semiconductor lasers operating at the same wavelength. QC lasers have been used to detect extremely small amounts (fewer than 100 ppb) of trace chemicals, opening the door to a new class of extremely sensitive, compact and portable chemical sensors.

- Unlike conventional lasers, a QC laser operates like an electronic waterfall: When an electric current flows through it, electrons cascade down an energy staircase with tens of steps; every time they hit a step they emit a laser photon, or light pulse. Each electron, therefore, generates tens of photons, as many as the number of steps, rather than a single photon as in conventional semiconductor lasers. The cascade effect is responsible for the high power of QC lasers.

4.9 Military/Aerospace Markets for Devices

The market value of devices in the military/aerospace sector was estimated to be US$240 million in 2000, falling to US$221 million in 2002. The period 2001–02 will see a fall of 8% as a result of the industry downturn. From 2002 this will transform into positive growth of over 15% per annum. As a result by 2005 the market will have grown to a value of US$388 million. The market estimate and forecast are summarized in Figure 4.13 and Tables 4.25 and 4.26.

Table 4.25 Worldwide Diode Laser Military/Aerospace Market by Region 2000–05 (US$ million)

	2000	2001	2002	2003	2004	2005
North America	101	93	106	122	141	163
Japan	38	35	40	46	54	62
Europe	74	68	78	90	104	120
Rest of world	26	24	28	32	37	42
Total	240	221	253	290	335	388

Figure 4.13 Worldwide Diode Laser Military/Aerospace Market 2000–05 (US$ million)

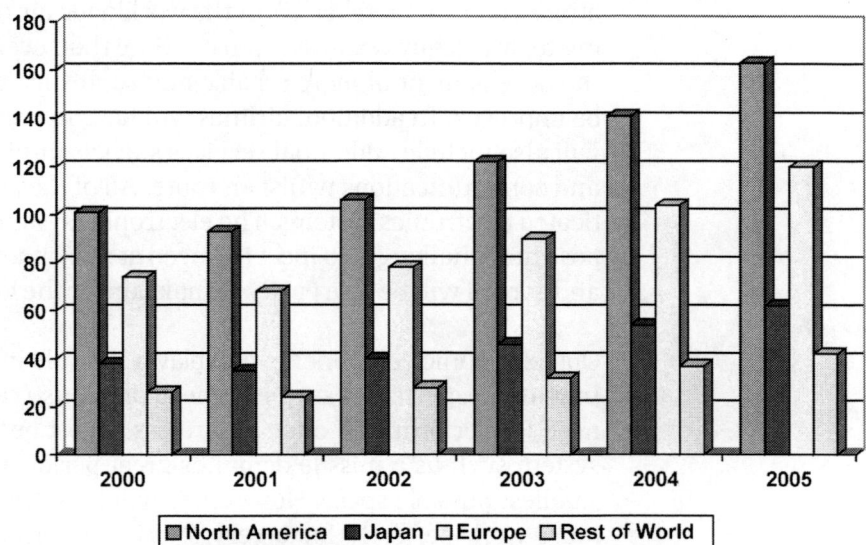

North America ■ Japan □ Europe □ Rest of World

Table 4.26 Worldwide Diode Laser Military/Aerospace Market by Type 2000–05 (US$ million)

	2000	2001	2002	2003	2004	2005
FOL	90	81	95	111	131	154
CD	55	50	52	55	57	60
VCSEL	31	28	34	42	52	63
HPDL	14	14	17	21	26	32
Red	43	41	45	50	55	60
Violet	7	8	9	12	14	18
Total	240	221	253	290	335	388

The industry sector has been affected by the industry downturn and fell by just over 8% in the period 2000–01. Thereafter the market is expected to pick up and will grow by nearly 15% per annum until 2005.

- Makers of equipment for the military and aerospace sector represent a relatively modest demand for components but one which carries amongst the highest unit prices in the marketplace.

4.9.1 Diode Lasers in the Military/Aerospace Market

Similar to military aircraft, civil aviation is making more use of electronics and this is boosting the total sales in this combined sector. For example, the European Airbus is a world leader in fly-by-wire control systems. Besides improved functionality and control, these systems provide significant weight savings and hence contribute greatly to improved overall economy of operation. On the drawing board are higher performance control systems based on fibre optics — so called 'fly-by-light'. Such systems will demand high numbers of optoelectronic components when they begin operational service in the next five years or so.

Thrown into the spotlight as a result of the events surrounding the terrorist attacks on 11 September 2001, the worldwide air transportation industry is having to drastically reassess security. More than ever it is going to have to invest in the development of more reliable and sophisticated systems to enable safety to be improved. In addition, airlines will have to regain customer confidence. This will also include additional services such as in-flight entertainment, navigation and communications whilst en route. All of these and more will require sophisticated electronics systems. The electronics systems on board these aircraft will pose new challenges to meet the need at reasonable prices. This sector will grow and overall will be able partly to make up for the loss of the military systems.

Optoelectronic components will play a key role in the defence/aerospace sector. In many respects this sector is ahead of mainstream electronics by virtue of the need for performance often regardless of cost but mindful of reliability. Often a system such as a missile demands great performance and functionality in the smallest possible space. However, it will be stored for as much as a decade but must work perfectly for a very short time prior to its complete destruction. Other military/aerospace applications require devices that can function over a wide temperature range. Rapid climatic changes can be experienced by a system and it must still deliver 100% operation for long periods with only rudimentary maintenance.

However, in a modern climate of regional conflicts, the military sector continues to slowly reduce demand as defence budgets shrink. Nevertheless, requirements for all involved, especially the forces of NATO, must be for secure communications, sensory systems and all forms of reconnaissance and electronic warfare.

The military operations of the past decade have reaffirmed an increasing reliance on electronics systems. In fact the fraction of all military systems devoted to electronics is increasing strongly and many types of high-performance components such as diode lasers will continue to be important for advanced aerospace systems of many kinds. In this sector there is also a requirement for more versatility from on-board systems and this is being achieved through micro- and optoelectronics.

It is expected that the move by the military to 'commercial off-the-shelf' (COTS) components will continue. The military is making more use of plastic-packaged components, for example. It can gain major performance/cost advantages through the use of these devices. The next five years is likely to see much more use of these which were formerly restricted to other application sectors. This has continued to be to the detriment of specialist component types and their suppliers and to the advantage of the larger commercial vendors. However, there is mounting pressure to reverse this trend from designers and manufacturers who have expressed dissatisfaction with COTS components. This will have repercussions for component suppliers who have closed or sold off mil-spec component lines or moved onto COTS lines.

In accord with the consolidation of companies that has been happening in the North American military/aerospace industry in the past few years, the European industry is also undergoing consolidation.

Space-based systems represent quite a large market for high-specification microelectronic components, a market which is expected to increase significantly. Whilst not all of these applications mandate the use of semiconductor components, these offer a very attractive mix of performance characteristics that should ensure their consideration in future systems.

Other key military/aerospace application sub-sectors include:

- Fibre optic gyros (FOGs). For improved accuracy in inertial navigation systems, the FOG is displacing mechanical systems. The reduced price differential is ensuring wider take-up for many key aerospace applications. However, the field has come under threat from lower-cost global positioning systems (GPS) which may over the longer term harm FOG sales for some applications. FOGs are likely to always be included in high-end systems given the possibility of jamming or other deterioration of GPS signals.
- Submarine communications. At present this is achieved using very low-frequency signals that need very long aerials — such systems are also very slow. A high-frequency blue-green laser would have high transmissivity through seawater, while being immune to jamming and immune from eavesdropping.
- Chemical and biological warfare agent detection. Availability of low-power portable monitoring equipment for combating these 'weapons of mass destruction' is ever more important in the post-Cold War period especially in view of the anthrax attacks in the USA in late 2001. Short-wavelength emitters and detectors promise to expand the range of diagnostics available in this field.
- Ultrahigh-speed data processing. Whilst today's microprocessors are working in the gigahertz speed range, there is need for much higher speed computational power for applications ranging from weather forecasting to stealth bomber design. Such speeds will require all-optical computing and very high-density optical data storage. High-speed information processing in real time is becoming critically important in defence and aerospace, for example in airborne radar systems.
- Laser-based radars. Today's radars (particularly those in aircraft) work using a mechanically rotated antenna. This has a number of drawbacks and is beginning to be replaced by fixed antennae called phased array radar (PAR) or active electronically scanned arrays (AESA). These comprise arrays of transmit/receive modules that electronically control the beam. An alternative or supplement to these systems is the laser-based radar where the microwave radiation is replaced by light wavelengths.
- Radiation hard systems. Despite the end of the Cold War, the possibility of nuclear warfare remains and so certain defence systems require immunity to the effect of intense radiation. Some semiconductors promise higher immunity than existing types and also offer improved power handling and higher-frequency operation. Resistance to radiation is also very important for equipment within atomic power stations, satellite equipment and space probes.
- Non-lethal weapons. A new trend in the counter-riot and security area is the requirement to immobilize opponents without causing permanent harm. Some of these weapons are based on the delivery of sudden, high-voltage electric shocks, which require high-voltage capable devices that would benefit from the special properties of wide bandgap semiconductors.

- Fibre optics are playing an increasingly important role in defence for hardened communications, missile guidance and towed decoys, to name but a few applications. All such systems demand robust, inexpensive, high-performance diode lasers.

4.9.2 Electronic Warfare

Electronic warfare (EW) systems such as electronic countermeasures, decoys and uncrewed vehicles, require-high performance systems to help to reduce the size, weight and power consumption and give increased reliability. Previously this was mainly concerned with microwave signals but now encompasses IR and visible wavelengths. Modern defensive and offensive EW systems rely more and more on a mix of sensors necessitating optical signal intelligence gathering and countermeasures. In extreme cases this also concerns the use of 'laser jamming' whereby the enemy optical sensor is 'blinded' by a laser beam.

In today's airborne warfare scenarios where a higher risk of interception of attack aircraft is encountered from advanced surface-to-air and air-to-air missiles fielded by previously third-world nations, more capable EW systems are required. Moreover, these systems must be more functional in the smaller volume of the aircraft. They must also be fielded in other aircraft such as for protection of the transportation used for high-ranking personnel.

However, all has not been stable in the EW marketplace. Much reliance continues on government budgetary allocations. In the USA (the largest market for such systems), several key programmes have been seriously affected by changes in budgets in recent years; for example, the cancellation of the B-1B's Defensive Systems Upgrade Program (DSUP). The DSUP programme was envisioned as a US$1.2 billion replacement for the ALQ-161 threat detection and jamming system to include the ALR-46M radar warning receiver and RF jamming system plus a fibre optic towed decoy. The ALR-46M would detect and display enemy radar systems, while the active countermeasures would have been used against both radars and missiles. The US Air Force decided not to fund DSUP.

As a result of the recent military actions in Europe and Middle East, the US armed forces have been forced to reconsider quite a few such decisions. Temporarily the war-footing requires equipment to be enhanced and emergency funds are found to carry these out. However, over the longer term these plans may have to be subjected to further reconsideration depending on the outcome of the newly announced 'war on terrorism'. Money will be found to fund essential projects but this could be at the expense of other projects. For example, while the JSF has been cleared to proceed it is deemed likely by industry observers that its future is not 100% assured. The USA is also funding the F22 and newer versions of the F/A-18 and Harrier plus the Osprey. Not all of these piloted systems will survive the next five years. This is partly due to continuing cutbacks in defence spending and also to the trend towards uncrewed aerial vehicles which fit in with the US requirement for minimizing friendly casualties. Nevertheless, crewed aircraft and other military vehicles are likely to always be required. It is expected that for the most dangerous missions 'robot' vehicles will be developed. Lack of pilots also obviates the requirement for expensive search-and-rescue forces.

4.9.3 Laser-based Weapons

Offensive weapons have long been important exploiters of lasers. Since the 1970s the precision munitions have become an important air-delivered weapon in all the world's air forces. While these systems are generally based on solid-state IR lasers, developments are underway for their replacement by semiconductor devices. These confer considerable savings in power and weight, permitting more compact units and room for additional fuel or weapons. Similar trends are underway in ground-based laser designators and related equipment.

This trend has percolated down to the lowest levels of weaponry. Today it is a simple matter to improve accuracy by adding a visible laser pointer to a pistol or rifle. The availability of very low-cost points has made this possible and has the added benefit of allowing the user to dazzle or even blind an opponent. Such non-lethal weapons have yet to be deployed by the world's security services but could, in time, become important with visible lasers playing a key role in this regard.

There are on-going programmes to develop higher-energy beam weapons for the defence of major installations from missile attack. The lasers that are expected to be used will not be diode lasers but these could, with development, achieve useful power delivery in compact form. Traditional lasers are bulky, fragile and consume high power. Alternative diode laser systems do not have these drawbacks and could be used in portable defence systems.

Because of the terrorist attacks on the World Trade Center and the Pentagon in late 2001, this industry sector has reached something of a crossroads. Two factors are at play here:

- The airline industry has suffered a severe decrease of passenger traffic. Lost revenues are over and above a general decline due to telecommuting but will result in deferment or cancellation of orders for equipment. However, it should boost sales of security systems, some of which are likely to rely on the advantages of handheld laser units over X-ray machines that are bulky and static.
- There is a likelihood that defence spending will increase in certain key areas. One of these is precision-guided munitions and reconnaissance systems so as to improve locating and targeting of enemies. It will also take steps forward with next-generation optical equipment and other systems developed since the end of the Cold War.

Overall, the trend will be for an upward growth for this sector. There will be a heightened awareness of threats that differ considerably in nature but not in seriousness.

Previously EW was concerned with the microwave region of the spectrum but today this is moving into the longer wavelengths of IR and even visible light. Modern weapon systems increasingly exploit a mix of sensors to overcome adverse climatic conditions. Countermeasures to such systems are required including compact diode laser light sources.

4.10 Automotive Markets for Devices

The market value of devices in the automotive sector was estimated to be US$126 million in 2000. This market has suffered from the general industry downturn and fell by nearly 7% to US$118 million in 2001. This will increase to U$138 million in 2002 representing an average annual growth rate of 18%, thereafter to reach US$228 million in 2005. The market estimate and forecast are summarized in Figure 4.14 and Tables 4.27 and 4.28.

Figure 4.14 Worldwide Diode Laser Automotive Market 2000-05 (US$ million)

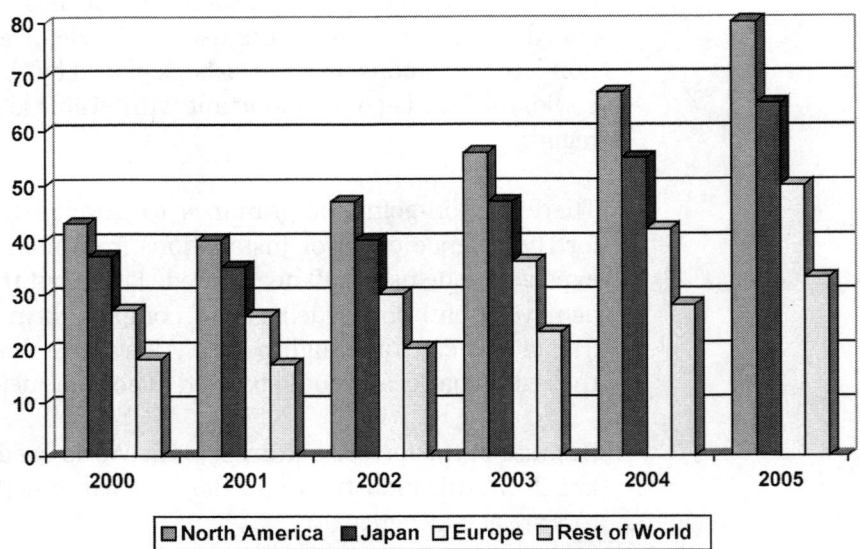

North America ▪Japan □Europe □Rest of World

Table 4.27 Worldwide Diode Laser Automotive Market by Region 2000–05 (US$ million)

	2000	2001	2002	2003	2004	2005
North America	43	40	47	56	67	80
Japan	37	35	40	47	55	65
Europe	27	26	30	36	42	50
Rest of world	18	17	20	23	28	33
Total	126	118	138	162	192	228

Table 4.28 Worldwide Diode Laser Automotive Market by Device Type 2000–05 (US$ million)

	2000	2001	2002	2003	2004	2005
FOL	18	16	19	22	26	31
CD	10	9	10	10	11	11
VCSEL	31	28	34	42	52	63
HPDL	37	35	44	54	67	83
Red	25	24	24	25	26	27
Violet	5	6	7	9	11	13
Total	126	118	138	162	192	228

4.10.1 Applications for Diode Lasers in Vehicles

Diode lasers are not found in most vehicles outside of in-car entertainment systems, i.e. CD players. They are likely to remain a fairly small contribution to the total market but when it takes off it will be characterized by robust components with fairly low unit prices in quite high volumes.

There is considerable potential for increasing application of these highly capable devices in a wide range of applications in vehicles. This is in particular regard to the safety, security, environmental compliance and information technology aspects of modern and prospective vehicles. The major trends in this regard are:

- Increased legislation for vehicle illumination.
- Legislation for improved environmental compliance, e.g. sensors for emission control.
- On-board networks based on fibre optics rather than copper wiring for reduced weight and higher bandwidth.
- Enhanced driver and passenger safety, e.g. occupant presence detection.
- Remote sensing and diagnostics via fibre optics.
- Security is playing an important role and optoelectronic components such as low-cost, intelligent sensors and warning systems will play a key role in this area.
- Handling of on-board information such as navigation and traffic hazards will require computer units with optical data storage for vehicle-to-vehicle and vehicle-to-roadside communication.
- Station keeping and collision avoidance in adverse weather conditions using laser radar-type sensing.

The general trend, which is expected to apply to future vehicle applications, is for initial penetration of high-end saloon cars and/or specialist vehicles, e.g. buses. Subsequently, once component costs have been significantly reduced, these systems will also cascade down into general vehicles.

Over the longer term, lasers will also become important for improved sensing and interconnection in vehicles. This will be contingent on the availability of suitably low-cost, robust devices. VCSELs and low-cost photodiodes will be promising candidates for this forthcoming market.

Today, the electronics content of vehicles imposes severe constraints on component manufacturers by virtue of the demanding price/performance and harsh operating environment. Moreover, although the total electronics content of a vehicle amounts to only a few percent of its total value, this is increasing year on year and the industry has many players wanting to offer an increasingly diverse range of products.

Suppliers continue to develop existing component devices and create new ones; they must also comply with ever more stringent requirements for quality, delivery and all round performance. In most cases the suppliers work closely with their automotive customers either through national/international programmes or through special relationships.

A range of technologies designed to help control road traffic, increase safety, reduce pollution and contribute to road maintenance costs are under way in North America, Japan and Europe. At key points in these systems devices are likely to play a role.

Electronic and other types of sensors for next-generation automotive systems are being developed at leading laboratories around the world. These include microwave, millimetre wave and laser systems. Functions required include:

- Collision warning.
- Collision avoidance.
- Adaptive cruise control.
- Blind-spot coverage.

In-vehicle serial communication has evolved strongly over the past few years and is providing system designers with a variety of benefits. Automotive systems are traditionally linked through wiring harnesses or looms strung throughout the vehicle. Over the years with ever more on-board systems, this has increased weight, and made it less easy to design automotive interiors. Moreover, diagnosing system failures has been made less straightforward. All these factors have had the consequence of substantially increasing costs to the consumer and manufacturer.

The main categories relevant to applications for diode lasers are shown in Table 4.29.

Table 4.29 Automotive Applications for Diode Lasers

Vehicle communications	Vehicle control
Automatic debiting (toll roads)	Collision avoidance
Mapping and orientation	Near-obstacle detection
Vehicle-to-beacon warning systems	Doppler road speed measurement

Most developments are being sponsored by a number of large government programmes such as IVHS in the USA, PROMETHEUS in Europe and AMTICS in Japan. Automatic debiting is being explored in Europe and the USA, in both the ISM band and at 4.7 GHz (in Italy and France).

With many defence and aerospace contractors leading the way, various technologies developed for the military could find commercial applications in smart highways. For example, Rockwell is promoting an automatic vehicle location system based on GPS. It includes a vehicle unit that receives and processes signals from GPS satellites to help dispatchers determine where the vehicle is.

The IVHS programme includes:

- Advanced traffic management systems that will manage traffic congestion and electronic tolls.
- Advanced traveller information systems that include vehicle displays and travel information.

- Advanced vehicle control systems, an effort that includes guidance systems and controls for autonomous driving.

There have long been potentially large market opportunities in the automotive sector. At present, collision avoidance radar (CAR) and near-obstacle detection systems (NODS) are receiving the most attention. A variety of microwave-based collision warning systems are under development. These include the Delco Electronics' Forewarn system that operates at 10 GHz and Vorad's CWS, currently available in North America for truck and bus applications. Automotive supply companies are developing low-cost collision warning technologies for the passenger car market. Diode laser-based systems are also under development to compete in this sector. It is expected that these systems will be available by the end of the decade.

CAR systems have not progressed as rapidly as first expected for a variety of reasons. Few of these are technical; for example, US bus company Greyhound fitted its fleet with radar-based collision-warning systems from Eaton Vorad and had to withdraw them. This was due in part to costly upgrades to a frequency that did not interfere with radar detectors. But the overriding concern is with safety and the failsafe requirement of systems in the light of potentially huge lawsuits against manufacturers should a system cause injury.

However, despite the technological advances, the main obstacles to the implementation of these systems are *legal and legislative* issues, rather than technological — for example, how much control will consumers be willing to hand over — and to whom. For vehicle-to-vehicle communication and/or vehicle-to-roadway communication for controlling the positioning and speed of the vehicles, control and decision-making will be handled by the roadway infrastructure. This requires keeping track of all the cars and helping to make decisions as to where they should move.

Bosch, Rockwell, Siemens, Nippon Denso and TRW are among those companies which have worked on smart-car systems. Delco Electronics has demonstrated some of the most advanced collision-warning systems, the Forewarn systems in production. One of these is used to alert school bus drivers to the presence of children in their blind spots when the vehicle is stopped, and the other is a side-detection system for heavy-duty trucks.

- Siemens' Ali-Scout field test in Oakland County, Michigan, allows two-way communication between vehicles and IR beacons at intersections. A central command system monitors how much ground the vehicles cover and, in the case of congestion, suggests alternative routes using in-car information displays.

The automotive electronics market continues to be very important for a growing range of compound semiconductors. Or, to put it another way, more and more compound semiconductor devices are finding their way into a wider range of vehicles. This is for two reasons. Firstly, that these devices offer functionality and performance which meet existing needs of automotive systems, and secondly, and most importantly in this very price-sensitive sector, they are available at competitive prices.

The uptake of CAR systems for vehicles largely remains an optional extra on high-end saloons. It is still a few more years away before CAR systems achieve significant permeation of the vehicle market.

A key factor in the successful penetration of devices in such systems will be competitive pricing of complete modules. All players in the area are adopting packaging so as to be compatible with automated chip assembly equipment. This will be a major contributory factor in the aim of achieving lowest possible system cost. These modules will consequently also be of very much smaller overall dimensions and this will facilitate installation within the tight confines of vehicles.

Automotive Adaptive Cruise Control (ACC), desirable so as to maintain driver-selected headway interval between vehicles, is a promising growth area. Mechanical cruise control has been standard on many North American vehicles where a set speed is sustained irrespective of ambient traffic conditions, while first-generation ACC monitors the headway interval providing audible warning or retards the vehicle when the inter-vehicle gap becomes potentially hazardous according to a preset distance.

Short-distance sensor functions such as park distance control systems use ultrasonic acoustic technology. This is a suitable opportunity for low-cost microwave sensors. These have several advantages: they are robust and can be mounted invisibly, e.g. behind plastic bumpers.

Follow-on generation smart systems automatically maintain the safe distance governed by throttle and brake control. ACC laser-based radar systems have been installed in a few Japanese executive saloons. Whereas in Europe companies such as Mercedes prefer radar-based ACC and in North America applications in saloons are catching up the early lead in haulage vehicles and Greyhound buses.

- The next stage in the evolution of this technology will be ACC integrated with antilock braking system (ABS) that has reduced highway accidents in the haulage sector.

The on-going successful penetration of these technologies is dependent on the further drastic reduction of unit prices. Cost is also important for the related automotive sector, that of roadside sensing, information presentation and signalling. Under consideration for some time has been the concept of embedded sensors at the roadside to monitor and control traffic flow. These are still deemed to be too expensive for use other than in special sections of highway such as urban roads or toll roads, e.g. bridges, etc.

Alternatively, the less costly route would be to employ the vehicle as the sensing platform and share the gathered information about weather, traffic flow, etc., with a central computer. All subscribing vehicles would thereby share out up-to-the-minute route guidance.

- This has been set up by Mercedes-Benz. Several models have been equipped with radio systems that gather and transmit data from on-board rain and light sensors coupled with satellite navigation systems, and stability sensors.

Awareness of the presence of objects in the vehicle's blind spots, to the sides of the vehicle and at the rear of the vehicle is important. A blind spot is the location along the side and roughly in the middle of a vehicle in which the usual combination of mirrors does not provide coverage. Blind spots are a common cause of sideswipe accidents during highway lane changes. By locating several sensors at various locations on a tractor-trailer is required.

In other sectors of the automotive marketplace, components are used to provide vehicle identification information for automated toll collection. High-performance microwave components offer one possible way of providing low-cost capabilities for assisted braking or autonomous collision-avoidance capabilities.

Pre-crash detection is foreseen to be able to increase passenger safety. One possible new function is the smart airbag that is inflated adaptively by taking into account manifold sensor information on the collision and occupant situation. While acceleration or pressure sensors trigger conventional airbag systems, the use of a radar pre-crash sensor could improve the reliability of the airbags, especially with respect to the side airbag, which is the most critical one. In the future, pre-crash sensors may also be used to activate reversible safety systems like pneumatic airbags, electronically activated safety belts, neck restraints and kneepads. In addition to the various radar sensors around the car, further in-car sensors (e.g. optical three-dimensional cameras for occupant detection) will support the safety and comfort of passengers.

For the foreseeable future there are good prospects for adaptation of existing technologies from other sectors within the automotive environment, e.g. further integration of sensors and systems with communications. The on-board systems will be moving on from a hand-carried cellular telephone and towards integrated wireless systems. These will enable vehicle owners not only to keep in touch at all times but also to check on the status and whereabouts of the vehicle. This has great ramifications for functionality as well as security. Already the on-line Mercedes-Benz Community has been set up. Upon purchase of a new model, the customer is given access to an on-line service from the car, at home or at the office. These services would include dynamic route guidance to avoid traffic jams, as well as on-line concierge services for making reservations, paying bills and even making mail-order purchases.

The technology is maturing rapidly and several key components of practical CAR systems have been demonstrated in the past year.

The automotive environment can be one of the most adverse experienced by components, and suppliers continually improve the robustness within price constraints. Electric vehicles will improve the markets for a wide range of components. They will increase take-up of power management modules and the electronics content and therefore component count will be higher than a conventionally powered vehicle. In between is the 'hybrid', which has both a conventional engine plus an electrical generator/motor. It will use the diesel engine on the motorway, for example, and be propelled by the electric power unit in towns.

In the automotive environment manufacturers are also concerned with preventing component overheating from external heat sources. This is often the case when, for example, the electronics is located close to the engine. Modern trends to ever more confined engine spaces or hotter running only aggravate these problems. It will to continue to be one of the major preoccupations of all designers and makers of automotive electronics systems to ensure that they reach the best compromise of all these factors within a price regime that satisfies everyone.

Traditionally, electronic circuitry has been located remotely from the adverse thermal environment of the engine compartment. However, there is interest worldwide in having more information and this can be achieved via fibre optic cabling using remotely located diode laser emitters. At present these command high insertion costs and are used only at critical areas in high-end systems.

It has been inevitable that with the spreading of electronics systems throughout the vehicle, moves are being made to link these systems with serial networks. This serial multiplex communication technology is commonly called multiplex or simply, MUX. MUX communication is typically achieved via single wire, twisted pair wires or plastic optical fibre. This method helps greatly reduce the size and complexity of the vehicle's wiring harness as well as providing other benefits such as improved reliability and redundancy.

Other key areas of automotive electronics that will have an impact on the supply of electronic components include:

- Vehicle security of even the most basic vehicles is a growing fraction of the vehicle's total cost. In most respects, the security measures deployed in all types of vehicle are increasingly reliant on electronics. For example, keyless entry via a hand controller. In fact, security has become one of the most sophisticated areas of vehicle electronics using advanced hardware and software to thwart the thief and reassure the owner.
- Electronic engine control (ECU) systems consisting of sensing devices that continuously measure the operating conditions of the engine to provide increased accuracy and adaptability in order to minimize exhaust emissions and fuel consumption while providing optimum driveability. These presently use microelectronic circuitry but increased use is considered of optoelectronic devices such as fibre optics to enable remote sensing within harsh ambients.
- Ignition systems have to provide precise control over the ignition of the injected compressed air/fuel mixture. The evolution to improved precision and reliability in ignition systems has come about due to the improvements in semiconductor devices providing the requisite function in an adverse environment. In particular, high-voltage power transistors, bipolar analogue ICs and microcontrollers have been optimized to withstand the adverse under-the-hood environment with respect to temperature, RFI etc.

The industry will also require the ability to manufacture cheaper components and yet retain functionality, maybe even improve upon this, and ensure that they will last for the life span of the vehicle without malfunction.

4.11 Other Markets for Diode Lasers

The market value of diode laser devices in the category under the generic title of 'other' was estimated to be worth US$130 million in 2000. This will decrease to US$122 million in 2001 as a result of the industry downturn, a fall of just over 6%. From 2002 to 2005, growth will recover and proceed at nearly 18% average annual growth rate, reaching US$234 million by 2005. This forecast information is presented in Figure 4.15 and in Tables 4.30 and 4.31.

Table 4.30 Worldwide Diode Laser Other Market by Region 2000–05 (US$ million)

	2000	2001	2002	2003	2004	2005
North America	33	31	37	43	51	60
Japan	32	30	36	42	50	59
Europe	33	31	36	42	50	59
Rest of world	31	29	34	40	47	56
Total	130	122	142	168	198	234

Figure 4.15 Worldwide Diode Laser Other Market 2000-05 (US$ million)

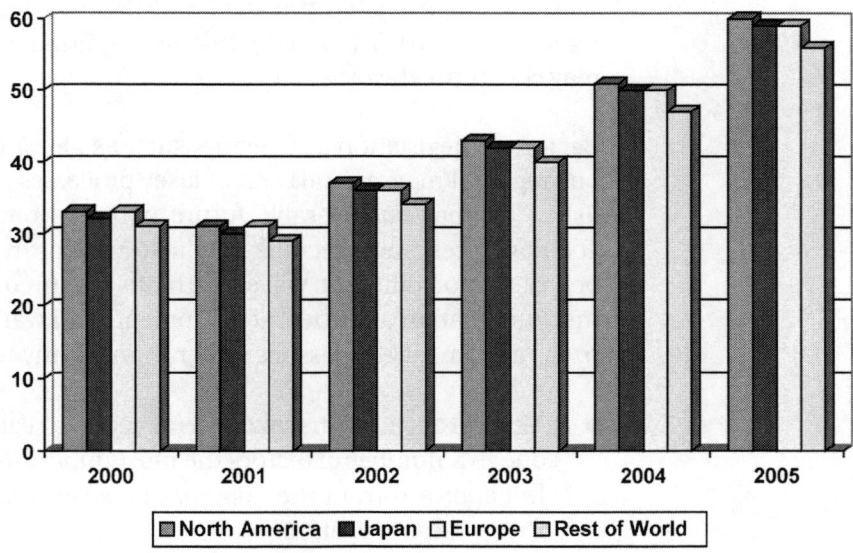

☐ North America ■ Japan ☐ Europe ☐ Rest of World

Table 4.31 Worldwide Diode Laser Other Market by Device Type 2000–05 (US$ million)

	2000	2001	2002	2003	2004	2005
FOL	14	13	15	18	21	25
CD	28	25	27	29	32	35
VCSEL	22	20	25	30	37	46
HPDL	34	32	40	50	62	77
Red	25	24	26	29	31	35
Violet	7	8	9	12	14	17
Total	130	122	142	168	198	234

The applications discussed above account for the majority of demand for devices. Other applications cover a wide range of niche applications, examples of which are given below. Generally, the next five years will see some interesting market activity in this application sector.

- Other is of course ranked as the lowest application sector in terms of value, but has in some respects future potential for component markets.

Diode laser applications for the 'other' section include:

- Medical.
- Security.
- Horticulture.
- Energy.

For example, diode lasers are penetrating more and more areas of the medical sector. Though a good proportion of this sub-sector belongs with the generic industrial sector, it is worthwhile highlighting a few specific applications that may constitute the other category.

Because of legislation and factors such as gaining approval for use in the various regional markets, take-up of laser processes has taken a while to get established but has considerable future potential once these have been achieved. Today devices have become available which are capable of handling higher power and/or different wavelengths as required for certain types of medical instrumentation and theatre equipment but await full-scale deployment owing to the aforementioned issues rather than for any technical problems.

- Medical applications take several years to reach commercial fruition. This is due to a number of factors the most important being the time taken to gain full approval from the appropriate governing body such as the US Federal Drugs Administration.
- There is also a certain amount of conservatism in this field and therefore it can take time for the medical fraternity to build up confidence in any newly applied technique.

A fairly low market forecast is made for this category given the immaturity of some of the diode lasers with respect to quality and legislation issues. The new field of diode laser devices could well come to exceed the estimate given here by an order of magnitude should the violet laser, for example, find particular application in some critical area of healthcare.

Medical applications for diode lasers include such areas as surgery (non-invasive as well as dental and cosmetic, e.g. hair and wart removal) and PDT techniques. There are a number of applications that utilize the shorter-wavelength light-emission (and detection) capabilities of these devices. Existing systems are based on gas or crystal lasers that do not have the optimum set of characteristics required for medical use.

Requirements include precise wavelength selection and longer stability of output. For example, in PDT (a process whereby accurately paced drugs are activated by intense light pulses) the availability of compact light sources in the green-blue to UV region expand the range of capabilities and choices for the treatment of sever medical complaints such as cancer.

In many respects, medical applications closely follow the breakdown of categories such as consumer and industrial, with strongest take-up for LEDs, lasers and associated components. Cost is important but functionality is paramount. Instrumentation is likely to find take-up first in private medical care centres where price is of secondary importance to the availability of a specific service. For example, a cosmetic technique may come within the 'price is no object' category of medical treatment. In other medical centres capital equipment costs are more important.

Another aspect that is growing in importance is the capability of the system being battery powered. Useful for mobile paramedic services or in the field, the higher performance of wide bandgap semiconductor-based components lends this group of devices to particular application in the portable equipment sector.

As has previously been mentioned, the prospective commercial availability of a compact, solid-state source of UV light also has important medical applications. UV light has long been used as a means to sterilize food and other products. Similarly there are many requirements for sterilization of equipment as well as, for example, in operating theatres or intensive care facilities. Low-power, compact UV sources available in competitively priced units are needed to preserve the integrity of ultra-clean environments in many medical and non-medical facilities. Such components would also be suitable for small, localized, portable sterile environments such as for the transport of drugs or organ transplants.

Those components based on diode lasers could make significant contributions to improving process control systems such as those within nuclear systems. For example:

- Electricity generating nuclear power stations. The construction of nuclear power stations is continuing but at a slower pace than in previous decades but the heightened safety requirement mandates higher performance monitoring and control systems.
- Military power sources. Systems such as submarines and aircraft carriers require compact nuclear power sources and weapon systems in close proximity to the crew and so safety systems able to withstand harsh environments are even more important.

- Spacecraft and satellites. There has been a reconsideration of the use of nuclear power sources aboard all types of space vehicles owing to the possibility of accidents at launch. This is also because of significant advances in solar cell and thermoelectric power sources. However, for deep-space probes these alternatives are not so useful owing to the remote distance from the Sun and so nuclear thermopiles are still being considered.

- Nuclear fusion. While all present nuclear power sources rely on nuclear fission, development of thermonuclear and other types of nuclear fusion powers sources continue. Handling very high power levels used in fusion reaction research requires very robust microelectronic components such as sensors and very rapid reaction control systems. Should thermonuclear fusion be achieved then there will also be a requirement for control systems able to handle even higher energy output than that from conventional power sources.

- Weapons research. While testing of nuclear weapons has become very restricted, this activity is still in progress and mandates some of the most severe environments for microelectronic components.

5 Technology Overview

5.1 Introduction

A background and overview of developments in the main technological R&D and production processes for devices is provided. The principal focus concerns the enabling technology for the production of the present and future generations of materials for diode lasers and related devices. This process involves the following sequence:

- Bulk growth of single crystals;
- Epitaxial growth of semiconductor single crystal layers;
- Device fabrication, i.e. gate and contact formation, etc.; and
- Packaging and testing.

This chapter looks at bulk growth and follows the process sequence from crystal synthesis to epitaxial growth with some coverage of device processing issues.

Earlier device types comprised semiconductor materials built upon identical substrate materials, e.g. GaAs on GaAs; virtually all of the new generation of devices require the use of other substrates. This process is termed 'heteroepitaxial' growth. Epitaxial processes that are covered in this chapter are as follows:

- Metallorganic vapour phase epitaxy (MOVPE); and
- Molecular beam epitaxy (MBE).

These are the preferred methods for the fabrication of most of today's diode lasers. Previously, liquid-phase epitaxy (LPE) and vapour-phase epitaxy (VPE) were the main approaches used for the mass production of diode lasers, and are still used for some basic devices. Newer generations of devices such as vertical cavity surface-emitting lasers (VCSELs) and high-power diode lasers (HPDLs) have been built using the more precisely controlled MBE or MOVPE processes. MBE is used to mass produce devices such as AlGaAs diode lasers, while MOVPE is very popular for LEDs.

Today, virtually all of the principal suppliers of epitaxial growth equipment manufacturing and supply systems are capable of precision growth of GaAs and related semiconductors. Each supplier has launched machines dedicated for the multiwafer mass production of these materials. In most cases this encompasses two types of reactor — R&D and production — each sharing common features so as to enable users to migrate upwards from R&D machines to full-scale production. Several of these notable commercial offerings from the major players are described here.

Also included is coverage of *in situ* monitoring (ISM) of epitaxial growth processes. ISM has become a very important tool in the preparation and control of epilayer growth for a wide range of materials but has particular application for a wide range of semiconductors. It is expected that within a few years this technique — which is really a family of techniques — will become standard practice throughout the electronics industry and especially in the manufacture of many types of optoelectronic devices.

GaAs is second only to silicon in terms of maturity and understanding of its properties. However, there are some fundamental issues such as defects, material constants, recombination mechanisms and so on that have yet to be fully resolved especially for the larger-diameter wafers. At the device level, reliability and other issues such as the role of dislocations and their effects are becoming important, hence the interest in alternative crystal growth techniques such as vertical gradient freeze (VGF).

Other compound semiconductors that are important in diode lasers are indium phosphide (InP) and gallium nitride (GaN). InP has properties making it important for a number of opto- and microelectronic applications. Its bandgap and lattice constant allow the fabrication of diode lasers with an emission wavelength well matched to the optimum transmission in fibre optics. Due to the high mobility of carriers in InP, this material is also important for the manufacture of high-speed and high-frequency transistors and related products, such as single-mode fibre optic transmission systems.

GaN violet and blue diode lasers have great promise for optical data storage and are finding interest in, for example, spectroscopic instrumentation. However, the new industry that has arisen around this semiconductor is based not on GaN substrates, which are a very immature technology, but rather on sapphire or silicon carbide (SiC).

On the open market, manufacturers of epitaxy equipment have developed their own proprietary processes that are an essential selling point for most customers. As per tradition, these processes were developed either in-house or via collaborative projects with universities and device companies. Such process technologies, which are basically recipes for specific devices, are in widespread use. There are certain patent issues however — from Rockwell in particular that relate to the pioneering work done by H Manasevit and others some two decades ago.

It is clear that certain devices can be produced by more than one route. For example, AlGaAs-based lasers can be produced equally well by either MBE or MOVPE. Preferences are more subjective than technical and one company will prefer one technique because of its longer-term familiarity with it. For devices such as visible lasers it is not generally agreed that one technique is preferable even though there are technical reasons for MOVPE over MBE relating to incompatibility with the required phosphorus compounds. Several companies have chosen to equip their fabs with both techniques in order to be best prepared for the challenges of future market supply. Other companies rely on merchant epiwafer supply rather than having installed capacity.

As a result, it is expected that neither technique will gain dominance for the next couple of years. However, by the end of the period covered by this report it may well turn out that MBE has lost out to MOVPE should phosphorus-based diode lasers continue to gain market share.

In conclusion, the new generation family of GaAs-based devices would probably not have been possible without the precision and versatility of the growth techniques described in this chapter. Epitaxy is likely to continue to be the foundation stone for all present and following generations of optoelectronic devices. Their importance is likely to continue to increase as even better process control and yields are established.

5.2 Crystal Growth

The materials supply tier of the III-Vs semiconductor industry is very well established and it is difficult for new entrants to gain a foothold so strong is the presence of existing suppliers. The competitive scenario is a moving target continually being refined to offer improved substrates at ever more competitive prices.

Specifications are beginning to settle down but industry-wide special parameters exist that are unique to one customer or another. This makes for a difficult supplier market. It is not straightforward to achieve volume orders of standard product. Not being able to produce to one or a few product specifications is in contrast to the silicon substrate industry that has been able to achieve economies of scale and very low-priced substrates. Achieving parity is likely to remain one of the major preoccupations of the III-V substrate business. Success is doubtful because the customer base does not equally share the interest.

Pricing is a key force in the optoelectronics business. It is absolutely vital to be able to produce devices at commercially desirable prices. The pricing can be broken down into several stages corresponding to the fabrication process:

- Substrates.
- Epitaxy.
- Device fabrication.
- Packaging.

The role of the substrate in pricing matters is a significant contributing factor. At the time of writing they are generally assumed to be as summarized in Table

5.1. Representative prices for other commercial semiconductors are also included in Table 5.1 for comparative purposes.

Table 5.1 Comparative Prices of Materials

Material	Price (US$)
Sapphire per wafer	32
Silicon carbide per sq in	425
Gallium nitride per sq mm	450
GaAs (SI) 50 mm per sq in	40
GaAs (SI) 75 mm per sq in	25
GaAs (SI) 100 mm per sq in	15
GaAs (SI) 150 mm per sq in	17
GaAs (SC) 50 mm per sq in	30
GaAs (SC) 75 mm per sq in	25
GaAs (SC) 100 mm per sq in	35
GaAs (SC) 150 mm per sq in	N/A
For LEDs by LPE (unpolished) 50 mm wafer	12
For LEDs by MOVPE (polished) 50 mm wafer	15
SOS per sq cm	50
Silicon per sq cm	0.5
Indium arsenide per sq in	110
Indium antimonide per sq in	110

N/A Not Applicable.

National programmes have been undertaken to improve substrate quality and manufacturability. For example, in the USA a number of DARPA-supported programmes have supported 100 mm and 150 mm diameter substrate development. Efforts to improve substrate quality are in progress worldwide and are expected to lead to lower prices and higher device yield. The general consensus is that the USA has been investing in technology for several years and is ahead, but the Japanese industry is now spending more and starting to catch up. R&D is also underway in Europe, but uncharacteristically there seems little interest in this activity in South-East Asian countries such as South Korea or Taiwan, the home of much optoelectronic device production.

Semiconducting GaAs substrates which make up a large proportion of source materials for diode lasers are grown as bulk crystals, known as ingots or boules, from which individual wafers are then prepared. Single-crystal GaAs can be grown by controlled freezing of a melt in a boat or ampoule but, more typically, by Czochralski techniques which allow the crystal to be grown with a free surface.

The most established methods are liquid-encapsulated Czochralski (LEC) and horizontal Bridgman (HB) techniques. The latter is now falling out of favour but is still relatively well suited for producing smaller, doped wafers used by the optoelectronics industry. Modifications of the LEC and HB methods, such as VGF, vertical Bridgman (VB) and vapour-pressure controlled Czochralski (VCZ), are also being developed. VGF, in particular, has emerged as a significant force in the market over the past five years.

5.3 Crystal Growth Technologies

It should be stated from the outset that the required single-crystal III-Vs have been difficult to grow compared with silicon. Purity is no longer a problem since both Group III, i.e. gallium and indium, as well as arsenic and phosphorus Group V elements, can be purified to at least a seven nines level (99.99999%). However, the differing vapour pressures of, for example, gallium and arsenic complicate the formation of GaAs with perfect stoichiometry — which also applies to InP. To avoid decomposition into its separate constituents, these III-Vs must be grown in an over-pressure of the arsenic or phosphorus vapour.

The various advantages and disadvantages of the crystal growth techniques used in R&D and manufacture are shown in Table 5.2.

Table 5.2 Comparison of Methods for Bulk Crystal Growth of III-Vs

	LEC	HB/HGF	VGF	VB	VCZ
Technical features					
Low EPD capability	Poor	Good	Very good	Very good	Good
EPD uniformity	Poor	Moderate	Good	Good	Good
Length scale-up	Good	Poor	Good	Good	Very good
Size scale-up	Good	Poor	Good	Good	Very good
In situ monitoring	Good	Good	Poor	Poor	Good
Commercial features					
Equipment cost	High	Low	Low	Low	High
Operator dependence	High	Medium	Low	Low	High
Process maturity	Very high	High	Medium	Low	Very low
Substrate price	Low	High	Medium	Medium	N/A

N/A Not Applicable

GaAs is also routinely subjected to post-growth ingot annealing. Annealing improves the uniformity of wafers, but since the process is difficult to control, ingot-to-ingot and wafer-to-wafer differences can be a problem. Furthermore, the issue of arsenic precipitation, which leads to non-stoichiometry and impaired properties, is not always satisfactorily resolved by conventional annealing. Wafer annealing is also important for perfecting VGF materials. Electrical properties are established through a multi-step post-growth annealing process developed and tailored specifically for this low-defect material. These epi-ready substrates are suitable for both MBE and MOCVD applications.

Other techniques appear in the literature from time to time. These are variants of the basic traditional processes. However, it is industry practice to keep the important improvements a closely guarded secret. It is a general truism that if a process change becomes common knowledge it may well have outlived its usefulness.

It should also be noted that for the preparation of violet diode lasers, hetero-epitaxial growth procedures are used exclusively. These devices are based on sapphire (crystalline alumina) or silicon carbide. The preparation of these materials is similar to the methods described here for the III-Vs but commensurately higher pressures and temperatures are required. While sapphire is a mature material in terms of its manufacture owing to its widespread use in conventional lasers and other applications, SiC is less so.

Devices based on sapphire dominate the short-wavelength light emitter market and are likely to do so for the foreseeable future. The technical difficulties of SiC wafers are being overcome but the wafer price is still an order of magnitude higher than that for the III-V compounds. This material could, however, become important for other applications such as microelectronic devices for high frequencies, high power and/or adverse environments. If so then the price could drop significantly and SiC could also become important for diode lasers.

5.3.1 LEC

The coverage here will be principally concerned with the LEC growth of GaAs. There are a number of distinguishing characteristics of the method that are common to all III-V materials. However, for the other important diode laser substrate material, InP, pre-synthesis is standard procedure. This entails chemical combination of indium and phosphorus so as to make polycrystalline InP. Subsequently, this material becomes the source material for the growth of single crystals.

GaAs is synthesized from gallium and arsenic together with the desired dopant, e.g. silicon, in a pressurized and heated crucible. The melt is covered with boric oxide (B_2O_3) to prevent arsenic from escaping. A schematic representation of this method is given in Figure 5.1.

Under computer control, a seed single crystal on a shaft is lowered through the boric oxide cap and then raised at about 50–100 mm/hour and rotated,

Figure 5.1. Schematic Representation of the LEC Method

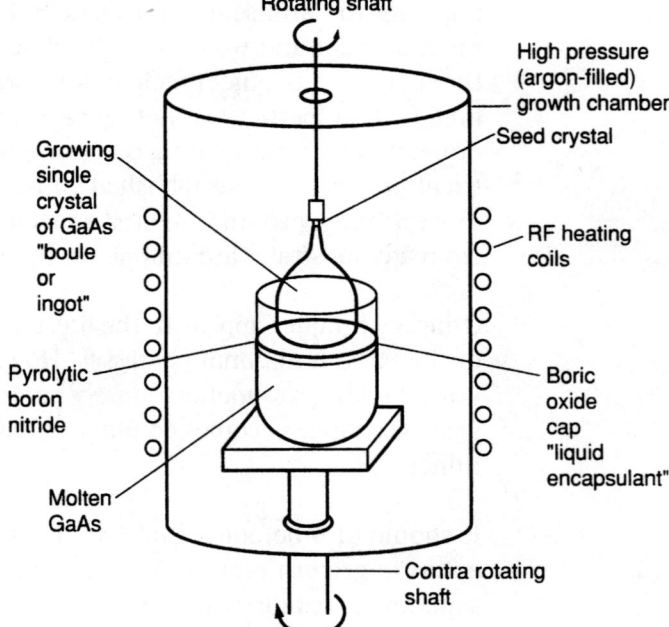

while the crucible is rotated in the opposite direction. A cylindrical ingot is pulled from the melt, the diameter being primarily determined by the rate of pull. LEC growth produces a semi-insulating (SI) crystal without added dopants.

The key drawback of the LEC method is that the thermal gradients present during crystal growth can cause dislocations and other stress-related defects. Therefore, it cannot produce wafers with very low defect levels and low stress, as required for some applications.

Recent years have seen major improvements in the quality of LEC-grown GaAs, driven largely by advances made in the elimination of residual contaminants in the starting materials. Another key area of development is to maximize the single-crystal yield per ingot by increasing melt sizes, so increasing manufacturing efficiencies.

For InP the procedure is very similar but different temperatures and pressures are required. The maturity of the process is advancing rapidly but is broadly speaking a few years behind that of GaAs, as are other indium compounds such as the antimonides and arsenides, which are also becoming important for diode lasers. For all of these indium compound crystals the resultant boule is commensurately smaller and thus yields fewer wafers.

5.3.2 VGF

VGF is a more recent development in commercial crystal processing. It is derived from the earliest crystal growth method, the Bridgman process. The drawback of this process is that crystals cannot be made SI because of contamination from the quartz vessels used. These can be replaced by boron nitride to prevent contamination and thus render the crystal SI. For semiconducting crystals, dopants can be intentionally added, e.g. zinc for p-type crystals. A schematic representation of the VGF method is given in Figure 5.2.

One of the initially touted attractions of VGF was, akin to HB, its significantly lower density of defects such as crystal dislocations. These defects cause device degradation, shortened lifetime and so on — acutely undesirable in susceptible optoelectronic devices such as laser diodes. However, it has so far proved to be the case that electronic devices such as monolithic microwave ICs (MMICs) are less vulnerable to defects than previously thought.

In common with LEC, VGF is capable of producing circular wafers requiring minimal post-growth shaping. It has also been shown to yield comparably good numbers of wafers per ingot.

As integration levels grow, epitaxy becomes more widespread and substrate diameters become larger, there is increasing demand for substrates with both lower dislocation densities (or etch pit density, EPD) as well as higher mechanical strength. Several methods can produce wafers with EPDs orders of magnitude lower than LEC (which ranges from 10 000 to 100 000 per sq cm) and without the limitations of HB. The four main methods all work by minimizing the temperature differentials that occur in crystal growth.

Figure 5.2. A Schematic Representation of the VGF Method

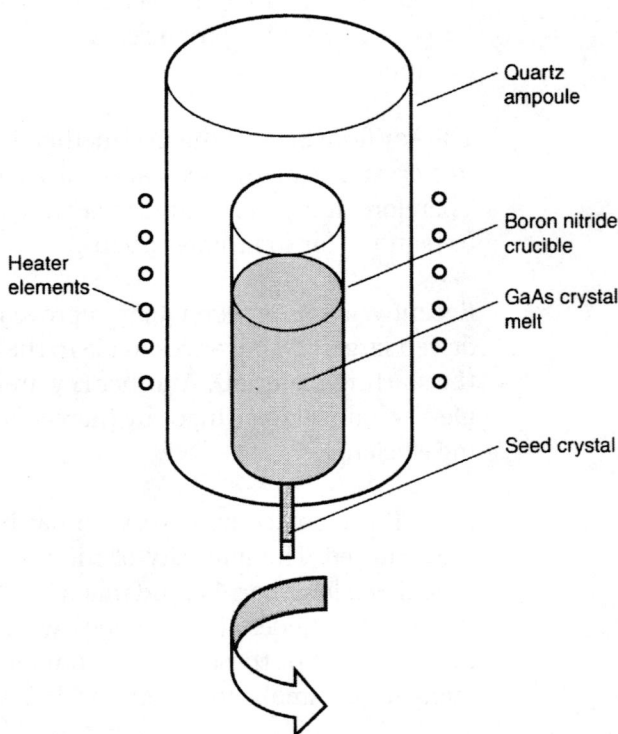

The greatest present disadvantage of VGF is its price premium over LEC, attributable mainly to the economies of scale achieved by the larger LEC suppliers. However, VGF suppliers have been making strenuous efforts to reduce this disparity by, for example, relocating production where overheads are proportionately lower.

In the USA, commercialization of VGF has been pioneered by AXT. To cope with demand, additional capacity is in place at its Fremont facility together with a new plant in China. AXT has seen demand for VGF low-EPD GaAs substrates increase as customers have been substantially increasing ordering patterns to match higher device demand. In May 2001 AXT announced its plan to approximately triple GaAs substrate capacity by the end of third quarter 2001. The expansion is aimed at relieving some of the previous short-term industry shortage, particularly for 150 mm substrates.

In VGF growth, the polycrystal charge in a pyrolytic boron nitride (pBN) crucible is heated in a multi-zone furnace. Crystal growth is initiated by melting the charge back to an appropriately oriented seed, then reducing the temperature gradient to cool the melt from the seed. The crystallization process is precisely controlled by electronically varying the temperature profile of the multi-zone furnace. Growth in low axial and radial temperature gradients is combined with diameter control imposed by a crucible, and does not require a large temperature gradient to freeze the crystal quickly. This reduces the strain, and dislocations in the crystals are typically an order of magnitude lower than for LEC material. The crucibles are made of pBN and chromium doping is not required. VGF has no moving parts and the temperature profile in the furnace moves the freezing point from seed to tail. The method was originally developed by AT&T Bell Labs in the late 1970s.

The main drawbacks of VGF are that the sophisticated monitoring and feedback processes in an LEC puller are not included (because diameter control is inherent in the process). However, the simpler mechanism results in reduced capital equipment and labour expenses, potentially leading to lower-cost wafers. The speed of crystal formation in VGF is rapidly improving; 100 mm substrates have been available for some time, and 150 mm substrates are in development.

VGF substrates from Freiberger Compound Materials' process for 100 mm SI material became available in 2000. These substrates have guaranteed state-of-the-art structural and electrical properties coupled with epi-ready surface. FCM also had a growth process for 150 mm diameter ingots in development. The furnace is designed to accommodate crucibles up to 350 mm long, and with a maximum length to width ratio of 2, thereby allowing growth of either 100 mm or 150 mm ingots. Other VGF crystal growers include Wafer Technology whose products are aimed at the optoelectronics market.

VB produces crystals with similar characteristics to those grown by VGF and is also potentially a low-cost technology. In VB growth, the seed is placed at the bottom of the crucible and the assembly is then sealed. The heating zones of the furnace travel in a vertical, as opposed to horizontal, direction. Since the heaters are placed very close to the crystal, HB-like temperature gradients are obtained giving low-stress, low-defect substrates.

5.3.3 Other Technologies

5.3.3.1 Annealing

Subsequent to crystal growth a number of manufacturers subject the materials to a further heat treatment. Called annealing, much like the metallurgical precedent, it serves to improve the structural perfection of the crystal.

The adoption of whole-boule annealing of GaAs did not occur until production shifted from 2- to 3-in diameter crystals. Recently, indium phosphide underwent a similar transition so it is likely that whole-boule annealing will become an important part of the process of InP wafer manufacture as well.

Many researchers have reported on the annealing of indium phosphide. Most concentrate on wafer annealing and even those that refer to ingot annealing are confined to crystal slabs with a maximum of 20 mm thickness. Wafer annealing, although effective, is not attractive as a production process. Consequently several crystal manufacturers have been studying the effects of whole-boule annealing.

5.3.3.2 HB

In an HB crystal growth system, pre-synthesized, polycrystalline GaAs is placed in a boat with a seed single crystal at one end. The boat, and a small amount of arsenic, is sealed in a quartz ampoule, which is inserted in a horizontal furnace with multiple heating elements. The elements are moved over the ampoule, and crystal growth follows the temperature front forming a solid ingot.

A key advantage of HB is that the shallow temperature gradient results in a very low level of defects in the crystal. While it is eminently suitable for semi-conductor substrates for optoelectronics applications and is still very important for lower-cost devices such as LEDs, HB has several drawbacks for SI production: to produce the SI properties, silicon impurities from the quartz ampoule have to be compensated for by adding chromium, which is generally (although not always) undesirable; the growth boat is not round, and to produce round wafers is a very low-yield process; and HB is not practical for growing wafers larger than 75 mm in diameter. HB therefore has limited opportunities for further development.

5.3.3.3 VCZ

The VCZ technique is essentially a modification of LEC. Crystal growth is performed in a gas-tight vessel maintained at a high temperature. The heaters are located outside the vessel and the system is sealed inside a water-cooled high-pressure chamber. The GaAs seed is placed in the melt, covered with B_2O_3 and pulled up while it is rotated, causing formation of a GaAs single crystal. Crystals of 100 mm in diameter have been grown by VCZ and 150 mm diameter wafers are in development. This technique is being developed almost exclusively by Sumitomo Electric Industries of Japan.

5.3.3.4 Horizontal Gradient Freeze

Horizontal gradient freeze (HGF) is a gradient freeze modification of conventional HB and suffers from similar limitations for production. It is mainly used to supply the market for lower-cost optoelectronics applications.

5.3.3.5 Summary

Of the main crystal growth technologies shown in Table 5.1, in commercial terms LEC is the most established method, and has the best development potential for high-volume production of larger substrates. VGF is now established as a significant commercial force, mainly due to the efforts of AXT. VB is used for large-scale production of doped wafers while the VCZ technique is still at the R&D stage, promoted only by Sumitomo Electric Industries.

GaAs crystal growth and wafer preparation has progressed significantly over the past five years, but still has a long way to go before it compares to standard silicon operational procedures. Polishing and related processes still require much development.

So-called 'epi-ready' wafers are rarely trusted to perform as such, and usually require pre-treatment. There is a lack of standardization, and wafers are customized to a large degree for each user. The electrical and mechanical properties can often vary considerably from batch to batch, although process consistency and reproducibility continue to make rapid progress.

5.4 Substrates

As a semiconductor material GaAs has several advantages over silicon including a high electron mobility and saturation velocity, and a large and direct bandgap. GaAs substrates can be prepared with SI properties, which allow active devices to be electrically isolated, reducing the presence of parasitics so improving speed and the power-delay product, and is also very useful for making bipolar devices. These properties enable the realization of ultra-fast logic, very low-noise amplifiers, highly integrated analogue ICs and a wide range of other devices.

However, offsetting these physical advantages are several practical disadvantages compared with silicon. GaAs substrates are relatively fragile and expensive, have poor thermal conductivity and stability, and passivation is difficult. As with all compound semiconductors, stoichiometry-related defects can cause problems at many steps in the manufacturing process. Substantial progress in the quality of GaAs and considerable reductions in prices have been key factors in enabling the commercial development of the industry.

In the market for InP substrates the industry is moving to larger diameters and to materials having better flatness and cleanliness. There has also been concern as to the longevity of the world's indium supply, which may expire within the next two decades.

A new factor has now entered the arena of substrate materials supply. This is not a new concept but rather a new technological means to provide the advantages of larger-area silicon wafers with the opto- and microelectronic properties of compound semiconductors such as GaAs.

Progress is also underway in 4-in InP as evidenced by the world's first multiple 4-in wafer-capable InP MOCVD reactor from AIXTRON. The industry is preparing the 40 Gbit/s generation of high-speed circuits that will be based on InP so it is important that a reliable MOCVD tool with 4-in capacity is available.

InP-based optoelectronics is also likely to move from 3- to 4-in. There will be a convergence of light-emitter technology and very high-performance electronics for which InP is the preferred alternative.

Other substrates that may become commercially important for diode lasers some time in the next decade include zinc oxide (ZnO) and aluminium nitride (AlN). In the USA, Cermet has received several contracts to develop ZnO and AlN as substrates for the growth of nitride-based devices. ZnO is a direct bandgap semiconductor so it could be used for light emission in the blue and UV. In 2001 Cermet received a US$1 million contract from the US Ballistic Missile Defense Organization (BMDO) to develop ZnO junction technology and demonstrate the fabrication of ZnO diodes.

Another substrate material is silicon carbide (SiC) — Cree of Durham, North Carolina, USA, introduced a 3-in n-type 4H-SiC wafer to complement its existing 3-in n-type 6H-SiC wafers and has demonstrated a 3-in SI 4H-SiC substrate, which it says will allow scaling of radio-frequency and microwave products for commercial production. The availability of 3-in wafers more than doubles the available device area per wafer compared with existing 2-in SiC wafers, and is expected to significantly reduce the cost of devices made from SiC.

Another important optoelectronic substrate material is germanium (Ge) for which the market is 95% dedicated to space solar cells; other current uses are for photodetector applications. However, there are also important new developments in the field of optoelectronics emitters especially LEDs where Ge can substitute for GaAs in the production of red LEDs as well as being a platform for integration of compound semiconductors on silicon. For example, Union Miniere, a leading wafer manufacturer based in Belgium, has demonstrated the use of Ge for red cavity LEDs and has also demonstrated production of Ge as a substrate for GaAs transistor manufacture.

5.4.1 GaAs *vs* Silicon

GaAs wafer sizes and prices will never match those of silicon, due to the inherent difficulties in the crystal growth of compound (as opposed to elemental) semiconductors, and the expense of the raw materials. But the gap between GaAs and silicon crystal growth technology continues to narrow.

One of the key advantages of GaAs and InP is that either semiconducting (SC) or SI substrates can be produced. It is the former that are the most important for optoelectronics. However, SI substrates are also important for certain applications and will become important for photonic ICs where electronic and optoelectronic functions are monolithically combined on one chip. This arises from the fact that the intrinsic carrier concentration of GaAs is about four orders of magnitude lower than silicon, so that true SI properties are possible.

Devices made on SI substrates are self-isolating, making the material ideally suited to IC fabrication. The reduced parasitics are essential to the high-speed capability and low speedpower product of digital GaAs, and enable the device isolation required to make analogue GaAs ICs or optoelectronic ICs.

To exploit these SI properties, two main requirements must be met. First, the bulk resistivity of the substrate must be high, typically greater than 10^5 Ω cm. Second, this resistivity must be preserved during all subsequent processing, especially after a post-implant anneal, which is carried out at 850–950°C.

One of the foundations of the emergence of the semiconductor industry is the use of epitaxial layer growth. Recently R&D has furthered the creation of working devices based on dissimilar epilayer growth on silicon substrates. Until this, development was proceeding at a slow pace suggesting that this approach would not be commercially significant in the timeframe covered by this report. However, the recent Motorola announcement reviewed in Section 5.4.5 seems to make a reappraisal of the whole substrate situation a matter of debate.

It would be impractical to review all substrate materials that have been used in R&D, since virtually all crystalline (and some non-crystalline) materials have been examined over the past five years. Only the more interesting materials that have some possible commercial impact will be looked at here.

5.4.2 Wafering

Subsequent to crystal growth the ingot is subjected to a process generically known as wafering. This refers to the sawing of the ingot into wafers and the polishing thereof through to packaging and shipment.

Substrate suppliers must ensure that during wafering the customer receives the most physically and chemically perfect surface together with exact conformity to other specifications such as mechanical dimensions, electrical behaviour and crystalline perfection. Wafer suppliers all have on-going process and quality control improvement programmes in this respect. In recent years, this process has been subjected to something approaching an international standard, via the quality conformance implicit in ISO9000.

In addition to various national programmes, there is an international programme, under the auspices of the Semiconductor Equipment Manufacturing Industry (SEMI) organization, to establish and maintain specifications for GaAs wafers. This important initiative is being delayed by the diversity of customer requirements and by confidentiality agreements. It would greatly assist the wafer business if a common standard, comparable to that used in the silicon industry, could be agreed upon. However, the industry has yet to concur on this standard and specifications are constantly evolving.

These considerations are important for all wafers but particularly the larger 150 mm GaAs and 100 mm InP wafers now reaching the fabs. Customers often refer to 'silicon-like' economics with respect to process yields, and it would greatly help the setting up and running of such a device manufacturing line. This is necessary in order to ensure that all participants achieve competitive cost control and profitability.

5.4.3 Aspects of Cutting

The process whereby wafers are created from a single-crystal ingot usually involves some form of mechanical saw. Cutting is usually preceded by an X-ray crystallographic measurement to assess and set the desired angle-off to customers' specifications.

The cutting equipment generally used in wafer factories is a diamond-coated saw. In contrast to the elegant, high-technology crystal growth process, the slicing environment is more akin to a mechanical workshop than to a semiconductor fab. Nevertheless, the necessarily high degree of precision is achieved routinely, together with high yields, low losses, timeliness and minimal wafer contamination or damage.

After slicing, the wafers are degreased and cleaned to remove process-related contamination such as wax. Next they are sent for stock removal, which is a wet chemistry etching process that removes several tens of micrometres from the front and back faces of the wafer. This is a more cost-effective route than the older method of stock removal on polishing equipment. The thinned wafers are then polished on a planetary bed chemo-mechanical machine, employing mechanical abrasion coupled with a chemical etch to remove precisely and uniformly the desired thickness and to produce a mirror-like surface. The resulting surface is particulate free, extremely flat and chemically stable. It is highly desirable to capture this pristine surface by some form of packaging for delivery 'as-is', 'epi-ready', so that the customer receives the wafer in just-polished condition. However, this is not straightforward and much work continues on perfecting this stage of production.

Samples from each process batch are taken and measured using a variety of sophisticated instruments. These data serve not only as a real-time process control check but also as a process monitoring record for any subsequent customer problems.

Today's wafering process is accomplished in ultra-clean (Class 10 in some stages) cleanroom conditions, with minimum human operator activity and the highest possible degree of automation. However, some stages still require a hands-on touch to achieve the necessary balance of throughput and yield, given the fragility of these wafers.

Of course, many of these processes are proprietary and this descriptive overview cannot go into detail. Wafering has reached a high degree of sophistication in order to meet the needs of device manufacturing customers as seamlessly as possible. All this has had to be achieved with a close eye on economics to yield as competitive a product as possible.

Laser marking is now a common procedure for wafer backing. Each wafer is coded after slicing and can be tracked not only through the wafer supplier's factory but also through the device manufacturer's fab.

5.4.4 Metamorphic Crystal Growth

While it shares many of the production characteristics of GaAs, InP in its present form has several shortcomings. These mainly derive from the substrate. Because it is a fairly immature technology, the crystal quality is less than that for GaAs. Wafers are more brittle and available in only a limited size range. Substrate sizes of 50, 75 and 100 mm are available but as yet these carry a relatively higher price (about US$ 50 per sq in). By contrast, GaAs is already available in 150 mm diameters at prices as low as US$ 15 per sq in. Also, processes such as wafer thinning and backside processing have become mature technologies for GaAs.

In early 2001 IQE of Cardiff, UK, said it was to begin supplying production quantities of metamorphic epiwafers on GaAs substrates for fibre optic and wireless communications applications. It is well known that InP-based electronic structures provide significant performance advantages over today's GaAs-based structures, and that the high-volume commercialization of such devices is imminent.

The metamorphic approach represents an attractive alternative to InP-based electronics and is potentially the lowest-cost solution to requirements for next-generation applications. Metamorphic HEMT epiwafers employ a buffer layer between the GaAs substrate and the active region to accommodate the difference in lattice constants. GaAs substrates are less expensive, mechanically stronger, of higher quality and available in larger diameters than InP The metamorphic approach enables the production of a wider range of epitaxial structures due to the flexibility of designing the buffer layer for a range of active region lattice constants.

Metamorphic InP epiwafers promise the realization of commercially effective high-performance InP devices previously unattainable using existing InP sub-strates. There exists, therefore, a potentially large market for GaAs substrates in this role. Conversely, it could result in the scaling back of development and pro-duction of equivalent large-diameter InP substrate technology. However, cer-tain devices have a higher temperature sensitivity that makes it more important to have a higher thermal conductivity substrate such as InP rather than GaAs.

5.4.5 GaAs on Silicon

In August 2001 Motorola scientists became the first to combine successfully the best properties of workhorse silicon technology with the speed and optical capabilities of high-performance compound semiconductors. The discovery, which solves a problem that has been vexing the semiconductor industry for nearly 30 years, opens the door to significantly less expensive optical commu-nications, high-frequency radio devices and high-speed microprocessor-based sub-systems by potentially eliminating the current cost barriers holding back many advanced applications. For consumers, the technology should result in smarter electronic products that cost less, perform better and have exciting new features. The technology will change the economics and accelerate the develop-ment of new applications.

Specifically, the discovery has an impact on the semiconductor industry by:

- Increasing substrate size and reducing substrate and processing costs for III-V manufacturing;
- Integrating the superior electrical and optical performance of III-V semi-conductors with mature silicon technology to create a new industry based on optoelectronic integrated circuits (OEICs);
- Extending the life of silicon and existing capital investments, improving cost effectiveness for higher-performance applications such as optical com-munications; and
- Enabling larger scales of integration.

This achievement has the potential, when fully commercialized, to transform the industry in a way similar to the transition from discrete semiconductors to ICs.

Most importantly, for the diode laser market, the new technology opens the way to OEICs, the monolithic integration of optical and electronic components. It is said to be analogous to Kilby's 1958 development of the first IC, which saw the birth of today's silicon industry.

IQE, based in Cardiff, UK, was involved in the commercial development that has resulted in the demonstration of the world's first 8- and 12-in GaAs-on-silicon wafers, the first to combine successfully the best properties of workhorse silicon technology with the speed and optical capabilities of compound semi-conductors. The technology enables very thin layers of III-V materials to be grown on a silicon substrate. Until now, this has been a virtually impossible task due to fundamental material mismatch issues.

Specifically, the underlying crystalline structures of silicon and the various III-V compounds do not match. As a result, previous industry attempts to combine them resulted in dislocations. The key to solving the problem was introducing an intermediate layer of material between the silicon and the III-V material. The solution was found in discovering exactly the right recipe for a material that would easily bond with both silicon and GaAs, reducing the strain between the two target materials in the process.

The idea was originally developed by Motorola scientist Dr Jamal Ramdani. Developing and proving the exact recipe and process grew out of work done by a broad team of scientists and engineers. Motorola is now working on developing the optimum intermediate layer for InP and other materials.

Motorola has already made working power amplifiers from GaAs on silicon wafers and successfully completed numerous wireless calls using those devices in several telephones. In addition, a light-emitting device was created to demonstrate the optical characteristics.

The underlying materials technology involves the MBE growth of strontium titanate (STO) on silicon wafers — a combination popular with developers of advanced DRAMs. This was initially carried out at Motorola's research laboratories in Phoenix, USA, and later at IQE's MBE production facility in Bethlehem, USA. The key to the process is that STO is well matched to the crystalline properties of silicon but also to those of GaAs. Subsequently, some of these wafers were shipped to IQE's MOCVD facility. There multiple epilayers of III-V semiconductors were grown to demonstrate that the concept could also be applied in future to optical devices.

GaAs-on-silicon is just the first step and has created a baseline technology for extending the research to other materials systems. The next goal is to complete the task of growing InP on silicon. This technology should support chip clock speeds of more than 70 GHz and long-wavelength lasers that are critical to fibre optic communications.

Until now, the industry has been dependent on more costly GaAs and InP wafers for opto- and microelectronics applications. Because of their brittle nature, no one has previously been able to create commercial GaAs wafers larger than 6 in or InP wafers larger than 4 in. Scientists have also been unable to combine light-emitting semiconductors with silicon ICs on a single chip.

Motorola has filed more than 270 patents on inventions related to this new technology (e.g. US Patents 5,904,553, 5,907,792, 6,022,410, 6,030,453, 6,110,840 and 6,113,690) and the company intends to broadly license the technology.

5.5 Epitaxial Growth

5.5.1 Introduction

The principal epitaxial growth process technology for the development and production of compound semiconductors relies on two principal approaches:

- The MOVPE process. This is also known as OMVPE (organometallic VPE) or MOCVD (metallorganic chemical vapour deposition). These terms are frequently used interchangeably but for precision the term MOVPE is used in this report. There are variants of the basic vapour-phase epitaxy technique — such as hydride VPE (HVPE), but in the main it is MOVPE that is the principal equipment type in use.
- The MBE technique is an ultrahigh-vacuum technology which has found particular application in the production of HEMT and HBT devices based on AlGaAs. Variants of this process are under development, which rely on gas sources (GSMBE) or metallorganic sources (MOMBE) plus other processes relying on special 'cracking' technology to provide phosphorus sources for the production of InGaP epiwafers.

Epitaxy has its origins in 1960, based on the work of Kleimack and others, as a new method for growing layers of silicon identical in structure with the silicon substrate itself. As the technique matured, epitaxy continued to be adapted for other multilayer semiconductors and semi-insulators. Subsequently, Cho, a US researcher, developed the first MBE system producing single-crystal growth one atomic layer at a time.

For GaAs and related compounds the MOVPE and MBE techniques are in widespread use. MOVPE is also the more important for the growth of wide bandgap semiconductors such as GaN and SiC while MBE is more important for SiGe.

Ion implantation is still the most important non-epitaxial growth technique but its application is becoming increasingly restrictive in the production of next-generation transistor devices. However, such is the production throughput — and hence resultant low die cost — of ion implantation that this is still unmatched by most epitaxial processes.

The basic techniques of epitaxy are examined here. These techniques are illustrated by reference to examples of commercial equipment from the principal suppliers. The wide range of approaches that are being deployed by MOVPE equipment suppliers for routine production of devices is illustrated. These examples are in the main developments of a basic reactor technology that has been adapted to the required growth regimes.

In every example discussed in this report users of the equipment have reported successful growth of devices and are therefore finding considerable take-up by research laboratories and production fabs of many device producers worldwide. There are a number of characteristics of the epitaxial equipment business that are worth noting as they help to create a picture of equipment provision and the operations of the suppliers.

In 2001 AIXTRON announced the development of III-V epi on silicon wafers for electronic devices but with the potential also for optoelectronic devices. The step is important because it uses standard silicon wafers in a commercial III-V LP-MOCVD system having additional proprietary features for the 'HeteroWafer' technology. AIXTRON owns a substantial number of patents and intellectual property rights protecting the key underlying MOCVD technology, and has further applied for additional patents in its field of business. It will strongly contribute to cost reductions and scaling of compound semiconductor manufacturing technologies in particular for electronic devices.

The AIX 3000 is a true multiwafer 12-in (300 mm) technology, with a capacity of three of these large 12-in silicon wafers. Detailed process development at this large scale continued, developing base processes for the required variety of III-V materials, from GaAs, AlGaAs, to GaInP and AlGaInP, and reaching into the InP material system. Following on, the first MOCVD epitaxial processes for these III-V materials were demonstrated on the multiwafer 8-in scale.

Advanced III-V MOCVD synthesis routes have further been developed on silicon, sapphire, SiC or germanium wafers in order to foster the understanding of the critical step to deposit compound semiconductor material onto wafers of different nature, thermal behaviour or conductivity. This contributed to the successful development of AlGaN/InGaN/GaN production on sapphire wafers, which is already in high-volume production for blue, green and white emitters.

The advantage of these materials is the enhanced thermal durability for high-power electronics, allowing prototype operation at 16 W/mm static heat dissipation, without degradation of performance. AIXTRON's 'HeteroWafer' epitaxy technology is likely to be attractive to the industry because it is being made available on a free-of-charge 'User Licence' basis along with all new MOCVD equipment purchases dedicated to such material systems.

5.5.2 Custom *vs* Standard Equipment

When the compound semiconductor device business began, device companies themselves traditionally developed the equipment used in many laboratories. A combination of circumstances has led to these practices becoming less commonplace. In particular, it has become necessary to ensure that any system is fully equipped to meet or exceed all safety regulations. So it is less common for reactors in research laboratories to be completely home-built and many are a basic apparatus that is locally modified. Equipment suppliers were once faced with the business problems of never building the same reactor twice.

Earlier there was no such thing as a standard system and to some extent this is still true. However, the leaders in the business have been successful in creating suitably versatile systems whereby customers can select specific features usually relating to the complexity of the materials growth they wish to research from a list of options. Nevertheless, for other areas of materials research, e.g. high-temperature superconductors, very specialized reactors are still being built from scratch so to speak.

It is now common to learn of specific sales of MOVPE or other epitaxy equipment to Japanese laboratories or companies. This has not always been the case.

Unlike Japanese equipment suppliers, home markets for other suppliers have not been sufficiently large for a usefully sized business. Western companies have therefore begun in their home country and then steadily expanded their business to other regions. In all cases considered here, whilst these companies have strong local connections with their origins they are truly international companies and operate as such.

In the following, examples of equipment from each of the principal Western suppliers are presented. In each case the background to the development of the equipment is given along with specific characteristics and examples of which companies have selected these systems. The discussion is largely confined to the key aspects of equipment but further information on the companies can be found in Chapter 6.

The coverage is not meant to be exhaustive because in certain cases customers exercise the right to confidentiality and suppliers have been unable to reveal placement of systems. This is especially pertinent in Japan and South Korea.

5.5.3 MOVPE

The particular technological requirements related to the deposition of semi-conductors have resulted in the need to re-examine reactor design by all manufacturers of such equipment. As a consequence this has created a level playing field across the entire reactor business for this application. The established epi-wafer manufacturers, both captive and merchant, once again have to look at the merits of the various MOVPE reactors offered by the major suppliers. A generalized schematic representation of the MOVPE system is shown in Figure 5.3.

Historically, the first developments in MOVPE were made by workers at Rockwell in the late 1950s. In particular, H M Manasavit and W I Simpson published the seminal papers on the use of metallorganic precursor materials. Previously, the main epitaxial growth methods in use were variants of the LPE and VPE techniques. MOVPE grew out of the latter process as a result of the difficulties encountered with the use of aluminium materials.

Figure 5.3. A Schematic Representation of the MOVPE Method

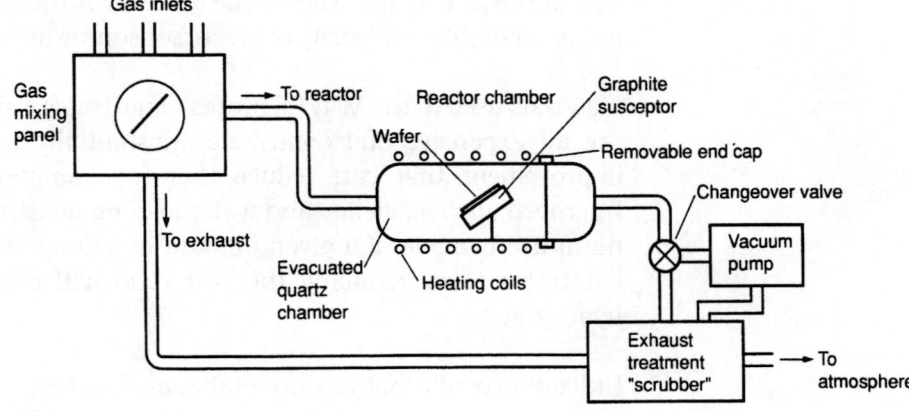

The production of devices took a few more years with initial results in electronic devices being reported by 1975. The earliest success with optoelectronic devices came with the growth of epitaxially grown diode lasers based on the AlGaAs alloy.

The MOVPE process began to be established as a production technique for photocathodes in the early 1980s. The first mass production of optoelectronic devices came in the mid-1980s. These were photocathodes and solar cells, which were exclusively produced by MOVPE techniques. MOVPE has only recently become a significant contributor to the mass production of emitter-type devices. The LED industry is built upon the earlier epitaxy techniques such as LPE and VPE but particularly the former. To put things into perspective, around 30 billion LEDs are manufactured every year and yet in 1994 virtually all were made by these techniques, not by MOVPE.

While MOVPE has certain advantages — such as ease of use with circular wafers that the older LPE technique normally does not — in due course it will have been developed to supersede LPE.

For laser diodes the situation is somewhat different. Up to 100 million laser diodes are manufactured every year and these are prepared by the three epitaxy techniques roughly in equal proportion. Long-wavelength LEDs use LPE, MOVPE and MBE, whilst AlGaAs lasers use either MOVPE or MBE. The AlGaInP laser types use only MOVPE.

Today there is much less to choose between each technique. This is largely down to the device application though even that is not clear and may depend a great deal on the preference and experience of the particular user.

A principal factor in the widespread take-up of the technique is the high degree of flexibility in the design of precursor molecules. Since the growth processes are far from equilibrium, stable and metastable compounds can be grown. Almost any combination of layers and layer sequences can be deposited on almost any crystalline substrate. Moreover, great progress has been made in the industrialization of MOVPE, and specific examples of the most popular MOVPE equipment in use today are examined in more detail below.

One of the principal drivers of the industry is the refinement in equipment and in the versatility and purity of precursor source materials, i.e. reactants.

Materials used in the MOVPE process substrates, source gases and carrier gases are all expensive and contribute substantially to finished wafer cost. Any improvement that can reduce their consumption can help reduce costs. Improved alkyl efficiency and wafer loading density will ensure that the maximum use is made of a given amount of source alkyls being deposited on the substrates. Also, reducing the V-III ratio will minimize the consumption of hydride gas.

Utilization of alternative sources that are less toxic such as tertiary butyl arsine (TBA) instead of the highly toxic arsine, is preferred for some circumstances. Use of TBA, etc., unfortunately has prohibitive impact on the finished wafer cost and has not found large-scale commercial take-up as yet. This may change should TBA costs fall as volume use comes along, but the price margin over arsine will remain high. Yet use of TBA can assist in the reduction of capital equipment and maintenance. This is because it has a lower vapour pressure and thus does not need high-pressure gas piping, etc. In all instances, therefore, users must take into account their specific requirements and balance out the often conflicting parameters for an efficient operation and minimization of the cost of the final product.

MOVPE machines capable of growing highly uniform layers over an area of hundreds of square inches are available principally for the solar cells industry whereby GaAs is grown on germanium substrates. Several efficiency, yield, safety and toxic waste disposal issues remain to be resolved.

The successful preparation of active devices also requires the incorporation of electronic doping materials from Group IV, e.g. carbon or silicon, and Group II, e.g. beryllium or magnesium. These, like other source materials the so-called 'precursors' must be of the highest quality that is commercially available.

5.5.4 Epitaxial Precursor Materials

III-Vs, such as GaAs, are direct bandgap semiconductors with applications in microelectronic and light emitting/detecting devices. The successful fabrication of devices based on this family of III-V compound requires high-purity source materials. Specifically:

- High purity — at least '5-nines', 99.9999% gallium, usually in the form of an organometallic (OM) such as trimethyl gallium (TMG).
- An arsenic-containing source. Arsine (AsH_3) is most commonly used as source of the Group V element for growth of GaAs (or phosphine (PH_3) for InP).

Precursors of various kinds have been investigated over the past five years each with particular advantages and disadvantages with respect to semiconductor growth.

Gallium is commonly in use for a fairly wide range of electronics applications, e.g. solders and electronic materials such as GaAs and to dope silicon. The metal and its highly purified compounds are much in demand for crystal and epitaxial growth. The availability of gallium has often been cast in doubt and hence by turn has its compounds and device markets based on them.

Unlike mainstream silicon, gallium is not plentiful and its extraction and purification, etc., are periodically considered uneconomic by the handful of providers. This has led from time to time to reappraisals of the future of any devices based on gallium. It is deemed unlikely that supplies of gallium will dry up but prices may have an adversely impact on device competitiveness. A number of companies (e.g. Recapture Metals and Eagle-Picher in the USA; III/V Reclaim in Germany) have successful operations based on recovery of valuable gallium from scrap gallium-containing compounds.

It should also be pointed out that restrictions in gallium supply may also adversely affect the prospects for single crystals compared with other single-crystal materials which could be considered as substrates. Aluminium, unlike gallium, is abundant. GaAs substrates use up orders of magnitude more gallium metal than epilayers, dopants, etc.

Other families of gallium -containing devices are also enjoying considerable market success. GaAs is now not only very successful in optoelectronics but also in microwave devices, etc. These rapidly expanding markets seriously affect the gallium market. On the one hand, they may make ore refiners reconsider their production plans and on the other they may cause the price of gallium to change. Either way this may adversely affect the manufacturing costs of GaAs-based devices at a time when great efforts are being made to make them even more competitive compared to silicon-based devices.

Whatever the outcome, the long-term availability of gallium and gallium-based precursors is likely to have an impact on the developmental success of GaAs-based microelectronics. In extremis, the very success of GaAs microwave ICs and optoelectronics may prove its undoing. There may be a gallium shortage and the whole market may be forced in new directions. Meanwhile, gallium recycling will see strong growth and further entrants or diversification in this sector may occur in due course.

There is also the prevailing trend within the microelectronics industry for leaner and greener manufacturing to be taken into consideration. For example, the industry has all but eliminated CFCs from its cleaning processes and now there are moves to eliminate lead from solders. Also important is the increased requirement for manufacturers to be responsible for their products from 'cradle to grave'. This has particular relevance to devices that contain potentially harmful ingredients such as arsenic in GaAs.

It is likely that gallium recovery will be extended to MOVPE exhaust products. Rather than being a nuisance, this will have higher value and hence be more economic to recycle.

Gallium price increases will necessitate improved efficiency of epitaxial processes.

5.5.5 Commercial MOVPE Reactors

5.5.5.1 Introduction

At this point, the principal MOVPE and MBE reactor types with application to optoelectronics are described. Recently, there have been some important developments in this area while at the same time there are some things that have changed very little. Merchant suppliers of epitaxial equipment have developed their own systems dedicated to answering the particular technological problems posed by multiwafer growth of GaAs in a commercial system available on the open market to all customers from R&D to production. This product line represents a substantial investment by these companies that has been rewarded by remarkable sales figures over the past three years. Makers of GaAs-based MOVPE equipment and MBE systems have enjoyed amongst the strongest growth rates in the market.

Another important recent business development has been the merger of three of the most important specialist players in the epitaxy equipment supply sector: AIXTRON AG has become a major shareholder in the Swedish SiC specialist Epigress AB and almost at the same time brought the UK MOVPE equipment company Thomas Swan Scientific into the AIXTRON group of companies. In contrast, the leading US supplier EMCORE Corp has continued to diversify into a variety of affiliated epitaxy-oriented businesses, not only Hall sensors but also wide bandgap semiconductor materials and devices.

Each of the reactor types is described here, together with some examples of the customers who are presently using them. By the end of 2000 it was estimated that worldwide the total number of epitaxial growth systems was well in excess of 200. It is not a straightforward matter to ascertain the precise number or locations of all of these systems because in many cases the customers have exercised their right for anonymity. Where information is publicly available, details of these systems are provided in the company profiles and in the listing of research institutes.

It is also worth reiterating that these systems are in common use for micro- as well as optoelectronic device manufacture. In the case of most machines, however, their use is exclusive to either one of the other device type. Few epitaxy machines are used for more than one type of device. Certainly it is rare for a user to grow epilayers for optoelectronic devices on a machine set up for the production of microelectronic devices. Perhaps towards the end of the forecast period certain special ICs will contain features common to both opto- and microelectronic devices and will be routinely fabricated on the one all-purpose epitaxial reactor, i.e. *in situ* manufacture of monolithic OEICs also known as photonic ICs (PICs).

5.5.5.2 The AIXTRON Planetary Reactor System

AIXTRON AG is based in Aachen, Germany, and has one of the broadest ranges of MOVPE and related equipment in the market. Nearly all are based on the patented Planetary Reactor MOVPE epitaxial process originally developed at the Philips LEP Laboratory at Brevannes, France, under an ESPRIT research programme. AIXTRON took out a long-term exclusive licence for the worldwide rights to manufacture the system. It has successfully further developed the basic process for a wide range of materials and scaled it up for high-capacity production machines. The AIXTRON portfolio includes a complete MOVPE line from R&D up to full-scale mass production.

For example, the AIX 2400/2600G3 has a flexible design for up to 5 × 6-in and 35 × 2-in wafers (a 7 × 6-in version has also been launched). The two-flow horizontal Planetary Reactor is recognized as the most widely used multiwafer MOCVD reactor for compound semiconductor applications.

The reactor concept allows laminar gas flow without turbulences for precise control of the material composition and achievement of ultra-sharp interfaces. Ultrahigh uniformity along with high growth efficiency on multiple 2-, 3-, 4- or 6-in wafers is achieved through dual wafer rotation with the patented Gas Foil Rotation.

AIXTRON offers a wide range of epitaxial systems for MOVPE, VPE and LPE from R&D-oriented systems up to large-scale Planetary Reactor production systems for up to 25 × 4-in or 9 × 6-in wafers. AIXTRON has sold over 500 systems worldwide and is represented in over 15 countries.

Table 5.3 Key Features of AIXTRON Series of Epitaxial Reactors

Series	AIX 200	AIX 2000	AIX 2400	AIX 3000
Load capacity/process technology	1 × 2 in, 3 × 2 in, 1 × 3 in, 1 × 4 in	7 × 2 in, 5 × 3 in	15 × 2 in, 8 × 3 in, 5 × 4 in	105 × 4.5 cm^2, 95 × 2 in, 40 × 3 in, 25 × 4 in, 9 × 6 in, 5 × 10 in
AlGaAs/GaAs	×	×	×	×
AlGaInP/GaAs	×	×	×	×
GaAsP/GaAs	×	×	UD	UD
GaInAs/InP	×	×	×	
GaInAsP/InP	×	×	×	
AlGaInAs/InP	×	×	UD	
InSb/GaSb	×	×	×	
InGaN/AlN	×	×	dev.	

UD Under Development.

Further industrial clients include names such as Agere, Alcatel, Anritsu, ATMI, AXT, Epistar, EPSON, Honeywell, IQE, ITT, JDS-Uniphase, Kopin, LumiLeds, Mitsubishi, Motorola, Nortel, Osram, Procomp, Samsung, Showa Denko, Siemens, Sumitomo, Thales, UEC, VPEC and numerous Japanese corporations. AIXTRON systems have also been installed in many research institutes including JPL, Sandia National Laboratories, Fraunhofer Institute, ITRI-OES and CNRS. AIXTRON is very active in advancing MOCVD technology through collaborations and joint R&D projects with major partners worldwide.

Overall, the AIXTRON Planetary Reactor is considered to have a number of advantages which makes it particularly suitable for GaAs materials for opto-electronic devices such as diode lasers and LEDs. It confers excellent interface abruptness and offers a rapid heating/cooling capability. The AIX2500G3 cassette-to-cassette system was the world's first 5-in MOVPE reactor, suitable for R&D systems for a wide range of different materials technologies. AIXTRON can supply complete MOVPE lines from R&D up to full-scale production, i.e. 1 × 2-in up to 95 × 2-in wafers, or equivalent 3-, 4- and 5-inch wafers.

Examples of AIXTRON's recent GaAs reactor sales include:

- In June 2000 AIXTRON announced that it had sold its 150th AIX 2500G3 multipurpose MOVPE reactor, the third-generation version of the Planetary Reactor. The customers which can be listed are: Kopin, EPI, OMMIC, Marconi, Showa Denko, UEC, Epistar, Procomp, VPEC, Arima Optoelectronics, Epitronics and Epiworks.
- Sumitomo Electric Industries of Yokohama, Japan, purchased an AIX2400 multiwafer Planetary Reactor for the production of epiwafers for communication network systems devices.

AIXTRON is also the only supplier of commercial HVPE systems for GaN including guaranteed growth processes. For wide bandgap semiconductors, AIXTRON machines are in production for blue, green and white LEDs (e.g. at Agilent Technologies, now LumiLeds Lighting). To date the AIXTRON reactor series has achieved world-class results for the growth of wide bandgap semiconductors. Numerous publications attest to the capability of these systems.

- AIXTRON was the first company offering GaInN multiwafer technology and demonstrating MOVPE epitaxial growth on full 2-in GaN and InGaN wafers. AIXTRON recently demonstrated device quality GaN with excellent PL uniformity better than 1 nm across a 2-in wafer and thickness uniformity typically better than 2%. Similar results were reported by Meijo University using the AIX 200/4HT system.
- The Institute of Electronic Materials Technology (ITME) in Warsaw, Poland, ordered an AIX 200/4 RF-S MOCVD system for GaN-based structures for lasers and LEDs. In addition to the existing AIX 200 MOCVD system that has been used for the fabrication of GaAs- and InP-based structures, the new AIXTRON reactor will play an important role in the Polish GaN project.

Other AIXTRON epitaxy systems include a single wafer HVPE system for III-nitrides, a single- or multiwafer VPE system, a single-wafer LPE system for III-V and II-VI materials as well as the AIXTOX scrubbing system for the decontamination of toxic exhaust gases.

The AIXTRON line of production reactors is based on a proven scaling design enabling all processes to be transferred relatively easily from one model/size to another. All systems display high homogeneity in film thickness and composition with excellent electrical and optical film quality.

One of the most important commercially used systems is the AIX 2400 Planetary Reactor with Gas Foil Rotation. This system has high up-time and throughput and has proven very popular for UHB-LED mass production. The average throughput for the AIX 2400 is up to 27 000 wafers per annum.

All Planetary Reactors have one advantage in common: the Radial Two-Flow reactors, which provide an inherent advantage for very low consumption of the speciality gases used in MOVPE. For example, in comparison with the rotating disc reactors from competitors, the material consumption (for the same amount of qualified epiwafer material in real production mode) is lower by a factor of foir in the AIX 2000 and 3000 reactor line. At the same time the V/III ratio is dramatically lower than in the rotating disc reactor concept. The consumption of the main carrier gas, hydrogen or nitrogen, is also much lower, and in fact many processes have already been qualified with nitrogen carrier gas based on a proprietary gas injection system so that the hazard from hydrogen is also eliminated.

The Planetary Reactor uses a rotating disc plate for wafer support. It uses gas bearing technology to provide smooth rotational drive and reduced particulate contamination. The unique feature of the system is that the centrally fed radial gas streams are used for both levitation and rotation of the main support and the individual wafer platter. The hydrogen carrier gas is fed through special grooves to provide rotation of the wafer carrier.

The rotating wafer holders are located in a support plate that will hold up to 7 × 50 mm, or 5 × 75 mm diameter satellite wafers. The individual rotation of the support plate and that of the individual wafers guarantees an extremely high degree of compositional and thickness uniformity for the growth of epilayers. This provides frictionless rotation of the wafer free from particulate generation and ensures outstanding wafer uniformity and extremely abrupt interfaces.

AIXTRON is playing a key role in the further development of MOVPE equipment for manufacturing purposes and introduced four years ago the first MOVPE system with robotic cassette-to-cassette wafer handling (AIX 2400/2500G3).

Of the top ten optoelectronic semiconductor manufacturers, seven are using AIXTRON production-scale reactors for their optoelectronic manufacturing.

- AIXTRON delivered two AIX-2400G3 MOCVD reactors to Princeton Lightwave (Cranbury, New Jersey, USA), the supplier of active components for the optical networking industry to enable it to satisfy high demand for its Wave-Power high-performance EDFA and Raman laser pump modules.
- Multiplex (South Plainfield, New Jersey, USA) has increased its production capacity for InP-based lasers by adding two AIX 2400G3 Planetary Reactors to its existing AIXTRON MOCVD tool.

5.5.5.3 The EMCORE Systems

EMCORE Corp, based in New Jersey, USA, offers two families of TurboDisc Tools for the production of epiwafers for GaAs, AlGaAs, InP, InGaAsP, GaN, InGaN, AlGaN and SiC: the Discovery and the Enterprise. Each of these families of tools is engineered to optimize performance and provide the highest-quality epitaxy. EMCORE focuses on the market demands for compound semiconductor applications such as LEDs and diode lasers as well as electronic materials for higher-performance transistors, solar cells and magnetic sensors.

By configuring these tools for each end market application, EMCORE claims to offer its customers unprecedented production tools with good ramping times from installation to material production. Features of the Discovery series are summarized in Table 5.4.

Table 5.4 Key Features of EMCORE Discovery Series of Epitaxial Reactors

Reactor	Semiconductor	Device
Discovery R&D	InGaAsP, InGaAlP, AlGaAs, AlGaN, InGaN, SiC	Customer specified
D180 LDM (Laser Diode Machine)	InGaAsP, InGaAlP, AlGaAs	VCSELs, diode lasers
D180 SpectraBlue and D180 SpectraGaN	AlGaN, InGaN	Blue LEDs, violet lasers

EMCORE has focused on the TurboDisc technology deposition equipment (see Chapter 6). This has proved itself both in the field with various customers worldwide and also in EMCORE's own epiwafer production facilities.

While applicable to a diverse range of devices including semiconductors and solar cells, EMCORE's TurboDisc technology is said to be uniquely suited for the large-scale epitaxy production that is required by the wireless industry to meet the projected cost requirements needed to compete with conventional ion implantation device fabrication methods.

EMCORE says that because of its scaleability of TurboDisc technology, growth systems range from 75 mm up to the mammoth 400 mm platter. This is claimed to be the only CVD deposition chamber in the world (silicon or compound semiconductor) capable of growth on a single 400 mm wafer.

The Enterprise family of TurboDisc Tools is the flagship fleet of EMCORE's high-volume production tools for epitaxial materials. Each tool in this series has been configured for specific end market applications for high process reliability and quality epitaxy.

Table 5.5 Key Features of EMCORE Enterprise TurboDisc Series of Epitaxial Reactors

Reactor	Semiconductor	Device
E450	GaAs, InP	HEMTs, HBTs
E400 GOLD	InGaAlP	High brightness red, orange and yellow LEDs
E400 M	InGaP	pHEMTs, HBTs, FETs
E300 GANZILLA	AlGaN, InGaN	Blue LEDs, blue lasers
E300 LDM	InP, InGaAsP, InGaAlP, AlGaAs	VCSELs, diode lasers, electronic materials

The TurboDisc systems are available in several platforms:

* Enterprise for volume production;
* Discovery for pilot production; and
* Explorer for research.

EMCORE also offers customers the EPIMETRIC *in situ* photo-reflectance system to monitor the growth rate and thickness uniformity of a broad range of materials.

In 2001 EMCORE announced the purchase of ten additional TurboDisc Enterprise production systems to United Epitaxy Company (UEC) in Hsinchu, Taiwan. UEC already had six EMCORE MOCVD systems, used for the production of high brightness LEDs. The additional systems are EMCORE high-volume Enterprise series MOCVD production platforms configured for the production of HB-LEDs and electronic materials.

EMCORE has been working with other US agencies to develop ISM of epitaxial growth. This is an important area of work and has potentially great impact on the viability of the system in a manufacturing environment. This is especially when such a system can be directly linked to real-time computer-controlled growth. It is no straightforward matter to simply add such a monitoring system to the growth chamber, as special ports are needed for optical access. Thus an ISM monitoring system has to be part of the overall design of an MOVPE reactor.

EMCORE views this as a 'paradigm shift' away from the traditional procedure of lengthy dummy runs to set up the growth system as close to perfect as possible. These calibration runs are time consuming and have a great impact on the economics of mass production processes based on MOVPE and the growth of GaAs is no exception to this rule. The move to *in situ* control will avoid this procedural approach and make MOVPE a much-improved economic solution for production (see below). EMCORE has developed and refined this technology in its own R&D laboratories working in collaboration with other agencies such as Sandia National Laboratories.

By 2001 reactors had reached 51 systems worldwide mainly for optoelectronics but also for microwave applications. To reach this total the company had secured an additional nine systems of which five represented multiple orders. The company declared that a single customer had no less than eleven Spectra-Blue systems that are dedicated to GaN optoelectronic devices.

The well-established Enterprise E400 MOVPE system has been proved in solar cell mass production and is also popular worldwide for a range of other optoelectronic devices. It was launched in 1995 and the Enterprise system was launched in 1995. Both systems are based on the proprietary TurboDisc process technology. These reactors are capable of deposition of semiconductor materials.

EMCORE has achieved a number of sales successes with the TurboDisc system in GaAs device manufacture. In late 1999, EMCORE announced the sale of an Enterprise 400 TurboDisc MOCVD platform to the NTT Optoelectronics Laboratories of Nippon Telephone and Telegraph of Tokyo, Japan. The E-400 EM will be used by NTT for the production of pHEMT and HBT electronic materials.

5.5.5.4 Thomas Swan Systems

MOVPE reactors manufactured by Thomas Swan, based in Cambridge, UK, have a broad applicability as regards materials systems but in recent years the company has focused on the wide bandgap semiconductors. Thomas Swan systems are based around the proprietary showerhead technology for nitride-based MOVPE (see Chapter 6). The showerhead technology was developed in collaboration with the IMEC research group based in Belgium. Both parties have a long-standing relationship for shared R&D; the successful Epitor range of reactors came out of this.

Thomas Swan has other collaborations with UK universities for equipment and software development, in particular with Cambridge University where it has six Close Coupled Showerhead (CCS) reactors for GaN process and optimization developments. It is also working with Salford University on advanced optical monitoring and with the North East Wales Institute on software for ISM. The company's CCS reactor is popular for optoelectronic R&D and pilot production at facilities around the world for GaAs and InP.

Thomas Swan's reactors have traditionally been very popular in the R&D and pilot production sectors but the newest reactor, the CCS reactor, is also aimed at multiwafer production. Thomas Swan's showerhead approach with its inherent scalability to deliver larger volumes has given the company a strong technology platform on which to grow its business. The company says that this technology can be developed to enable it to compete with success in multiwafer systems for GaN as well as InP/GaAs applications.

Following the success of its 3 × 2-in GaN system, the company introduced 5 × 2-in reactors for InP/GaAs and further expanded its GaN-related offering with a 5 × 2-in wafer reactor, dry ammonia scrubbing system and an *in situ* interferometer during 1999. Also, it planned to deliver its first 12 × 2-in (5 × 3-in or 4 × 4-in) system for InP/GaAs in the first quarter of 2000, emphasizing the company's push to scale its products.

Improvements in growth uniformity and reduction of dependence on growth conditions have included the development of improved reactor geometry. These developments mean insensitivity to such growth conditions as carrier flow, reactor pressure and susceptor temperature. This facilitates the optimization and leads to better run-to-run reproducibility.

- In collaboration with IMEC, a novel showerhead type of attachment for a vertical tube, rotating susceptor, multiwafer MOVPE reactor has been made.

Thomas Swan has traditionally worked very closely with a number of universities, but particularly the University of Ghent in Belgium, to develop new products. In the future it plans to maintain these relationships but has created its own in-house R&D unit to permit it to focus on developing systems for mainstream manufacturing. However, the announcement in 1999 of Thomas Swan's move into the AIXTRON group meant that the company would have access to that company's extensive R&D and sales/support organization worldwide.

5.5.5.5 Other Systems

The Japanese company Nippon Sanso has developed a novel horizontal MOVPE system in which 5×3-in wafers or 18×2-in wafers are mounted on a 10-in diameter susceptor. The MHR-8000 reactor has two special features: a heating system and wafer setting. The heater unit is set above the susceptor surrounded by a radiation shield consisting of three main parts controlled independently to achieve uniform temperature at the growing surface. A face-down setting has been developed for the wafers so as to minimize thermal convection and attraction of particulates. The wafers are placed downwards and source gases flow under the growth surfaces. The carbon tray that inversely holds the wafers is transported through the load-lock chamber for each run and hung on the susceptor. The wafers are rotated with the susceptor to attain uniform growth.

The European MOVPE company dedicated to the manufacture of crystal growth equipment for SiC (and also SiGe) is Sweden's Epigress AB. The announcement in 1999 of Epigress' move into the AIXTRON group meant that the company would have access to that company's extensive R&D and sales/support organization worldwide. The company supplies two systems for wide bandgap semiconductor materials:

- The Epigress SiC crystal growth machine that is based on seeded sublimation growth, where SiC is transported from a solid source to a seed crystal.
- The Epigress SiC MOVPE system which has a single-wafer capacity but can also accommodate a pair of wafers with a diameter of up to 2 in. Separate cells are used for growing ultra-pure and doped material.

Epigress is also one of the licensees of the IBM SiGe UHVCVD system. These are included here for completeness. Some time ago Epigress re-directed itself into SiC and SiGe epitaxial products exiting the GaAs and InP areas. However, it does have a number of GaAs units in the field that it continues to maintain.

Epigress has a collaboration agreement with the Industrial Microelectronics Centre (IMC), Kista, Sweden, to carry out the design evaluation and process development work for the new multiwafer reactor incorporating rotation being developed for the Epigress VP508 hot-wall, SiC production system. After the completion of the thermal and mechanical tests to verify the expected performance, a version of the new reactor design was installed in the Epigress SiC CVD system at IMC in Kista.

The VP508 system for IMC is configured and designed as a production unit including two separate reactors for un-doped and n-doped material and p-doped material, respectively. The system will be equipped with the latest version of the Epigress GUI system that is based on an in-house developed software package called Visual Process.

5.5.6 Commercial MBE Reactors

5.5.6.1 Introduction

MBE is used in the fabrication of many optoelectronic devices. MBE is especially appropriate for the fabrication of diode lasers and is a key tool in the mass production of transistors and integrated circuits for applications such as mobile telephones and satellite receivers. MBE frees molecules of an element by heating the element in an effusion cell. A fraction of the molecules escape into a chamber with a ultrahigh vacuum (UHV) so that the molecules are drawn into a linear beam. A wafer is mounted in the centre of this beam and the freed molecules can then be deposited one layer at a time directly on top of each other. A generalized schematic diagram of the basic MBE configurations is shown in Figure 5.4.

In today's climate of high-volume demand, the basic issue is cost per wafer. In the case of MBE systems, depreciation is only a small component of the entire epiwafer cost structure. However, the trend to reduce the total cost is irreversible. This is especially important when consumer products are driving market requirements. Another key question is how demand for devices will be met by MBE (and MOVPE) and how it will meet this challenge following on from its success in AlGaAs.

The bulk of AlGaAs production is presently based on solid-source MBE. For future needs, InGaP, which is well positioned with MOVPE, will present technological challenges that are being addressed by the makers of MBE equipment and their customers. At present this situation is yet to be resolved but upon it may well hinge the future success of one of the techniques against the other in the global marketplace.

Figure 5.4. A Schematic Representation of the MBE Method

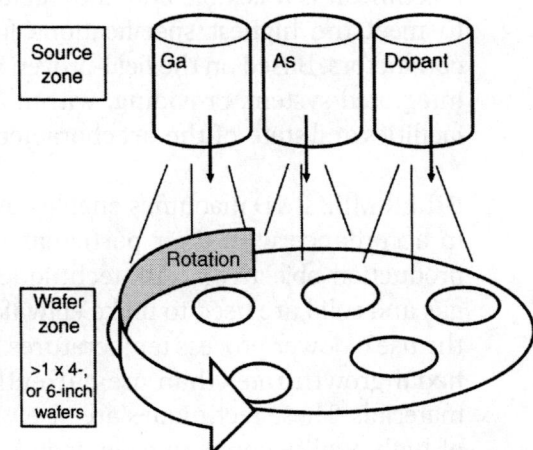

There is a consensus that MBE continues to perform extremely well in the HBT market as evidenced by RFMD's entire HBT output being sustained by solid-source (i.e. traditional) MBE. It has been estimated that RFMD's HBT production alone accounts for more than half of the world's total HBT market.

Through the use of such methods the quality of epitaxial material has steadily improved to the point where it is not possible to discriminate clearly between MBE or MOVPE — it depends on the application. Neither approach is as yet clearly universal for all device types.

In the following, each MBE system is described giving an explanation of key distinguishing features together with (where available) examples of operators of these systems.

5.5.6.2 RIBER Systems

French company RIBER offers a complete range of solid-source MBE, gas-source MBE (GSMBE) and CBE (or MOMBE) systems for fundamental and applied research applications, including single- and multi-substrate models (see Chapter 6). RIBER is a leading supplier of MBE products and services to the compound semiconductor industry, and offers a full range of MBE systems, from R&D reactors up to high-volume 4 × 4-in and 4 × 5-in production units, answering present and future MBE requirements (see Table 5.6). RIBER MBE systems are also suitable for the deposition other materials such as metals, SiGe and SiC.

Table 5.6 Examples of RIBER Equipment for the R&D and Manufacture of Optoelectronic Devices

Material	1-in	2-in	3-in
CdHgTe	MBE32	MBE32	–
InSb	MBE32	MBE32	–
ZnSe	MBE32	MBE32	–
GaAs	Compact 21	Compact 21	Compact 21
InP	Compact 21	Compact 21	Compact 21
GaN	Compact 21	Compact 21	Compact 21

RIBER say the Compact 21 is the new baseline MBE system. A versatile machine, it is a flexible and affordable system with features carefully designed to meet the highest specifications for the research of all compound semiconductors. Based on the field-proven Vertical Reactor technology, it is a 2-/3-in integrated system, providing, within a small footprint, all the necessary MBE facilities and state-of-the-art characterization capabilities.

RIBER MBE R&D machines enable customers to use either solid or gas sources in accordance with their particular research and development goals or pilot production objectives. MBE techniques using gas sources or a combination of gas and solid are used to make epiwafers for the production of devices through the use of lower process temperatures, to increase the possibilities of higher epitaxial growth rates than are currently possible with MBE using solid source materials. These techniques are also used to make epiwafers for the production of high quality compound semiconductors made up of four elements, such as

GaInAsP. Since 1995 RIBER has shifted its focus away from the manufacture of MBE research machines to pursue a strategy that gives priority to the manufacture of MBE production machines.

The newer generation of MBE machines can grow seven 6-in wafers per run instead of four, which is the current standard. This will allow a 75% increase in productivity. These machines can also grow 4-in wafers with the same gain in productivity.

The next-generation, high-throughput, fully automated 4- and 5-in multiwafer machine is designed for today's large-scale production of electronic and opto-electronic device structures. The unique capability of fully automated performance allows for 24-hours-a-day, 7-days-a-week operation, thus minimizing the per-wafer cost of production.

With 11 materials ports surrounding a large vacuum chamber where the epitaxial layers are grown, the machine is said to dwarf earlier MBE systems. The amounts of gallium, arsenic and aluminium and other materials placed in the effusion cells of the Model 5000 are measured in kilograms rather than the grams of earlier machines.

RIBER's MBE machines and parts are standardized to a significant extent to allow customers to start with an MBE R&D system and, with minimum learning time and expense, scale up to commercial production by purchasing one of the RIBER MBE production systems. While the research systems primarily are for R&D relating to compound semiconductors, in limited circumstances customers use MBE research systems for low-volume production. According to the customer requirements, RIBER's adaptable technology also enables the use of either solid or gas source materials in the MBE process.

RIBER's continued involvement in R&D programmes also allows it to follow closely developments in evolving technologies and to detect opportunities for future industrial applications of these technologies. For example, the Compact 21, the latest model in RIBER's line of MBE research machines, was developed within the framework of the Advanced Nitrogen Source for Epitaxy of Thin Film project. The project was funded by the EC in collaboration with several partners with the goals of developing a more cost-effective MBE R&D machine and of researching the potential applications of gallium nitride-based epiwafers in the production of blue laser diodes.

Since the introduction of their first R&D machine in 1975 (MBE 500), as of year-end 2000 RIBER had designed, manufactured and shipped a total of 342 research machines. The features of each model are adapted to accommodate the compound semiconductor source materials used, which depend on the research objectives of the user.

5.5.6.3 Thermo VG Semicon Systems

Thermo VG Semicon is a business unit of Thermo Electron (Waltham, Massachusetts) based in East Grinstead, UK, and has been in the business of MBE for many years (see Chapter 6).

The V100 is recognized as one of the key, enabling technologies in the success of GaAs-based microwave electronic devices and is emerging as a preferred route to certain high-performance phosphide-based optoelectronic devices. Coherent, the world's largest manufacturer of lasers, relies on a Thermo VG Semicon V100 MBE system for its laser diode production. The epiwafers are produced by Coherent Tutcore, located in Tampere, Finland. The company offers a wide range of MBE machines for R&D and production of a wide range of semiconductors, in particular the V100 and V150 multiwafer high-volume production machines.

The company has said that orders for its MBE systems for the first nine months of 2000 were a record US$42.9 million, more than double the total for the same period during 1999 (US$19.4 million) and already exceeding orders for all of 1999 (US$38.3 million). The company has installed more than 60 multiwafer GaAs production systems worldwide.

Recently Thermo VG Semicon received the award in the innovation category for its V100 production MBE reactor. The award recognized outstanding and continuous innovation from substantial improvement in business performance and commercial success.

The company has achieved considerable sales success with its V150 which is designed as a high-volume 4- or 5-in MBE system. The platen size accommodates either 9×100 mm or 4×150 mm wafers. Depending on the version of the V100 used for comparison, the V150 provides up to three times the 100 mm wafer throughput based on the platen size alone and four times the 150 mm throughput. The V150 addresses the industry's need for 100 mm wafers and also the rapidly emerging 150 mm market. It was specifically designed for the pHEMT and HBT business, based on VG's knowledge of the requirements — but is currently less important for diode laser manufacture.

Introduced in 1997, the V80H (III-V) system for GaN is the most advanced MBE system available dedicated to III-V research and prototype production. The V80H is designed for growth onto wafers of up to 3 in in diameter. The system consists of three chambers with independent pumping and control. The deposition chamber incorporates all the necessary components for the controlled growth and doping of III-V materials. The source flange may be equipped with up to ten effusion cells. A monitoring ionization gauge, RHEED system and residual gas analyser are used for growth and environment monitoring.

The V90 was the world's first MBE system configured for 4-in GaAs wafers. The V90 is ideal for MBE research and pilot production of electronic devices. It is also well suited to full-scale, high-volume production of laser diodes and other optoelectronic device structures. The system can be operated manually or is available in a fully automated version.

Another key area is process stability. VG has invested a great deal of development resources into a new range of sources called ThermoCells. These sources provide flux stability (i.e. long-term stability of growth rate) improvement of about a factor of ten over the previous generation of sources (from any vendor). All of the design changes come from specific knowledge of user requirements from the V100 user group.

Agilent Technologies has placed orders for three V150 MBE systems which will be used to equip Agilent's new 6-in wafer fabrication line in Fort Collins, Colorado, USA.

Reliable generation of phosphorus has historically been a problem for MBE. VG Semicon's 'Compound Phosphorus ThermoCell' offers a simple and reliable solution. It provides 100% P2 species by evaporating P2 from the solid compound GaP. The small percentage of gallium atoms that are co-evaporated are removed from the flux by a simple baffle arrangement. The source is designed for rapid thermal response, so that flux control is easily achieved by simply adjusting the temperature of the source. In contrast, the traditional approach requires a multistage combination of evaporation, condensation and re-evaporation of the phosphorus into a high-temperature thermal cracker zone. Although the conversion in a thermal cracker is high, in practice 100% P2 production is never achieved, leading to a background of P4, which can lead to problems.

5.5.6.4 Applied Epi Systems

Applied Epi, a privately owned company, is a global supplier of epitaxy equipment, solutions and services that enable the manufacture of compound semiconductors. In September 2001 Veeco Instruments signed a definitive merger agreement with Applied Epi. Applied Epi is a profitable company with year 2000 revenues of US$25 million, and forecasted 2001 revenues of approximately US$50 million. Their product leadership is demonstrated by their large installed base of MBE equipment (>5000 deposition cells and 200 research/ production systems).

The company is an innovator in MBE technology, having developed such products as SUMO effusion cells and Valved Crackers. It offers a complete line of MBE production systems, research systems and components, including its industry-exclusive silicon-style GEN2000 and GEN200 systems. In September 2001 Applied Epi reported that a leading silicon semiconductor manufacturer had purchased a GEN200 system in order to grow novel memory devices requiring the integration of MBE growth technologies with silicon manufacturing techniques.

Applied Epi introduced its revolutionary multiwafer production MBE system, the GEN2000. As with its other products, it has taken a fresh approach to designing production MBE systems. RF Micro Devices (RFMD) has qualified its Applied Epi GEN2000 MBE system for production. The GEN2000, which is being used by RFMD to grow HBTs (heterojunction bipolar transistors), can be configured for either 4- or 6-in wafers. This is the second GEN2000 to be qualified and growing high-quality wafers. The GEN2000 being used by RFMD and the previously qualified IQE GEN2000 are currently the highest-capacity MBE systems in the world.

Applied Epi's GEN2000 (7 × 6-in) and GEN200 (4 × 4-in) are the first systems in the compound semiconductor industry to integrate MBE and UHV technologies with cluster tool wafer-handling architecture from the silicon manufacturing industry. The systems' silicon-style designs deliver higher throughput, longer growth campaigns and excellent quality wafers in footprints 40–60% smaller than any other 4- or 6-in multiwafer production systems.

It has modelled its system after equipment commonly used in the processing of silicon semiconductor devices, utilizing a 'cluster tool' wafer-handling system. As in the silicon semiconductor industry, the cluster tool addresses reliability and throughput. Coupled with Applied Epi's patented high-capacity SUMO and Valved Cracker designs, to maximize uptime, the GEN2000's combination of reliability, throughput and capacity translate directly to low per-wafer cost. The GEN2000 is produced at Applied Epi's 55 000 sq ft manufacturing facility in St Paul, Minnesota, USA.

Applied Epi is also a manufacturer of MBE sources, with more than 3000 effusion cells and crackers delivered to more than 350 different facilities. These sources are in use in virtually every type of commercially available MBE system, as well as many customized systems.

There are four types of MBE sources: effusion cells, solid-source cracking effusion cells, gas crackers and injectors, and special-purpose sources. In addition to MBE systems and sources Applied Epi also manufactures effusion cell controllers, substrate holders, substrate heaters and viewports.

If the shift towards silicon-style manufacturing is as successful for the compound semiconductor industry as it was for the silicon industry, Applied Epi represents the future of production MBE.

The company has system orders from companies such as epiwafer vendor IQE and MMIC companies such as RFMD (one GEN2000 machine) and Filtronic Compound Semiconductors (two GEN2000 machines).

5.6 *In Situ* Monitoring

The world of semiconductor optoelectronics component manufacturing is on the verge of a revolution in productivity. ISM will enable a higher than ever degree of control, reduce costs, improve existing devices and enable new ones. European researchers and companies are at the leading edge of this new technology.

Modern device manufacturing processes are exceedingly complex. At the same time they are under pressure to drive up yields and thereby create cheaper devices. Key to the success of this is a precise understanding of what is happening with the process as it is actually happening. This must be in real time so that the process can be adjusted and productivity maximized. For the development of ISM as a practical commercial product a holistic approach has been necessary and many parties have been involved in its on-going development.

Two of today's key process steps are epitaxy and etching. Epitaxy is the process whereby precise thicknesses of semiconductor are used to build up a device. Dry etching involves selective removal of material. Knowing as much as possible about how things are going is crucial. This applies not only to the R&D of new devices but also to mass production, the latter being particularly acute in today's booming optoelectronics device business.

ISM is especially important for MOCVD mainly centred on a collaborative project in Germany but also on a French effort in a related field. For example, Lay-Tec, based in Hardenbergstr, Germany, is a spin-off from the Technical University of Berlin, Germany.

ISM is becoming one of the most important adaptations for MOCVD layer growth analysis. These methods are very important procedures for layer growth and previously were not possible at the typical higher pressures used by MOCVD. Importantly, ISM techniques have also rapidly closed the process monitoring advantages that high-vacuum processes such as MBE layer growth processes enjoy. MBE is commonly used for the production of diode lasers for data storage applications. The fundamental problem of window depositions, which has been solved for MOCVD by window purging, remains to be solved for MBE.

Optoelectronic component manufacture relies increasingly on advanced epitaxy systems such as MOVPE. These systems rely on precision control while at the same time improving production yields. Key to this trend is the use of ISM. Co-operation has been important in the development of ISM. To build on existing knowledge, a pan-European ISM project has been commenced that is likely to have a big impact on future development. Funded by the Fifth Framework Programme (FWP), the Online Monitoring and In Situ Control of Epitaxy During Metallorganic Chemical Vapour Deposition project will combine real-time optical diagnostic techniques with next-generation software to improve growth of advanced optoelectronic devices such as VCSELs and detectors. The Fifth FWP is an umbrella research funding programme of the European Union, where different topics of material science, nanotechnology, etc., are addressed and European consortia (also with US participation) can compete for projects.

Milestones of this shared cost, three-year project are combining the MOVPE with the embedded optical and X-ray diffraction sensors and then making *in situ* measurements. It is led from the University of Linz, Austria. Data will provide process control information on stoichiometry, roughness, voids, interface and surface quality, growth rate, homogeneity and doping. The partners will correlate *ex situ* data with *in situ* data to develop a closed-loop control system. Over the course of the project, there will be three different sensors attached:

- Spectroscopic ellipsometry (SE) from Sentech Instruments (Berlin, Germany).
- Reflectance anisotropy (RAS) from LayTec.
- X-ray diffraction (XRD) system with a Johannsson Monochromator from Philips Analytical (Eindhoven, the Netherlands).

This set-up has been made to work for GaN and has also been shown for GaAs and GaP in at least ten publications. It is undergoing trials for installation in a production environment. The set-up has a reactor equipped with a LayTec RAS system in the laboratory, which has already shown very good improvements in process control for a wide range of growth processes and materials.

ISM has received a great deal of interest in the past couple of years and will be a critical component of MOCVD R&D and production systems in the future. Optical ISM is coming of age and being introduced for production MOVPE systems rather than being just a retrofit. This because of the improvement in techniques — particularly interferometry — to provide information needed for a production system and is easier to implement. Also, growth of nitrides for blue-green LEDs, etc., is not a process as well behaved as for other materials. For the other III-Vs it was the real-time sensitivity of RAS to doping levels, its patented anti-wobble techniques and its direct combination with reflectance monitoring that finally caused the breakthrough even in the necessarily rather conservative growth community.

As the pan-Euro ISM project has done, ISM is conveniently split into *in situ* sensors and in-line (post-growth) sensors. In general for MOCVD and MBE, RAS and reflectometry will do the industrial ISM job. This is also based on *in situ* research with SE but SE, RS and XRD will be attached for certain in-line applications.

ISM techniques presently under consideration include: SE, RAS and reflectrometry and Raman spectroscopy (RS). Also under consideration is the non-optical technique of XRD. However, this is less convenient and equipment cost is somewhat higher than for the optical techniques.

Throughout the semiconductor industry, and not just in III-V epitaxy, optical probing methods are being developed to measure such parameters as temperature, surface composition, layer thickness and refractive index on the wafer during growth. ISM methods can also be used to measure and control the chemical composition and flux of chemical vapour source beams used for deposition of semiconductors and other materials.

ISM is not just required for R&D. Device makers and epiwafer manufacturers such as laser diode producers are in dire need of ISM. It has become vitally important to be able to measure the actual temperature on the wafers as they are grown. For example, in the manufacture of lasers such as DFBs or VCSELs, after visual inspection manufacturers presently have to reject as much as half of the epitaxy as being unacceptable for further processing. If temperature control was available the reject rate would, it is said, be reduced to near zero. Thus ISM has potentially a great impact on yields and thereby costs.

Optical diagnostics have the advantage that they can be used at atmospheric pressure. They can also be used at reduced pressure even in the presence of plasmas. Because they are located outside the reactor vessel they therefore do not interfere with the on-going process. Also, since they have low incident energy and are non-contact methods they have minimal effect on the growing epilayer. They can be used for a variety of reactor susceptor and heater configurations so that they can be used for either fixed or rotating wafer systems.

At present, optical diagnostics are the most popular ISM methods. Some commercial reactors have already been equipped with them. Ellipsometry (phase measurement) is sensitive to monolayers. With RAS using the symmetry effect, sub-monolayers can be detected with reflectometry when the technique is properly used. A layer thicknes of less than 50 nm is needed for growth rate and composition measurements.

The main scope of the group at the Institut für Festkörperphysik (IFP) of the Technische Universität Berlin (TUB) is the combination of optical analysis methods with MOCVD, which allows *in situ* growth studies. Techniques under study include RAS, SE and elastic light scattering. Detecting the difference in reflectance for light polarized along two perpendicular crystal orientations, RAS is highly surface sensitive for cubic materials whereas SE gathers information from the entire light penetration depth.

Based on the experimental set-ups at the TUB, a commercial LayTec EpiR-AS(TM)-200 sensor RAS system is being used at the Ferdinand-Braun-Institute (FBH) to study device-related issues like exchange reactions at heterointerfaces, control of doping and GaInAsP composition in dependence on growth parameters. Close collaboration between the partners is bringing a valuable research tool into a production environment where it allows for improved process control and shorter development cycles.

Another important contribution to the development of ISM is underway at the North East Wales Institute (NEWI) where work has focused on making quantitative measurements that are useful to the epilayer grower and to provide a basis for *in situ* control. There is a need for more *in situ* diagnostics due to the non-classical nature of much of these growth processes. A modified system to cope with GaN-onto-sapphire situations has been tested at the University of Ghent.

Notably, the well-known suppliers of equipment to the epitaxy market generally offer ISM systems. These systems have been available from specialist suppliers of peripheral MOVPE equipment for some time. For example, the NTM1 system from the Israel-based company CI Systems is a dual-channel electro-optical monitor for temperature measurement for MBE or MOVPE.

Another requirement is available computing power, especially for a closed-loop ISM system. Data gathering will require fast computers able to handle a great deal of data in real time. However, this volume of data is really only a side issue. Complex real-time analysis is now integrated into the sensors. Also, intra-networking of the MOCVD reactor's own computer with other sensor systems is now possible. A software user interface adapted to the needs of process engineer has now become a standard feature of such products. The next step in the evolution of ISM is to provide closed-loop feedback control of the growth process.

It is now not unusual for an epiwafer manufacturer to be faced with the loss of an entire load of expensive wafers. Modern MOVPE can handle 8×3-in wafers per run and each run might last several hours. Since each wafer can produce thousands of dice, a lot is at stake. Anything that can help avoid run failures will receive serious consideration from everyone in the business.

ISM can provide process control feedback for temperatures, fluxes (for chemical composition and concentration), growth/etch rates, surface composition and particulates. ISM will become critically important for new devices such as VCSELs. It will aid development of newer devices previously impossible to make without the control provided by ISM. However, ISM is not going to be an inexpensive addition to a reactor. If the need is there, cost-of-ownership will not be a problem, as ISM is also a yield-enhancement tool and is likely to recoup initial costs within the first year of operation.

While ISM will add significantly to the capital cost of MOVPE reactors, it will provide a rapid return on investment, say its proponents. The first systems to fully install and exploit ISM will be those that are highly optimized for mass production of optoelectronic and other devices. They will find most significant take-up with manufacturers of devices where capital cost is of less concern than cost-of-ownership and, most importantly, cost per die. This applies most rigorously to the makers of optoelectronic devices such as blue-green LEDs.

5.7 Device Processing

5.7.1 Introduction

In this section some of the process steps that are required in order to turn epi-wafers into functioning devices are described. Those described here are just some of the steps needed to manufacture diode lasers. These are broadly similar for the generic device but there are likely to be major differences according to device type. This derives in part from the diverse range of semiconductor materials required for light emission over the complete spectrum. So far there is no one material that can be made to emit coherent light over the entire spectrum (IR through visible to UV). Different sub-types of semiconductor compounds are required, e.g. AlGaAs for IR, InGaAsP for red and InGaN for violet-UV. Whilst some processes may be adapted for each type they require specific chemistries and processes to be developed so as to extract maximum performance.

The process steps required for diode lasers include some but not necessarily all of the following:

- Photolithography — defining patterns for follow-on process steps using resists and light exposure similar to a photographic development process.
- Ion implantation — placing dopants or changing electrical/optical characteristics of photolithographically defined areas by means of high-energy ions.
- Thermal annealing — repairing structural damage caused by crystal defects or ion implantation so as to improve device performance.
- Dielectric deposition — coating selected areas with insulating materials.
- Etching — selective removal of areas of metal, semiconductor or dielectric from areas; also to define optical facets, which can be achieved via wet chemicals or vacuum dry processes.
- Metal deposition — coating of selected areas so as to emplace pathways for electrical connections.
- Encapsulation — enveloping the device for environmental protection.
- Bonding — linking the electrical pathways to the external package.

These steps are carried out in an order depending on the device type and some such as photolithography may be carried more than once. In some instances further process work is required to optimize the optical output of the laser, which in some cases is not ideal for a particular application. This may be achieved by adding an external lens and so on.

Diode lasers differ from electronic devices in that they require additional attention to the optical part of the devices, in other words, the means to ensure highest possible electrical-to-optical efficiency of conversion. As a result these can in some respects be at least as complex as the steps required for the manufacture of an integrated circuit. A subtle difference is that the electrical and optical processing has to be compatible and in some respects a compromise has to be reached in order to effect a commercially viable device, i.e. with respect to yield and therefore cost. The makers of these devices therefore have had to develop their own means to manufacture the devices. These developments are usually patented but specific detail is often proprietary and so anything of significance seldom reaches the open literature.

Basically, much of the engineering of a diode laser takes place in the epitaxy stage. Some devices have dozens of epitaxial layers that are of an electrical as well as optical function. The very nature of the materials often requires the careful build-up of many subtly different layers so as to match two dissimilar semiconductor types. This engineering is a precise art as well as a science to achieve a working commercially viable device. Through such knowledge the business of independent epiwafer supply came into being and is now an important competitive sector in the business. Increasing numbers of today's diode lasers have their origination in these facilities. They are later taken into full-scale commercial production in the factories of the companies that placed the initial order.

However, the epiwafer engineering takes the diode laser only so far and much depends on the subsequent fabrication. This achieves the following:

- Electrical supply.
- Optical output.
- Thermal environment.
- Rugged durability.

An example of the device processing requirements of a topically important new diode laser is illustrated by the VCSEL. These are implicitly simpler and therefore cheaper to produce than the well-established edge-emitter laser. Nevertheless, precision is required during the epitaxial growth stage. All the work is in the growth, which makes laser manufacturing easier because there is no need to cleave up the individual die.

The edge-emitting laser requires the manufacturer to cleave the process wafer into chips and then package the chip before testing. Manufacturing VCSELs should be implicitly cheaper than the manufacture of edge-emitting lasers due to the availability of on-chip testing of the devices. However, manufacturing tolerances on the growth are much tighter requiring control over the thickness of layers to better than 1%. Another challenge is to maximize the uniformity of deposition over a 3- or 4-in wafer so as to maximize yield.

Like most LEDs, VCSELs emit light perpendicular to the wafer, which allows them to be tested in wafer form. The entire electro-optical characteristics of a VCSEL wafer can be mapped before forming individual devices, which allows automated die sorting and leads to application-specific sorting of VCSELs from the wafer. The vertical emission from the VCSEL also enables the use of traditional LED-style packages for use with VCSELs, such as surface mount and plastic encapsulated leadframes. The uniformity of VCSEL growth has increased substantially from 1996 and has led to a more reliable and cost-effective product.

Many VCSELs are for the IR wavelength range (780, 850 and 980 nm wavelengths) and are thus based on GaAs or AlGaAs. They can be fabricated in either oxide or implant form and longer-wavelength VCSELs are currently in development.

VCSELs also have the attraction by virtue of their structure to be formed into arrays of one- and two-dimensional emitters which is another virtue of the wafer-level processing:

- One-dimensional arrays, with ribbon optical fibre, look promising for 'outside-the-box' communications, e.g. server networks.
- Two-dimensional arrays are under consideration for free-space interconnects, e.g. communications 'inside-the-box' from microprocessor-to-memory, etc.

Finally, VCSELs are also among the first of the diode laser family to exploit micro-electro-mechanical systems (MEMS) in their construction. The telecommunications market is demanding tunable diode lasers and one of the routes taken to achieve this has been taken by US-based company Bandwidth9. Their tunable VCSEL has a fixed bottom mirror, a multi-quantum well active region, and a top mirror fabricated into a cantilever structure. Applying a small voltage to the top mirror causes the MEMS cantilever to move up or down, which changes the length of the laser cavity and hence the wavelength of operation.

It is expected that MEMS will play a key role in the further development and market penetration by VCSELs and other laser types over the next five years. The additional process technology is already in wide use in associated industry sectors, which is likely to see an adaptation into optoelectronics and thereby further expand the capability of the ubiquitous diode laser and related devices.

5.7.2 Ion Implantation

Whilst epitaxial growth dominates the R&D and fabrication of devices based on GaAs and related semiconductors, researchers are also developing alternative methods. The reason for this is two-fold:

- Cost — epitaxy can be an expensive process.
- Patents — alternative processes are needed so as to obviate those patented by the pioneers in the industry.

In mainstream semiconductors many important devices such as microprocessors and memories rely on processes such as ion implantation. This technique has been optimized so as to provide economics that are almost unbeatable by any other process. The challenge is to develop processes with equivalent throughput and economics. At present the main thrust is to perfect the epitaxy techniques of MOVPE and MBE but it cannot be ruled out that at some future stage the implant process will not be reconsidered. This particularly applies to other materials systems such as SOI, SiC and GaN that are becoming more important for electronic devices.

Implantation is a physical process, i.e. no chemical reactions are involved, and so in essence it is a 'cold' process as it does not involve intentional heating of the substrates. The reaction chemistry involved in epitaxial growth, however, relies on elevating the substrate to some of the highest temperatures ever used in the field. Such a thermal budget does not come cheap and is proving to be a major obstacle in the route to cost-competitive devices. It would be a major achievement if the industry could adopt cheaper processes such as implantation and this could have a great impact on the market acceptability of next-generation devices.

In the USA, Implant Sciences, based in Wakefield, Massachusetts, demonstrated in May 1998 the first gallium nitride blue LED created by ion implantation. Implant Sciences developed a procedure for implanting magnesium and silicon into adjacent regions of a GaN film that makes this diode structure possible. However, the LED is created using a film of GaN grown by MOVPE. After implantation, the sample is annealed to activate the dopants.

Ion implantation is widely used throughout industry as part of the technology for fabricating electronic devices and integrated circuits. Its use is due to several important advantages it has over other doping techniques. For example, the concentration of dopant atoms in the semiconductor can be controlled and varied between wide limits. A schematic representation of the ion implantation method is shown in Figure 5.5.

Also, the technique enables the doping to be performed uniformly and reproducibly over large areas, and by varying the energy of the incident ion beam, it is possible to vary its penetration into the semiconductor. Hence, the distribution of carriers with depth can be modified easily. With the aid of a suitable mask, it is also possible to implant selective areas to form, for example, the source and drain contact regions of a field effect transistor.

Thus ion implantation has enabled ICs to be fabricated in GaAs with better yields, better controllability and improved reproducibility compared with the

Figure 5.5. A Schematic Representation of the Ion Implantation Method

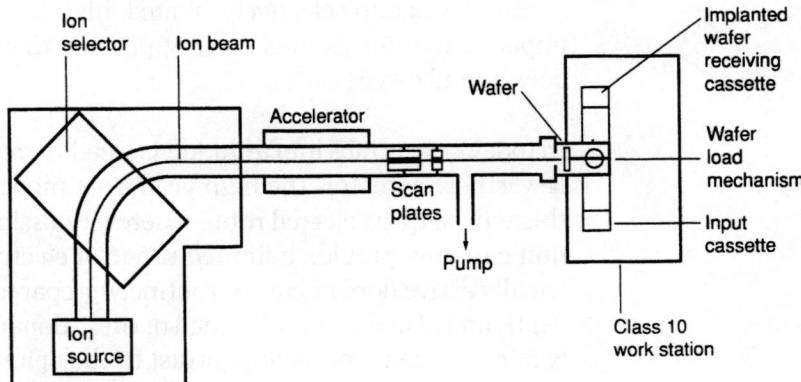

use of epitaxy. This is certainly so for silicon ICs, but for GaAs the use of ion implantation had a troubled gestation. Nevertheless, these problems were overcome and the technique is a key stage in the production of many millions of devices per annum. For example, being a compound, GaAs decomposes at the temperatures required to remove radiation damage and to activate electrically the implanted ions. This problem still requires an ideal solution, although many groups have developed adequate methods to prevent dissociation during annealing.

The technique of ion implantation entails the bombardment of a material with high-velocity, positively charged ions produced in a source held at a high DC potential. After extraction from the source, the beam is mass analysed and allowed to accelerate to the target (sample) that is at earth potential. When the ions impinge on the target, the majority penetrate some distance and slow down by random interaction with the nuclei and electrons of the target. The ions come to rest at a depth that is a function of the ion energy and of the mass and atomic number of both ion and target atoms. Perhaps the main disadvantage of ion implantation is the structural set damage caused to the substrate due to the dissipation of energy by the incoming ion. The amount of damage depends on factors such as the mass and energy of the ion — the heavier the ion, the greater the damage. In extreme cases, a high dose of a heavy ion can produce an amorphous or finely crystalline layer. To remove this damage, annealing is required. This not only repairs the lattice but provides the energy required to return the atoms to the correct lattice sites and for the implanted ions to act as dopants.

The post-implantation anneal once posed serious problems to the application of ion implantation to device fabrication, since GaAs readily decomposes above 540°C. It was found in the early days that thin-film dielectrics such as SiO_2 and Si_3N_4 were able to suppress the decomposition of the GaAs and allow the electrical activation of implanted ions to occur.

Recently, rapid thermal annealing has been used successfully to limit the diffusional broadening of profiles that occurs during long-time anneals in a furnace. In this way it is possible to implant higher doses and obtain a higher peak hole concentration.

In addition it is important that various parts of an IC are electrically isolated. As the complexity of ICs increases, so it is necessary to place circuit components closer together making adequate electrical isolation more difficult. To obtain the required isolation selectively created, high resistivity layers are made using ion implantation; for example, using protons that create defects that compensate both p- and n-type GaAs.

In today's GaAs fabs implantation is steadily making way for epitaxial processes. Nevertheless, such is the high yield from modern implantation techniques that this will be the preferred route wherever possible. The problem is that implantation can only provide a limited range of electronic devices. Both types of electrically active dopant can be routinely prepared by implantation but in terms of depth and also the lateral spread of these dopants, device structure engineering is fairly limited. This is in contrast to the epitaxial growth of many — possibly several dozen — thin or thick layers so as to build up very complex microelectronic and optoelectronic devices.

Implantation will continue to be important in the device fab, not for the preparation of active layers but rather the isolation between devices.

5.7.3 Rapid Thermal Annealing

Another key step in the mass production of diode lasers was the successful utilization of annealing processes. Most devices have in common the fact that they all must have electrically active regions. These regions that are prepared usually by epitaxy, have to be made electrically active via a subsequent thermal treatment, called annealing. This thermal treatment is a critical measure of the quality of the substrate. The substrate must retain its electrical characteristics, i.e. resistivity and mobility, throughout processing but it is most important that it survives the first anneal.

As with ion implantation where annealing is a routine process step, this treatment also improves the crystal structure of the wafer by removing defects and so on. Early on in the development of GaAs and other devices, mixed results were too frequently obtained when trying to activate the n-region via a conventional furnace anneal. This furnace process was derived from diffusion processing in the silicon industry and was less appropriate for GaAs owing to the dissociation tendency of the arsenic at anneal temperatures.

While the electrical qualities of substrates have been improved over the past two decades, at the same time alternative anneal techniques have been developed. Most importantly was the commercial development of the rapid thermal annealing (RTA) process. Today RTA, or more generally speaking rapid thermal processing (RTP), is in widespread use throughout the semiconductor industry including GaAs and silicon.

The essence of the process is that it is rapid. Not only is the duration of the actual thermal processing much briefer but also the total time taken from wafer loading to wafer retrieval is only a few minutes. Conventional furnace anneals used to require as much as an hour per wafer. Other advantages of the RTA process include:

- It is very cost-effective — RTA is very efficient in terms of energy usage (it is basically a high-intensity flash lamp which is on for only a few moments).
- It is compact — a cassette-to-cassette RTA system occupies much less room than most other equipment in a typical semiconductor fab, some models are truly 'desktop' in appearance.
- It is a uniform process which fully activates the dopant through the entire structure and yields little variation across the wafer and from run to run.
- It is a minimal impact process in that it does not adversely affect any other part of the device structure.
- It is essentially free of side effects although some care needs to be taken to ensure that the wafer experiences no dimensional distortion, i.e. warpage, or crystallographic slip.
- It is safe and easily operated via a standard programme or process recipe.

Since its inception in the 1980s, RTA technology has seen numerous process refinements and improvements. These include the ability to process even larger numbers of larger-diameter wafers automatically delivering even better uniformity of thermal treatment. Furthermore, virtually every device fab has at least one RTA station that not only performs implant activation but also contact alloying.

It is likely that RTA will further establish itself as a highly useful, low cost-of-ownership process in all future fabs. This will also be the case when next-generation devices begin to come on stream. RTA has already become an indispensable anneal tool in SiGe and SOI devices and the latest high-brightness GaN/SiC LEDs and diode lasers and will undoubtedly perform a similar role for microelectronic devices based on these materials.

5.7.4 Etching of Compound Semiconductors

The technical and business trends in etch processes for manufacturing opto-electronics are overviewed here, i.e. the dry etching for the fabrication of optoelectronic devices covering GaAs and InP as well as GaN and II-VI materials.

In order to manufacture optoelectronic components such as diode lasers, dry process systems are becoming the established tool of choice. The total III-V material equipment market now exceeds US$0.5 billion. Dry processing (etch and deposition) represents nearly half of this (US$220 million), with optoelectronics accounting fir nearly two-thirds of that (roughly US$145 million in 2000).

Optoelectronics places varying demands on manufacturers, largely because it still relies on discrete devices made on 50 mm (2 in) wafers rather than the larger wafers favoured by MMIC fabs. Devices are small (e.g. 500 × 400 μm) giving several thousand for every 50 mm or 75 mm (2 or 3 in) diameter wafer.

Nonetheless, demands for fibre-related telecommunications equipment provide a strong driver for the development of more capable etch processes and throughput. There is a clear trend away from low-scale R&D to production tooling. In fact, there are many more optoelectronics companies than there are for MMIC manufacture. There are currently frequent announcements of new start-ups with the larger companies (e.g. Nortel, Lucent and Marconi) having to add capacity or build new factories. In recent months this activity has seen a slower pace.

Emphasis on discrete devices rather than ICs means that dry processing for optoelectronic devices is some way behind the manufacture of electronic devices such as MMICs. However, it is catching up fast and has its own set of technical challenges to meet.

To achieve the optimum economics demanded by the price-sensitive handset market, GaAs MMIC manufacture is having to move to the largest available wafers (150 mm). GaAs manufacturers are in transition from 75 mm or 100 mm to 150 mm, whereas optoelectronics manufacturers seem content to use 50 mm and 75 mm wafers for the foreseeable future. This is the direct result of the emphasis on discrete devices for optoelectronics rather than MMICs for the analogue market. The two paths are also diverging: going to single-wafer processing for larger wafers for MMICs while optoelectronics looks more likely to pursue the batch approach. So, process tool manufacturers can only share so much technology across both these markets.

Indeed such has been the impetus for MMICs that companies which have chosen to dedicate themselves to this market have, to some extent, had to turn away from optoelectronics. Nonetheless, this manufacturing experience should prove invaluable when optoelectronics moves to larger wafers.

Now that today's compound semiconductor device-processing market is approaching silicon-like scales, a half-hearted approach is inappropriate. Companies can therefore not spread themselves too thinly, so others have been able to pick up business from the optoelectronics market. However, that is also showing signs of spectacular growth (25% per annum for key components such as LEDs and lasers).

The past two to three years have seen many announcements of start-ups and expansions within the optoelectronics field. These all demand more process tools, of which more will be dry process equipment (such as reactive ion etching (RIE) and, in particular, inductively coupled plasma (ICP) etching).

Also, wet etch processing is having to make way for dry etching. However, even dry processing techniques like ion beam milling and RIE are now making way for ICP etching.

There are several suppliers of dry process equipment, including:

- Oxford Instruments Plasma Technology.
- SAMCO.
- Surface Technology Systems.
- Tegal.
- Trikon.
- Unaxis (the company that includes Plasma-Therm, BPS, Nextral and ESEC).

Indeed, SAMCO has seen some success in Japan for etching applications in optoelectronics. It recently announced that it has sold more than ten of its Tornado ICP etch systems to key manufacturers of optoelectronic devices for InP, GaAs, GaN and related materials. Similarly, the 6520 plasma etch system which uses a dual-frequency plasma reactor for device-side and back-side etching of GaAs and InP films from US company Tegal has proved popular owing to its low-ion-energy etch process. These are tailored for damage-free etches of III-V films but still use simple process chemistries.

In the manufacture of InP-based lasers (e.g. for visible laser applications) the user requires a process which must not only be able to etch vertically with low damage but also have non-selective characteristics through the multilayer stacks. This used to be achieved by ion beam processes but, since this introduces ion damage, ICP is now the more common approach. ICP has much reduced damage and so is rapidly becoming the tool of choice in optoelectronic as well as MMIC device manufacture.

MMICs are principally being driven by the demands of the telecommunications industry, which is also the main driver for the optoelectronics industry. But, in contrast to the MMIC industry, which is pursuing wireless technology markets, optoelectronics industry's key driver is fibre optic telecommunications and data communications. The massive demand for more bandwidth for Internet and multimedia is driving infrastructure deployment, which in turn boosts demand for optoelectronic components. These are also in demand from several other market sectors.

In particular, optoelectronics is also driven by evolving demands for data storage, for which, in an increasingly optical-oriented marketplace, diode lasers and detectors are now the components of choice. In general, the telecommunications market is the greatest in value but the data storage market is the biggest in terms of unit volume of devices produced.

The optoelectronics market is also distinguished by demand for incoherent emitters (i.e. LEDs for status indication and information display). These are mainly simpler devices than lasers and so are of less importance to dry etching. Nevertheless, in the past five years this has seen an enormous resurgence because of the SiC/GaN-based blue-green LED.

Although these are also principally discrete devices, they demand drastic cost reductions to be competitive with older red and amber devices. Improved manufacturing techniques based on refined dry processing will play a key role in ensuring this.

This technology will be even more important for the development of the next big step in data storage. This key market has already gained the attention of the process tool manufacturers, who are quietly at work refining the process chemistry and wafer handling to be ready for the next wave. But at present most of the market is taken up with DVD lasers for movie players, games consoles and PCs.

New laser markets include higher-power units for CD-R/RW and DVD-ROMs. There is already a huge market for CD-R/RW archiving, but within the next five years the recordable DVD for home use will become established. This requires a new laser unit for reading and writing. At even higher powers, the demand for laser diodes for pumping or direct heating is growing strongly. All of these new laser markets demand more precise and lower-cost manufacturing techniques, leading to greater demand for dry processing tools.

At the moment the most significant driver for dry processing is the need for lower-cost components. This is seen as a determinant in the often discussed success of the prospective 'fibre-to-the-home' market. Increasingly, process tool companies are looking to achieve lower costs for the users via direct etch on wafer.

Another key application area for lasers is dense wavelength division multiplexing (DWDM), which requires a move from relatively low-volume discrete devices to high-volume, wafer-scale, integrated optical sub-systems (e.g. laser arrays, detector arrays, multiplexers, etc.). These are placing more stringent demands on the manufacturers of dry processing tools. Machines are having to be optimized for new etch chemistries to give higher etch rates, while also giving cleaner processing, fewer process steps and higher tool utilization. These products must provide considerable reduction in overall cost-of-ownership for optoelectronic device manufacture. This is similar for GaAs MMIC processes, but optoelectronics needs selective etching with better control of selectivity in tools that have a smaller configuration for smaller wafers. As a result, optoelectronics can use batch processing rather than single wafers.

In the etching of InP using ICP systems (i.e. shallow etching for waveguides and deep etching for mirrors) a hot electrostatic chuck allows a fast etch rate and vertical profiles. It achieves this with a simple chemistry. Also important is dielectric etching on InP substrates such as hard masks and for use in the fabrication of contacts.

Trikon's high-temperature electrostatic chuck targets plasma etching of compound semiconductors, especially InP. Accurate temperature control during plasma etching of compound wafers is critical and this has been a contentious issue for the earlier generation of chucks that could not effectively clamp III-V wafers for hot processes, leading to slower plasma etch processes and limited choices of etch chemistries. The only alternative was mechanical clamping, introducing particulates and causing wafer breakage that can drastically reduce yields and throughput. By adding reliable and uniform heating to a chuck that has been field proven in production, Trikon has brought a production-worthy solution to a major optoelectronic device manufacturing problem.

As was the case for other devices, damage remains an emotive subject as its quantification is highly dependent on the device type and etch chemistry. Typically, the required etch depth is shallow (<100 nm), which allows low-power (and hence low-ion-energy) RIE systems to be used for low-damage etching. There has been a transition to ICP source technology as this offers an increased level of control within the low-ion-energy parameter space, which satisfies some front-side, low-damage, gate-etch applications. However there are still sufficient concerns among some device manufacturers that they resort to a dry selective etch stop (GaAs/AlGaAs) followed by a finishing wet etch.

It looks certain then that more demands are being made on etch systems for optoelectronic devices in terms of productivity. With the telecommunications market driving up throughput and driving down price etch tools are helping provide the desired yields of high-performance devices. Moving on from R&D-type systems capable of handling only smaller wafer sizes for discrete devices, these etch systems are having to meet the needs for volume, wafer-scale, integrated optical sub-systems.

Automated handling is just around the corner and specialized wafer clamping improvements such as ES chucks are being developed. The focus of the tool supplier is changing. Previously, it was acceptable to order hardware, develop processes and maintain equipment in-house, but today higher levels of customer support from the tool suppliers — with all the consequent associated financial commitment — is mandatory.

This accumulated market will gather even more momentum over the next five years, ensuring that the optoelectronics market stays ahead of the III-V microelectronics market by an order of magnitude. However, this will involve mostly discrete devices, so it is likely that optoelectronics will retain smaller wafer processing and batch processing.

Though bigger wafers will become available in due course and the industry will steadily grow, it will be at a slower pace than for MMICs. Discrete devices do not grow much in size and require none of the passives that take up so much area on MMICs. Nevertheless, optoelectronics is moving towards more integration (albeit only for specialists applications at the moment). This involves arrays of detectors and emitters for a variety of wavelengths but mostly in the IR for DWDM and related applications such as instrumentation.

The way is beginning to open up for monolithic integrated optoelectronic circuits (also including MEMS) — the so-called photonic ICs or OEICs. Tools developed for the manufacture of higher-performance discrete devices at lower prices will also be adopted for these applications. Depending on how well this market expands over the next five years, this will demand larger wafers to accommodate larger-area optoelectronic devices as well as more per wafer.

R&D examples of OEICs are already being demonstrated whereby monolithically integrated multiplexers have been incorporated with diode lasers. These are in effect the first of a new generation of product opportunities — photonic ICs — and a key factor in the successful commercial realization of such complex devices will be the availability of the right dry processing tools with optimized cost-of-ownership.

6 Company Profiles

6.1 Introduction

This chapter contains an alphabetical listing of profiles of major merchant manufacturers of materials, devices and related equipment supplying the diode laser industry. There are presently around a dozen suppliers of substrates active worldwide. These include companies such as Freiberger, AXT and IQE. There are also several epiwafer suppliers such as ATMI, IQE, and Japan Energy. It is in this latter area of business that new entrants are most active. There has hardly been any change in the complement of substrate suppliers in the past decade whereas this period has seen the appearance of quite a number of epiwafer suppliers. Included in these new entrants is a mix of companies from university department spin-offs and equipment vendors supplying limited numbers of test wafers to full-scale start-up businesses.

GaAs wafer manufacturers are not restricted to supplying electronic device companies. Many of these wafer suppliers are also active in the optoelectronic business. This market sector is comparable in size to that of the electronics sector but is almost exclusively based on discrete devices rather than on ICs. Companies that specialize in more than one sector relating to optoelectronic devices include:

- AXT.
- EMCORE.
- Japan Energy.
- Picogiga.

In terms of geography, the key players in the wafer business are located in the three main regions: Europe, Japan and North America. However, there are a few additional suppliers in the rest of the world region, particularly in South-East Asia.

Equipment companies are one of the groups that tend to specialize in one specific area. Companies profiled here that manufacture equipment for the processing of diode laser devices include:

- AIXTRON AG.
- RIBER.
- Thermo VG Semicon.

The most important companies in terms of overall turnover are the long-established device manufacturers. Most of these companies, especially those in Japan, operate in several tiers of the industry from substrates through to modules. These include:

- Agilent.
- Alcatel.
- Coherent.
- Furukawa.
- JDS-Uniphase.
- Nortel Networks.
- ROHM.
- Sharp.
- Sony.
- Thales.

Finally, the optoelectronics industry has seen an unusually large number of start-up companies in the past decade. Examples of this type of company include:

- Alfalight.
- E2O Communications.
- Princeton Lightwave.

6.2 Agere Systems

Agere Systems Inc
Central Campus
555 Union Boulevard
Allentown
PA 18109
USA

The former Microelectronics Group of Lucent Technologies was renamed Agere Systems at the end of 2000; it completed an Initial Public Offering (IPO) in March 2001 for US$6 per share, and will either initiate a second public offering or be spun-off from its parent in early 2002. Agere manufactures analogue and digital lasers. Lucent is the world's largest manufacturer of optical networking equipment for dense wavelength division multiplexing (DWDM) applications.

- Agere reported FY 2000 sales of US$3.7 billion, excluding its captive sales to Lucent of US$1 billion.
- At end-year FY 2000 Agere employed 17 400 people. However, since then, owing to the industry slowdown experienced by electronics companies, and in particular telecommunications, Agere has been forced to reduce its workforce by 6000 people worldwide.

Acquisitions

Ortel Corporation (Alhambra, CA, USA) was acquired in FY 2000 for approximately US$2.95 billion worth of Lucent shares. Ortel's products include 10 Gbps

Figure 6.1 Agere Four Year Sales (US$ billion)

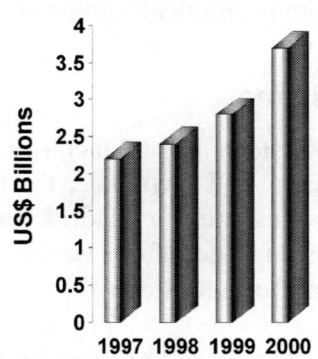

digital receivers and lasers, 1310 nm and 1550 nm analogue lasers and DWDM lasers.

R&D

R&D Expenditure (Lucent)
FY 2000: US$4.0 billion
FY 1999: US$4.2 billion

Bell Labs performs R&D for Agere. Bell Labs is one of the world's leading industrial R&D organizations; its inventions include the LED, transistor, laser, solar cell and the first 32-bit microprocessor. Quantum cascade (QC) lasers were invented and demonstrated in 1994 at Bell Labs. QC lasers are the first mid-IR semiconductor lasers that operate at room temperature with operating power as high as 0.5 W and feature unequalled performance in the mid-IR portion of the spectrum.

● In 2000 Bell Labs announced a QC semiconductor laser that emits light in the far reaches of the IR spectrum; it is the longest wavelength semiconductor laser ever made from standard semiconductor materials used in photonics. Intended applications include chemical analysis and diagnostic applications.

Agere has recently recorded two US patents:

● US Patent No 6,272,163 for a 'method for fabricating high speed Fabry–Perot lasers for data communication'. A method for fabricating a Fabry–Perota laser having a cavity length and facet reflectivity product tuned to cause the laser to exhibit a relaxation oscillation frequency of at least 10 GHz, wherein the laser is capable of being modulated at 10 Gb/sec at drive currents between about 20 and 40 mA.
● US Patent No 6,275,513 for a 'hermetically sealed semiconductor laser device'. A wafer assembly that includes a wafer substrate. A plurality of micro-optomechanical or micro-optoelectrical devices is positioned on a surface of the wafer substrate. Each micro-optomechanical or micro-optoelectrical device has a seal surface. A plurality of seal caps is coupled to the micro-optomechanical or micro-optoelectrical devices. Each seal cap

has a seal ring. The seal cap seal ring is coupled to a seal surface of the micro-optomechanical or micro-optoelectrical device to form a hermetic seal.

Facilities

Agere Systems has US facilities in Reading and Breinigsville, PA; Orlando, FL; Alhambra and Irwindale, CA (Ortel); as well as in Pathumthani, Thailand; Matamoros, Mexico; Madrid, Spain; and Singapore. The company has design centres and sales offices worldwide. All of these facilities have ISO 14001 environmental certification.

Agere has updated its optoelectronic manufacturing operations with a US$30 million expansion in Breinigsville, a US$6 million investment in the Reading facility, and a US$40 million project to expand the two Ortel facilities in California (new 20 000 sq ft fab at Irwindale plus renovation of existing plant, and an upgrade to equipment at Alhambra).

Agere's Optoelectronics Products unit produced a new manufacturing process 'Laser 2000', a low-cost platform that targets high-volume manufacturing and tight product distributions on all optical subassemblies.

Products

Agere Systems has a broad portfolio of integrated optoelectronic components including analogue and digital lasers, photodetectors, passive components, transceivers and fibre amplifiers in the telecommunications, CATV, data communications and undersea markets. It was the first company to offer a complete range of high-power FP, fibre Bragg grating-stabilized and DFB lasers for Raman and EDFA applications.

Agere's range of laser diode products includes:

- The 269-type DFB pump laser module designed as a CW optical pump source for EDFAs. It incorporates an InGaAsP/InP high-power, strained MQW laser chip that achieves fibre powers up to 280 mW.
- In May 2001 Agere announced the industry's most powerful 14xx nm DFB Raman pump lasers used to boost signal power in ultrahigh-capacity optical networking systems. These pump lasers support Raman amplification in both co-pumping and counter-pumping configurations with excellent

Figure 6.2 Lucent FY 2000 External Sales (US$ billion)

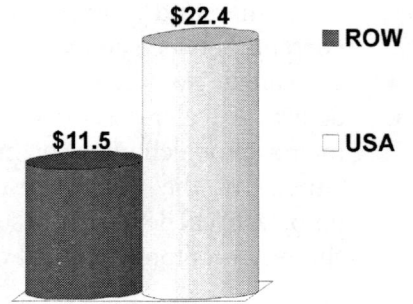

wavelength stability. The InP devices feature operating power of 280 mW, and drive currents comparable to high-power FP lasers.

- The new A1612A/B 1310 nm forward-path DFB laser modules are designed for both broadcast and narrowcast analogue applications. The devices feature up to 31 mW output power with superior distortion performance over an enhanced operating temperature range of -40°C to $+85^\circ$C, increasing the laser's reliability even in harsh environments.

- The D1861A is manufactured by Agere's subsidiary, Ortel. D1816A is a 10 Gb/s 1310 nm DML module designed as a cost-effective solution for digital transmission applications up to 50 km using traditional intra-city SMF 28 single-mode fibre links. The lasers are easily integrated into lower-cost, high-performance SONET/SDH metropolitan network and high-speed data communications applications.

- Agere introduced a 10 Gbit/s tunable transmitter containing lasers with monolithically integrated EA modulators. Inside the chip, the Fabry–Perot mode can be tuned to up to 20 channels by applying a current to the Bragg mirror. The wavelength is stabilized using Agere's on-board LambdaLock device (a patented system that locks the laser emission to 50 GHz channel spacings with a wavelength stability of 20 pm). Applications include terminal equipment in DWDM. Volume production is scheduled for the end of 2001.

Alliances

In July 2000 Lucent issued a licence to Applied Optoelectronics giving them the right to manufacture and distribute its QC laser for applications in industries other than telecommunications. The agreement was the first time Lucent had licensed this technology.

Lucent, Nortel and Hitachi have an agreement on the packaging of diode lasers for telecommunications, achieving a common size and pin-out structure for diode lasers (13.2 × 7.6 mm) and pin-compatibility with 14-pin DIL packages.

Table 6.1 Lucent Financial Highlights (US$ billion)

	2000	1999	1998	1997	1996
Net sales	33.8	30.6	30.1	26.4	15.9
Net income	1.2	4.8	1.0	0.5	0.2
Working capital	10.6	10.1	3.7	1.8	2.1
R&D	4.0	4.2	3.7	3.0	1.8
Total assets	48.8	35.4	25.1	21.0	20.2

6.3 Agilent Technologies

Agilent Technologies
370 W Trimble Road
San Jose
CA 95131
USA

Agilent Technologies was created in July 1999; it was formerly Hewlett Packard's semiconductor business, prior to that company's strategic realignment. An IPO of 72 million shares at US$30 per share was effected in November 1999.

- Agilent reported FY 2000 net revenue of US$10.8 billion, an increase of 29% on its FY 1999 figure of US$8.3 billion. The company employs 47 000 people worldwide.

Acquisitions

During FY 2000, Agilent acquired the Optical Technology Center (OTC) from Telecom Italia's Central Research Laboratory. OTC provides critical IP and design expertise for the next-generation optical transceivers market.

R&D

R&D expenditure
FY 2000: US$ 1.3 billion
FY 1999: US$997 million

Agilent's Technologies Laboratories (ATL) are located in Palo Alto, CA, USA and were formerly part of Hewlett Packard. In December 2000 the company announced a new research group, Agilent Labs China, located at its existing site in Beijing. Agilent also has a semiconductor R&D facility in Italy (see Acquisitions).

Agilent researches into semiconductor lasers and VCSELs, as well as LEDs and photodetectors. The company is also conducting research into novel photonic devices such as tunable lasers and filters, microcavity lasers, optical gain blocks and planar integration.

Agilent is the co-ordinator of the Parallel Optical Link Organization (POLO), which intends to find faster and less expensive ways to link computers for high-speed optical communications.

Figure 6.3 Agilent Three Year Sales (US$ billion)

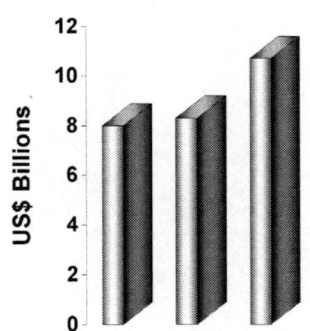

Figure 6.4 Agilent FY 2000 Sales by Business Segment (%)

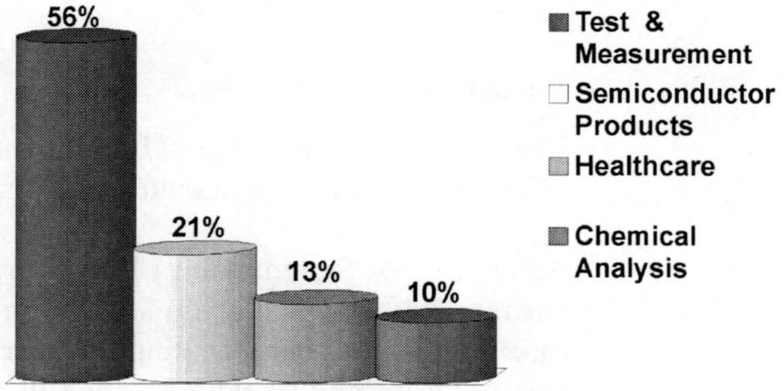

Facilities

Agilent's semiconductor operation has eight manufacturing sites, located in CA and CO, USA; Malaysia; Singapore; and the UK. Most of its silicon and GaAs wafer fabrication is carried out in the USA and Singapore.

Products

Agilent's semiconductor products business is a leading supplier of semiconductor solutions, focusing on the provision of high-performance optical, mixed-signal and digital ICs for networking, wireless, imaging and computing applications. This segment was responsible for US$2.2 billion (approximately 21%) of net revenues in FY 2000.

Agere's range of laser diode products includes:

- Agilent's HFCT-5215B/D 155 Mb/s single-mode laser transceiver is suitable for ATM, SONET and SDH applications. It is a high-performance module for serial optical data communications specified for a signal rate of 155 MBd and is designed to provide a SONET/SDH compliant link for 155 Mb/s long-reach links. The module is designed for single-mode fibre and operates at a nominal wavelength of 1300 nm. The transmitter section uses a MQW laser, whilst receiver section uses a planar PIN photodetector for low dark current.
- The LST2525 (200 µW coaxial laser) and LST2825 (1 mW coaxial laser) laser modules feature a high reliability SMQW Fabry–Perot laser diode and rear facet monitor photodiode. These are electrically connected to four pins in an industry-standard configuration. The devices are designed for use in short-, medium- and long-distance networks with bit rates up to 622 Mb/s.
- The Agilent XMT5370622 is a 622 Mb/s high-performance un-cooled optical laser transmitter. Suitable for applications such as CCITT SDH and ANSI SONET, it is designed with an ECL/PECL logic interface for 622 Mb/s transmission and can be operated with either a +5 V or −5 V power supply. The compact transmitter module contains a pigtailed laser, data interface, bias and modulation control circuitry.
- In May 2001 the company announced the augmentation of its range of DFB lasers with the launch of the 81663A family of high-power (+13 dBm) DFB laser source modules for the L-Band. The 81663A DFB laser modules

offer users high-wavelength and unsurpassed power stability, which ensure the highest test accuracy.

Alliances

The company is a member of IrDA (Infrared Data Association), which promotes interoperable, low-cost IR data interconnection standards.

Agilent, Cisco Systems and CIENA Corp created the Optical Internetworking Forum (OIF), an open forum focused on accelerating the deployment of optical internetworks. Other founding members are AT&T, Bellcore, Hewlett Packard, Qwest, Sprint and WorldCom. The OIF provides a venue for equipment manufacturers, users and service providers to work together to identify and resolve issues and develop key specifications to ensure the interoperability of optical internetworks.

Agilent is the world's largest supplier of visible LEDs; it has a joint venture company with Philips, LumiLEDs Lighting.

Table 6.2 Agilent Financial Highlights (US$ million)

	2000	1999	1998
Net sales	10 773	8331	7952
Net income	757	512	257
Working capital	2897	1857	1476
R&D	1258	997	948
Total assets	8425	5444	4987

6.4 AIXTRON

AIXTRON AG
Kackertstr 15–17
52072 Aachen
Germany

AIXTRON AG was founded in 1983, as a spin-off from the University of Aachen. The company is a leading supplier of epitaxial reactors for R&D and mass production of optoelectronic and other devices. AIXTRON is the largest manufacturer of MOVPE systems in Europe and has a large market share abroad. It offers a wide range of epitaxial systems for MOVPE, VPE and LPE from R&D-oriented systems up to large-scale Planetary Reactor production systems for up to 25 × 4-in or 9 × 6-in wafers. AIXTRON has sold more than 500 systems worldwide.

- AIXTRON reported sales for FY 2000 of €157.9 million, an increase of 87% over its FY 1999 figure of €84.7 million. The company employs over 300 people worldwide.

Acquisitions

AIXTRON acquired a 70% controlling interest in Epigress AB, of Lund, Sweden, in 1999. Terms of the acquisition were not disclosed.

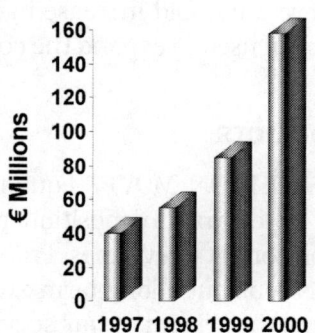

Figure 6.5 AIXTRON Four Year Sales (€ million)

Also in 1999, AIXTRON acquired Thomas Swan & Co Ltd, Cambridge, UK. Thomas Swan became a wholly owned subsidiary of AIXTRON. Again, terms were not disclosed.

AIXTRON has another wholly owned subsidiary in the USA — AIXTRON Inc of Buffalo Grove, IL.

R&D

R&D Expenditure
FY 2000: €9.6 million
FY 1999: €6 million

The company is involved in many collaborative projects with research centres worldwide.

AIXTRON employs 10 people at its Application Laboratory in Aachen, which is dedicated to the continuous improvement of hardware and processes.

- Development work is carried out on visible and IR lasers.

AIXTRON is also involved in several EC BRITE/EURAM projects, including LAQUANI, for laser quality III-V nitrides; LASBE, for layers and structures for blue emitters; and MIRIAD for microwave and IR industrial applications for diamond.

AIXTRON systems are installed in many of the world's research institutes including Fraunhofer Institute, JPL, Sandia National Laboratories, ITRI-OES and CNRS.

Facilities

Apart from its manufacturing facilities at its headquarters, the company has a worldwide service, sales and distribution network, including service centres in Taiwan, Korea and Japan. AIXTRON holds DIN ISO 9001 certification.

Phase I of the construction of AIXTRON's additional facility at Herzogenrath, Germany, was completed ahead of schedule in 2001. This facility is for the final assembly of semiconductor manufacturing equipment and will be built in three phases on a 10 000 sq m site; when completed, it will enable AIXTRON to achieve a ten-fold increase in its current capacity. AIXTRON's original facility has been used to expand the company's R&D activities.

Products

The AIXTRON MOVPE equipment range is based on the patented Planetary MOVPE epitaxial deposition process originally developed at the Philips LEP Laboratory at Brevannes, France, under an ESPRIT research programme. AIX-TRON took out a long-term exclusive licence for the worldwide rights to manufacture the system. It has successfully further developed the basic process for a wide range of materials and scaled it up for high-capacity production machines.

- AIXTRON can supply complete MOVPE lines from R&D up to full-scale production, i.e. 1 × 2-in up to 95 × 2-in wafers, or equivalent 3-, 4- and 6-in wafers.

AIXTRON is the leading supplier of MOCVD equipment for long-wavelength laser production at Agere, Alcatel, Anritsu, ExceLight, IQE, JDS-Uniphase, Samsung and Thales. Other customers include ATMI, AXT, Epistar, Epson, Honeywell, ITT, Kopin, LumiLeds, Motorola, Nortel, Osram, Showa Denko, UEC, VPEC and many Japanese corporations.

- In June 2001 the company announced that it had sold its 100th MOCVD system with automated wafer handler.

The AIX 200/4 RF-S GaN MOCVD system is based on the successful AIX 200/4 reactor series. The 200/4 RF-S is a stainless steel version and offers high flexibility in the choice of *in situ* measurement methods. Various optical ports allow the usage of reflectance spectroscopy with the Filmetrics Reflectance Monitor, RAS, pyrometry, etc. The robust design of the MOCVD system ensures reproducible and reliable operation.

- In June 2001 AIXTRON announced its 4-in Multiwafer MOCVD technology for InP, featuring 5 × 4-in and 8 × 4-in wafer size and throughput.

The AIX2600G3 cassette-to-cassette system was the world's first 6-in MOVPE reactor, suitable for R&D systems for a wide range of different materials technologies.

Figure 6.6 AIXTRON FY 2000 Geographic Sales (%)

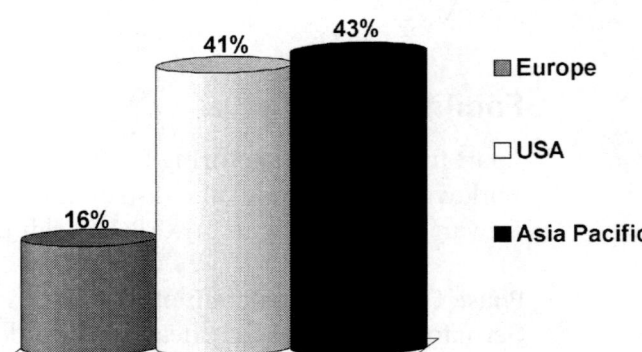

The AIX 2400/2600G3 has a flexible design for up to 35 × 2-in and 5 × 6-in wafers. The two-flow horizontal Planetary Reactor is the most widely used multiwafer MOCVD reactor for compound semiconductor applications. The reactor concept allows laminar gas flow without turbulence for precise control of the material composition and achievement of ultra-sharp interfaces. Ultra-high uniformity along with high growth efficiency on multiple 2-, 3-, 4- or 6-in wafers is achieved through wafer rotation with the patented Gas Foil Rotation.

In August 2001 Princeton Lightwave Inc (PLI), Cranbury, NJ, USA, took delivery of two AIX-2400G3 MOCVD reactors. The added production capacity will enable PLI to satisfy the high demand for its WavePower high-performance EDFA and Raman pump modules, WaveRider broadband GainChips and Wave-Harp tunable DFB semiconductor laser modules.

In April 2001 PLI reached the 1 W power level with a single narrow-stripe pump laser chip grown in an AIX 200/4 MOCVD reactor, positioning the company as a market leader in 1480-14xx pumps.

In May 2001 Marconi Optical Components ordered two more AIX 2400 (8 × 3-in) systems for its Caswell, UK site. These will be used for the production of InP-based lasers.

Mitsubishi Electric placed a repeat order for an AIX 2400G3 MOCVD Planetary Reactor with an integrated EpiRAS advanced *in situ* characterization module in May 2001. The machine will be used by Mitsubishi's Laser R&D Group to develop new device structures for high-quality lasers.

Opto Speed ordered an AIX 2400G3 Planetary Reactor to increase its production capacity for InP-based lasers and detectors, enabling it to grow advanced epi layers for 10 Gb/s products.

In June 2001 the Institute of Electronic Materials Technology in Warsaw, Poland, ordered an AIX 200/4 RF-S MOCVD system for the growth of GaN-based structures for lasers and LEDs, adding to its existing AIX 200 MOCVD system that is used for the fabrication of GaAs- and InP-based structures.

SDL in summer 2001 ordered two additional AIXTRON AIX 2400G3 MOCVD systems for InP applications.

Multiplex of South Plainfield, NY, USA, increased its production capacity for InP-based lasers, detectors and several other optoelectronic devices by adding two AIX 2400G3 Planetary Reactors to its existing AIXTRON MOCVD tool.

Alliances

In July 2001 the Department of Materials Science and Engineering and Electrical and Computer Engineering at Virginia Tech, USA, announced it had signed a research co-operation agreement for an AIXTRON 200/4 RF-S GaN MOCVD system. The group will focus on the growth mechanisms of Group III nitride alloys and heterostructures and on the application of these materials in optoelectronic devices.

AIXTRON also collaborates with:

- CNRS Valbonne for blue laser research.
- The University of Erlangen-Nürnberg, Germany, for research into blue and green solid-state laser technology.
- RWTH Aachen, Germany, in the investigation of Group III nitrides for blue LEDs and lasers and AlGaInP, UHB LEDs. AIXTRON grows the layers for RWTH.

It is also a member of the LAQUANI Project (see R&D section); AIXTROMN is the only industrial partner of this project which is lead by the French university, CRHEA-CNRS.

Table 6.3 AIXTRON Financial Highlights (€ million)

	2000	1999	1998
Net sales	157.9	84.7	55.2
Net income	18.5	10.4	5.4
Working capital	78.7	64.5	25.6
R&D	9.6	6.0	3.5
Total assets	226.7	136.0	58.1

6.5 Alcatel Optronics

Alcatel Optronics
Route de Villejust
F-91625 Nozay
France

Alcatel Optronics is a subsidiary of French company, Alcatel. In FY 2000 Alcatel subsidiary Alcatel Optics restructured its optical component activities within a single business unit called Optronics. The Optronics division comprises Alcatel Optronics (Nozay/Lannion, France), Optronics USA (Plano, Texas, USA) and Innovative Fibers of Quebec, Canada.

Alcatel Optronics designs, manufactures and sells high-performance optical components, modules and integrated sub-systems for use in terrestrial and submarine optical telecommunications networks.

- Alcatel Optronics reported sales for FY 2000 of €432.3 million, an increase of 144% over FY 1999's figure of €177.1 million. It employs 2200 people.

Figure 6.7 Alcatel Optronics Three Year Sales (€ million)

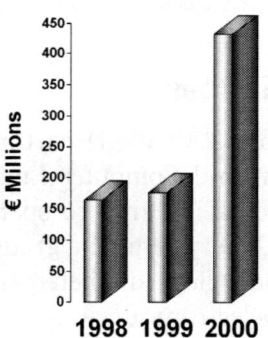

- The company is the world leader in submarine pump lasers and gigabit lasers and has a 25% share of the market for fibre Bragg gratings.

Acquisitions

Alcatel acquired privately owned Innovative Fibers of Gatineau, Quebec, Canada, in July 2000 for US$175 million. Innovative Fibers is the world leader in DWDM optical fibre Bragg grating (FBG) technology.

In May 2001 Alcatel acquired Thales' 48.83% stake in Alcatel Space for €795 million, making Alcatel Space a wholly-owned subsidiary of Alcatel. This reduced Alcatel's stake in Thales from 25.3% to approximately 20%.

Alcatel Optronics acquired UK company, Kymata, of Livingston, UK, for an undisclosed sum in September 2001. Kymata is a key player in mastering planar technology for high-end passive optical components, producing arrayed waveguide gratings and multiplexer/demultiplexers. The acquisition strengthens Alcatel's planar waveguide expertise and passive component product lines.

R&D

R&D Expenditure
FY 2000: €57.6 million
FY 1999: €24.9 million

The company's research centre is based at Marcoussis, France. Alcatel has conducted research into:

- A manufacturable process for multicolour laser diodes for all-optical networks.
- A breakthrough in broadband tunable laser technology, allowing rapid progress towards the release of a broadband tunable laser module, the 1905 TLM broadband CW tunable laser module.

Facilities

The company has manufacturing plants in North America and Europe, including facilities at its headquarters in Nozay, France. All facilities are ISO 9001 certified.

Figure 6.8 Alcatel FY 2000 Geographic Sales (%)

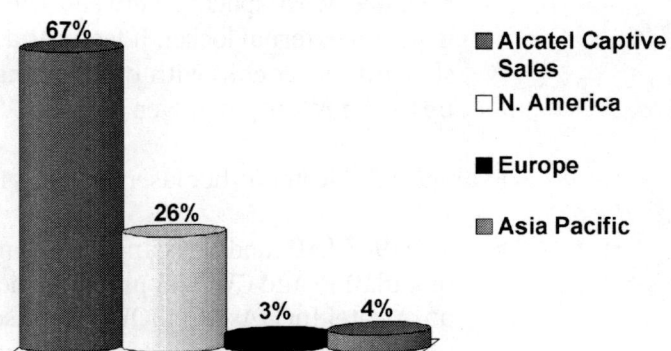

- Alcatel has doubled capacity at its Nozay facility, with the addition of an €93 million (US$100 million) expansion, increasing the site to more than 6800 sq m of manufacturing area. The existing InP product lines have been increased and added a new line for GaAs processing for 980 nm pump products.
- In January 2001 the company opened a €15 million, extension of its Lannion, Brittany production facility. The extension includes cleanrooms for the manufacture of laser sources and pump laser modules.
- The company opened a new 3000 sq m optical components facility in Plano, TX, USA, in March 2000.

Products

The Optronics Division's product range has four lines of active components: discrete modules, primarily DWDM lasers, sensors and optical routing modules; pump modules designed for terrestrial and submarine networks; optical amplification sub-systems; and optical interface sub-systems. Some 67% of its products are for captive use.

- The range of laser products includes high-performance FP lasers, DFB lasers and pump lasers manufactured using InP.
- Alcatel Optronics was first to market with Raman pumps for ultra long-haul applications, and now offers a complete Raman product line.

In March 2001 Alcatel introduced a new line of 980 nm submarine pumps (for high bit-rate, DWDM submarine systems), delivering higher output power than previously available products.

- The Alcatel 1992 SGP is suitable for a wide range of pump wavelengths between 972 and 984 nm, and the low drive current, at 450 mA for 200 mW, reduces the total power consumption of the system.

Alcatel introduced its four-wavelength tunable laser with wavelength locking for DWDM applications in August 2000. The wavelength locking function has been integrated into the 1946 LMM (2.5 Gb/s EA integrated laser), 1945 LMM (10 Gb/s EA integrated laser modulator), Alcatel 1945LMI (2.5 Gb/s DFB laser) and the Alcatel 1935 LMI (CW DFB laser). The lasers provide up to 30 mW of output power.

- The 1935 TLI for high-power DWDM applications contains an SLMQW DFB laser and is designed for use with external modulation. This module is designed with spacing between the channels down to 50 GHz without using an external locker. It features a high reliability InGaAsP buried ridge structure laser chip with outstanding long-term wavelength stability, and up to 20 mW output power.

Examples of Alcatel's other laser products include:

- The 1905 LMI module is an up to 20 mW WDM L-band version for external modulation and CW 1.55 µm laser module with optical isolator; it contains an Alcatel InGaAsP SLMQW DFB laser and is designed for use with external modulation optimized for high-power DWDM systems.

- The Alcatel 1900 LMC module family includes field-proven, reliable, un-cooled 1310 or 1550 nm DFB or 1310 nm FP laser diode, with a backfacet InGaAs PIN photodetector, in an hermetically sealed coaxial package.
- The Alcatel 1902 LMC consists of an un-cooled, Strained-Layer MQW InGaAsP laser and a backfacet InGaAs PIN photodetector in a hermetic window cap TO CAN coaxial module.

Alliances

Alcatel Optronics and Laser 2000, a specialist distributor of components and instrumentation for telecommunication networks, signed an agreement in March 2001 to develop the distribution of Alcatel's optical modules and sub-systems in Europe. Alcatel Optronics will be able to extend its services to clients: Laser 2000 operates in several European countries including Belgium, France, Germany, Holland, Italy, Sweden, Switzerland and the UK.

Alcatel and JDS Uniphase announced an agreement in March 2000 that determined a consistent standard for an internally wavelength-locked WDM source laser.

JDS Uniphase and Alcatel Optronics extended their agreement (which dates from 1999) to 2003, for the supply of 980 nm laser chips and pump stabilisers manufactured by SDL (now part of JDS Uniphase) to Alcatel Optronics. Alcatel uses the chips in pump modules for use in undersea fibre optic networks.

In March 2001 Alcatel Optronics, Agere Systems and OpNext expanded their multi-source agreement (MSA) for 10 Gb/s transponder modules used in optical networking systems to include Agilent, Ericsson Microelectronics, Mitsubishi Electric, JDS Uniphase, NEC and ExceLight.

Table 6.4 Alcatel Financial Highlights (€ million)

	2000	1999	1998
Net sales	432.3	177.1	164.1
Net income	37.5	16.1	24.7
Working capital	86.2	25.4	36.3
R&D	57.6	24.9	18.5
Total assets	576.2	140.8	129.3

6.6 Alfalight

Alfalight Inc
1832 Wright Street
Madison
WI 53704
USA

Alfalight was spun-off from the University of Wisconsin in 1988. It designs and develops high-power laser diodes for the optical fibre communications market. The company has secured financing during early 2001, totalling US$ 16.1 million.

R&D

Wisconsin Alumni Research Foundation (WARF) has an equity position in Alfalight. Consequently, WARF granted three exclusive licences to Alfalight for pending laser patents:

- High-power narrow spectral width lasers.
- High-power single-mode lasers.
- High-power short-wavelength lasers.

Facilities

The company has facilities in the USA and Canada, with a combined area of more than 50 000 sq ft. Alfalight opened its packaging facility in Vaudreuil-Dorion, Québec, Canada, in September 2000.

Products

Alfalight specializes in the development of high-power diode lasers using phosphide-based ternary and quaternary compound semiconductor materials. These aluminium-free active (ALFA) lasers permit the fabrication of advanced high-power laser structures. The company uses device designs that permit higher single-mode powers and greater mode stability than conventional diode lasers.

- Alfalight is in the process of developing 980 nm pumps for EDFAs.

6.7 American Xtal Technology

*American Xtal
Technology
4311 Solar Way
Fremont
CA 94538
USA*

Formed in 1986, American Xtal Technology (AXT) is a developer and supplier of a range of compound semiconductor substrates. AXT designs, develops, manufactures and markets high-performance substrates for the communications industry. The company produces VCSELs for fibre optics and has a proprietary VGF crystal growth technology for the production of low-defect, semi-insulating and semiconducting GaAs, InP and Ge substrates. It is also a manufacturer of LEDs for the display and lighting industries.

- AXT reported FY 2000 revenues of US$121.5 million, as opposed to 1999's figure of US$75.4 million.
- The company announced that it will discontinue its laser diode and consumer product lines in 2001 in order to focus on its core businesses of substrates and visible emitters.

Acquisitions

Lyte Optronics of Torrance, CA, USA, is an AXT subsidiary, which was acquired in 1999 for approximately US$75 million. Lyte Optronics manufactures laser

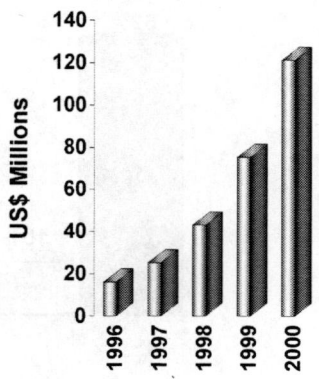

Figure 6.9 AXT Five Year Sales (US$ million)

diodes and LEDs and designs and markets laser-pointing and alignment products for the consumer, commercial and industrial markets.

R&D

R&D Expenditure
FY 2000: US$8.8 million
FY 1999: US$2.6 million

R&D projects include research into the development of GaN and high-purity GaAs epitaxy substrates. The company also funds part of its R&D through contracts with the US government and customer-funded research contracts.

AXT first began manufacturing VGF-grown InP substrates with support from a Small Business Innovation Research (SBIR) programme.

- The company has 10 Gb/s 850 nm VCSELs and long-wavelength VCSELs (1.5 mm and 1.3 mm) currently under development.
- AXT was one of the contractors awarded a DPA Title III Office programme for improving the quality of semi-insulating InP substrates and scaling up production to more than 50 KSI per year. Under this programme, TRW is a subcontractor to AXT carrying out analytical studies of the material and comparing it to other materials.

Facilities

AXT has received ISO 9002 certification for its manufacturing facilities. These are found in the USA at Fremont, Torrance, Monterey Park and El Monte, CA. The company also has plants in Beijing and Xiamen, China.

The Fremont plant includes 80 000 sq ft of production space and houses the company's proprietary VGF crystal growth operation (Phase I — 50 000 sq ft was completed in 1996; and Phase II — an additional 30 000 sq ft was completed in 1998). Phase II is used mainly for new product development, such as GaP and InP. During 1998, AXT also purchased an additional 58 000 sq ft facility in Fremont.

Figure 6.10 AXT FY 2000 Geographic Sales (%)

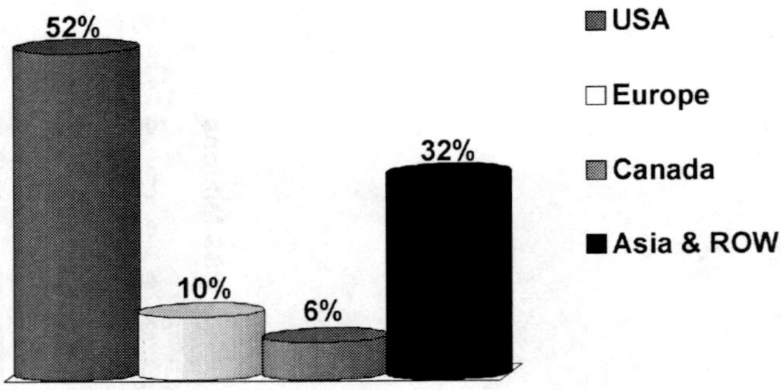

- USA
- Europe
- Canada
- Asia & ROW

The company also has a new LED manufacturing facility in El Monte, which began volume production during second quarter of 2001.

Products

AXT's product range includes the manufacture of 2-in to 6-in GaAs substrates and 2-, 3- and 4-in InP substrates. AXT has also been developing GaP and GaN substrates and has developed 100 mm VGF Ge substrates for solar cells used in satellites.

The company claims to produce the industry's lowest defect semi-insulating and semiconducting GaAs and InP wafers. The crystals are grown in-house using AXT's proprietary VGF process.

- AXT announced in February 2001 volume production of ion-implant VCSEL devices designed for fibre optic applications such as CWDM data communications. The devices are available with custom-matched wavelength spacing from 780 to 870 nm and can be customized to meet specific product designs. The devices have stable power output of up to 13 mW and can be modulated at frequencies in excess of 2.1 GHz.

AXT has been supplying 780 and 850 nm VCSEL wafers grown using MOVPE since August 2000. These wafers are available in two types: one is suitable for ion-implantation processes and the other for oxidation processes. The wafers have a thickness variation of less than 0.5% and are suitable for one-dimensional and two-dimensional VCSEL arrays for applications such as Gigabit Ethernet, Storage-Area Networks, laser printers, high-speed optical input/output and WDM. Available in 2- and 3-in diameters, the VCSEL wafers have an AlGaAs epi structure on a GaAs substrate. For the ion-implantation process, the VCSEL wafers have demonstrated a threshold current as low as 3.0 mA and a peak output power of 12 mW with approximately 18 μm aperture devices. The wafers can be used to create a range of arrays, such as 1 × 4 and 1 × 12 and two-dimensional matrix devices with speeds up to 2.5 Gb/s for each device.

- In April 2001 AXT began volume production of oxide confined VCSEL devices designed for higher data-rate fibre optic transmission. Applications for this product include VSR Ethernet, fibre channel for SANs, optical interconnections and other optical transceiver applications. The 850 nm devices have been successfully modulated at data rates in excess of 3.125 Gb/s with typical power output of 1.0 mW and peak power output exceeding 5.5 mW. The oxide VCSELs are highly efficient, with output slope efficiencies of 0.45–0.50 W/A obtained.

- In May 2001 AXT announced volume production of oxide confined VCSEL arrays in 1 × 12 and 1 × 4 configurations. These arrays are suitable for use in short-reach optical interconnections and optical transceiver applications and can include InfiniBand (SM) designs. Each device within the 850 nm wavelength array can be modulated at data rates in excess of 2.5 Gb/s, permitting optical designs with very high data rate products through the use of parallel data channels. The 1 × 4 array allows for modulation speeds reaching 10 Gb/s data rates (OC-192) and the 1 × 12 array supports 30 Gb/s rates. Typical power output for each VCSEL is 1.0 mW with output slope efficiencies of 0.45–0.50 mW/mA.

- AXT's VCSEL/LD Technologies Division announced volume production of 980 nm FP edge-emitter lasers in February 2001. The 980 nm lasers exhibit stable single-mode operation with output power up to 200 mW, threshold currents less than 10 mA and slope efficiency better than 0.9 W/A. Initial reliability tests showed favourable results and long-term tests are underway.

- In April 2001 AXT announced volume production of 1310 nm Fabry–Perot edge-emitter lasers designed for high data-rate telecommunications applications such as Gigabit Ethernet, short-reach SONET and other optical transceiver and data link applications. The lasers are available in production volumes, epi and processed wafers, as well as in bare die and TO-56 packaged forms, with both ball lens and flat window. 1310 nm EE lasers have operating powers up to 10 mW at 85°C in a stable single mode. The packaged lasers have typical threshold currents of 9 mA and slope efficiencies in excess of 0.42 W/A. Operating currents for the packaged part are typically 22 mA.

- The company announced availability of its 1550 nm edge-emitting FP lasers for medium- and long-haul transceiver applications in July 2001. The lasers exhibit excellent performance characteristics with high-temperature power output of 10 mW at 85°C and peak linear power of 20 mW. The devices are highly efficient with typical threshold currents of 10 mA and slope efficiencies of up to 0.5 mW/mA. The lasers can be driven to modulation speeds in excess of 2.5 Gb/s, for high-speed transceiver applications, satisfying OC-48 requirements. Volume production is scheduled for 3Q 2001.

Alliances

AXT has an agreement with Inner Mongolia Mining and Nanjing Germanium to mine germanium in Xilin Gol League, 250 miles from Beijing, China. Inner Mongolia Mining owns the mine property and Nanjing Germanium is one of China's largest germanium refiners. The agreement, which supplements AXT's source of supply, gives the three partners exclusive rights to the germanium for 25 years. AXT also has the right to purchase refined germanium from the joint venture at competitive prices.

AXT also has an agreement with MBE Technology of Singapore; it was announced in September 2000 that they had signed a one year, multi-million-dollar contract for the supply of 6-in GaAs wafers during 2001. MBE also joined AXT's new Supply Guarantee Program.

Table 6.5 AXT Financial Highlights (US$ million)

	2000	1999	1998	1997	1996
Total revenue	121.5	75.4	43.3	25.3	16.2
Net income	21.6	0.2	6.3	3.3	2.0
Working capital	140.4	40.5	41.1	14.2	5.5
Total assets	250.2	115.8	75.0	30.6	17.4

6.8 ATMI Epitaxial Services

ATMI Epitaxial Services
3832 East Watkins Street
Phoenix
AZ 85034
USA

ATMI Epitaxial Services is the new name for the former Epitronics Corp subsidiary of ATMI. This division also includes the former ATMI Diamond Electronics Division and epitaxial thin film manufacturer Lawrence Semiconductor, which was acquired by ATMI in 1997. The Epitaxial Services division, which employs 125 people, is divided into two groups: silicon and III-Vs.

ATMI Epitaxial Services is a one-stop shop for advanced semiconductor wafers. It was the first to provide GaN epi expertise to the open merchant market and became the second source of merchant 4H-SiC and 6H-SiC starting wafers.

- ATMI Inc reported net sales for the year ended December 2000 of US$ 300 million, an increase of 48% over FY 1999's figure of US$ 202.5 million.
- Owing to the industry slowdown, its second quarter 2001 revenues were US$ 55 million, down by 23% on the corresponding period in FY 2000.

Acquisitions

ATMI announced at the beginning of 2001 that it had purchased the remaining 30% minority interest in its South Korean chusik hosea joint venture with KC Tech for US$ 5 million. The joint venture manufactures, sells and distributes thin film materials to the semiconductor industry in South Korea. ATMI holds approximately 2.5% of KC Tech's shares and KC Tech will manufacture ATMI chemical delivery systems for the Korean market on an OEM basis for five years.

Figure 6.11 ATMI Inc Five Year Sales (US$ million)

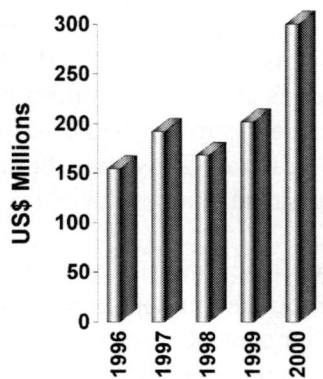

R&D

R&D Expenditure
FY 2000: US$28 million
FY 1999: US$18 million

ATMI holds more than 250 patents for its intellectual property.

ATMI is developing GaN, AlGaN and InGaN epitaxial layers and device structures on sapphire and silicon carbide substrates with customers and under contract with the US government. ATMI is also developing a high-quality, strain-relaxed, highly conductive HVPE GaN substrate. Use of the HVPE GaN substrates eliminates the need for a MOVPE- or MBE-grown buffer layer and permits straightforward homoepitaxy of device structures.

- In October 2000 the company announced a significant expansion of an ongoing venture to manufacture high performance optical materials used in the development of GaN blue laser diodes. ATMI has developed and patented processes that allow the manufacture of high-purity GaN wafers 50 mm and larger in diameter. Based on its results, ATMI committed significant additional funding for continued development and pilot manufacturing capacity. It was also awarded contracts totalling almost US$4 million from the Office of Naval Research (ONR) and the Ballistic Missile Defense Operation (BMDO) to further develop GaN wafers for electronic and optoelectronic devices.

In February 2001, a joint development agreement was announced between ATMI and SC Fluids of Nashua, New Hampshire, to develop and introduce supercritical fluid materials into semiconductor cleaning processes. The two companies are working on innovative technology with environmental benefits for future chip generations where current systems are unlikely to be able to function effectively. Under the joint development agreement, ATMI and SC Fluids are developing supercritical fluid chemistry and manufacturing processes for cleaning partially processed semiconductor wafers and removing photoresist.

Facilities

The company's 15 000 sq ft Phoenix facility specializes in AlGaAs and InGaAP HBT structures for HBTs and also offers PHEMT and FET structures. Products obtained through distribution agreements such as with Nippon Steel Corp for SIMOX (separation by implanted oxygen) enhance the product line.

In June 2000 the company announced that it would expand its silicon epitaxial technology at its Mesa, AZ, USA, facility to enter the power discrete semiconductor market; it added several Mattson EpiPro 5000 batch tools (adding to the existing 36 single-wafer tools already in place), which are capable of depositing thick, single or multiple, epitaxial layers of single crystal silicon. With the added machinery, ATMI will be able to process more than 100 000 wafers per month. Both facilities hold ISO 14001 environmental certification.

Late last year ATMI announced the expansion of its Epitaxial Services business into the Taiwan and Asia-Pacific semiconductor regions, building a local, Taiwan-based, epitaxial services facility. At the time, ATMI qualified wafer specifications for major Taiwan customers at its Mesa facility.

Products

The Epitaxial Services III-V Division provides III-V epitaxial products used in wireless communications, satellites and optoelectronics. Its major products include electronic devices, including AlGaAs and InGaP HBTs. It also supplies a wide range of custom epitaxial structures including solar cells, Hall sensors, waveguides, varactors, PINs, HEMTs and FETs.

The Silicon Division in Mesa, AZ, provides silicon epitaxial services for thin-film growth of single layers, on buried layers and patterns and advanced device structures on wafers up to 8-in.

A leading developer of advanced semiconductor thin film processes and semiconductor materials and a leading provider of point-of-use environmental equipment to the global semiconductor industry, ATMI develops products on and around its proprietary and patented CVD technologies.

ATMI offers a broad line of compound semiconductor wafers and epitaxy. Customer device requirements for wireless and optoelectronics products become custom and semi-custom InGaP, AlGaAs and GaAs epitaxial structures. Heterojunction bipolar transistor (HBT) structures with InGaP alloy emitters are ATMI's latest product, producing more reliable device performance in new wireless systems applications.

The company is exploring new markets by developing a thick GaAs on GaAs process. Its distribution of Nippon Steel SIMOX is focused on low-power electronics CMOS designs for portable computing and wireless handsets. The product line supports customers' advanced semiconductor wafer needs at each stage of their product life cycle, from R&D (SiC, GaN), production (SIMOX), to commercial insertions and volume production (III-V epi).

Figure 6.12 ATMI FY 2000 Geographic Sales (%)

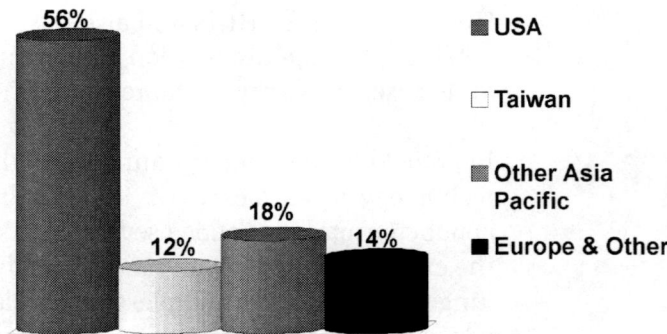

Alliances

ATMI Epitaxial Services has an agreement with Applied Materials which dates back to 1999. The collaboration provides customers with a qualified facility in Mesa for demonstrating and developing SiGe epitaxial processes on Applied Materials' Epi Centura system. ATMI also uses the Epi Centura system to manufacture SiGe epi wafers for customers.

ATMI has had many partnerships (beginning with Millipore in purifiers for CVD) and now with nearly a dozen major partners who span the semiconductor industry. The company also has partnerships with Candescent Technologies Corp (CTC, formerly known as Silicon Video), IBM, TI and Micron for commercialization of leading edge thin-film technologies. ATMI also works closely with Lucent Technologies on barium strontium titanate (BST) capacitors and with Infineon on BST-based non-volatile memories.

ATMI has a joint development program with CTC, developing and manufacturing cold cathodes for thin CRTs.

Table 6.6 ATMI Financial Highlights (US$ million)

	2000	1999	1998	1997	1996
Net sales	300.0	202.5	168.1	192.4	154.4
Net income	43.7	10.8	4.3	6.3	13.1
Working capital	203.0	121.3	104.2	49.0	37.1
Total assets	350.8	228.8	210.6	154.5	125.9

6.9 Coherent

Coherent Inc
5100 Patrick Henry Drive
Santa Clara
CA 95054
USA

Founded in 1966, Coherent is a leading manufacturer of laser-based solutions for medical, scientific and commercial applications. Coherent's subsidiary, Lambda Physik, produces NovaLine excimer lasers (using deep-UV microlithography) and the StarLine diode pumped solid-state laser.

- Coherent has restructured its operations, merging the Auburn Group, Laser Group and Semiconductor Group to create a new Photonics Group.
- For the year ended September 2000, Coherent reported net sales of US$568.3 million, an increase of 21% on FY 1999's figure of US$468.9 million.

Acquisitions

In 1998 Coherent acquired an 80% stake in Finnish company Tutcore OY for US$16 million. Tutcore specializes in the growth and processing of aluminium-free epitaxial wafers for the production of semiconductor lasers.

It also acquired Micracor, a manufacturer of semiconductor-based solid-state microchip lasers for the telecommunications market.

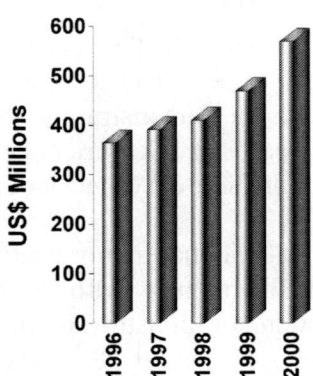

Figure 6.13 Coherent Five Year Sales (US$ million)

Coherent acquired solid-state laser manufacturer, Microlase Optical Systems of Glasgow, UK, for US$ 3.2 million in 1999.

It followed this in 2000 by acquiring Laserlec BV, Barendrect, The Netherlands, Star Medical Technologies of Pleasanton, CA, USA, and Crystal Associates Inc (CAI), of East Hanover, NJ, USA, in a cash transaction. The CAI acquisition enabled Coherent to accelerate deliveries into several key sectors including FreD lasers for fibre Bragg grating manufacturing for telecommunications and Avia lasers for the advanced packaging and interconnect markets.

R&D

R&D Expenditure:
FY 2000 US$57.4 million
FY 1999 US$46.8 million

In January 2001 Coherent demonstrated a prototype of its Azure second-generation CW diode-pumped laser, with a 266 nm laser source. Coherent began shipment of its first-generation CW 266 nm solid-state laser system in 1999 and intends that the new Azure system will be available by the end of 2001.

The company has developed and patented a new facet-coating technique which significantly increases catastrophic optical damage levels and improves lifetime.

Facilities

Coherent has a state-of-the-art laser diode facility in Santa Clara, CA, USA, bringing its total manufacturing space to 275 000 sq ft. During FY 2000 Coherent allocated US$ 36 million to upgrade its facilities in Auburn, Göttingen and Tampere.

- In 1995 Coherent obtained proprietary technology for laser diode production from Uniphase.

Products

The Coherent Semiconductor Group is part of the newly-created Photonics Group, which accounts for almost 50% of Coherent's net sales.

Figure 6.14 Coherent FY 2000 Geographic Sales (%)

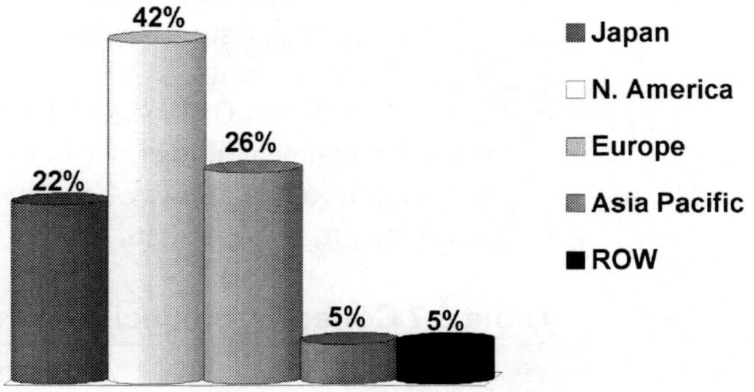

The Semiconductor Group's products include high-brightness red laser diodes and aluminium-free InGaAsP semiconductor laser material. The InGaAsP, aluminium-free technology shows an improved performance of InGaAsP (as opposed to AlGaAs) in areas such as resistance to dark-line defects, sudden failures and gradual degradation.

Coherent produces fibre-coupled, single-stripe LDs for illumination, medical diagnostics and reprographics applications.

- In May 2001 it announced a 500 μm broad-area emitter device rated at 5 W, CW. An expansion of Coherent's existing family of aluminium-free NIR (790 to 813 nm) OEM components, this new single-stripe diode laser is for integration into direct diode thermal medical and industrial applications as well as a pump source for Nd-based diode-pumped, solid-state lasers. Typical drive current and compliance voltage for the 5 W device is less than 6 A and less than 2 V. Spectral width is typically less than 2.5 nm, FWHM. Beam divergence is less than 35° by 10°, FWHM.

Coherent's Laser Division introduced the new AVIA Ultra series in January 2001. An addition to its solid-state, 355 nm, Q-switched, diode-pumped 'AVIA' product line, the AVIA Ultra series has been tailored for high repetition rate applications such as stereolithography, marking and materials processing.

- The AVIA 355-4500 is a frequency-tripled, Q-switched, diode-pumped, all-solid-state UV laser, producing 4.5 W of average power at 25 kHz and is suitable for a broad range of materials processing applications, including micromachining and micro-via drilling for the electronics packaging industry.

Coherent's range of laser products include:

- The 990–1000 nm LD linear array CW bars based on aluminium-free active area technology give high power and long-life performance in the 995 nm wavelength range. Featuring narrow bandwidth, low divergence and long life, they are suitable for printing, micromachining and marking applications.
- The Coherent range of 800–820 nm single-stripe, CW laser diodes are based on the company's aluminium-free technology. An AR-coated, fast-axis collimating lens can be included on each single-stripe device. The lens

collimates the fast-axis beam divergence from ‹35° to ‹1° FWHM, making applications in optical pumping and illumination, or integration into more complex optical systems, easier. Applications include laser pumping, printing and process control.

- The 790–830 nm QCW stacked arrays are available in 1, 2, 5 or 7 bar formats. Typical peak power per 1 cm long bar is 100 W at less than 1:15 A and can support applications where the typical pulse width is 00/s with a maximum duty factor of 400/m.

Table 6.7 Coherent Financial Highlights (US$ million)

	2000	1999	1998	1997	1996
Net sales	568.3	468.9	410.4	391.0	364.4
Net income	69.9	11.8	18.8	26.3	30.3
Working capital	422.9	194.9	173.7	152.7	123.6
R&D	57.4	46.8	44.5	39.4	44.5
Total assets	744.8	495.5	390.8	361.7	311.5

6.10 Cree Inc

Cree Inc
4425 Silicon Drive
Durham
NC 27703
USA

Cree Inc was founded in 1987 and is a world leader in the development, manufacturing and marketing of SiC-based semiconductors for certain optoelectronic, power, high-temperature, RF, microwave and non-volatile memory applications.

Although it has not as yet commercialized a blue laser using SiC technology, Cree is second in importance only to Nichia; it has created the market for and had great success with, blue LEDs (from which it derives the majority of its income). It is only a matter of time before Cree accomplishes a SiC blue laser.

- For the year ended June 2001, Cree recorded revenue of US$ 177.2 million, an increase of 63% on FY 2000's figure of US$ 108.6 million. The company employs 250 people.

Figure 6.15 Cree Five Year Sales (US$ million)

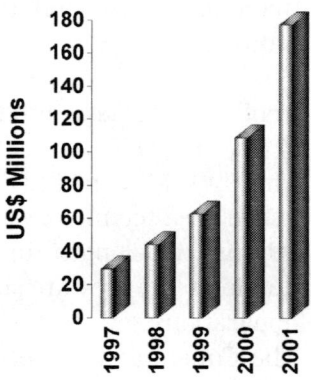

Acquisitions

Wholly owned subsidiary Cree Lighting Co, of Goleta, CA, was formerly Nitres Inc, which Cree acquired in May 2000. It develops solid-state technology for next-generation lighting and electronics devices using SiC and GaN materials.

Cree acquired UltraR in December 2000. UltraRF designs and manufactures LDMOS and bipolar RF power semiconductors.

R&D

R&D Expenditure
FY 2000: US$13 million
FY 1999: US$7.1 million

Cree is researching into new product programmes based on its experience in SiC- and GaN-based semiconductors, including blue laser diodes for optical storage applications,

- In October 2000 Cree reported operating results for its blue laser device on SiC. It announced CW lifetime in excess of 100 hours with an operating output power of 1–3 mW.
- Cree also demonstrated up to 100 mW of CW power from a single device, dramatically exceeding the 30–40 mW level presently required for read/write applications. This power level is believed adequate to meet the market requirements for virtually all consumer optical storage applications. Cree's commercial target is to further increase lifetime and reduce costs.

Cree introduced the world's first commercially viable blue LED chip in 1995 (the DH-85) and has since developed several generations of blue LEDs.

Facilities

All of Cree's manufacturing activities are undertaken in Durham, NC, USA. Cree Lighting has an R&D facility in Goleta, CA, USA.

Products

Cree has yet to commercialize a blue laser diode. Its main products are LEDs manufactured using silicon carbon.

Alliances

Cree has an alliance with ROHM Co Ltd dating from December 2000. The agreement includes a licence agreement under which ROHM granted Cree a five-year exclusive licence to several US patents and a non-exclusive licence under a Japanese patent for optoelectronic devices. The companies have also signed a non-binding memorandum of understanding for the co-operative development of a packaged blue laser diode for consumer applications. In addition, Cree and ROHM have entered into an annual supply agreement for the purchase by ROHM of LED chips manufactured by Cree.

Legal Disputes

Cree has various patent infringement lawsuits in process against Nichia Corp and Nichia America for GaN-based semiconductor devices.

- In September 2000 Cree alleged that Nichia was infringing the US Patent No 6,051,849, entitled 'Gallium Nitride Semiconductor Structures Including a Lateral Gallium Layer That Extends from an Underlying Gallium Nitride Layer'. This patent was issued to North Carolina State University (who are co-plaintiffs) and is licensed to Cree. Cree alleges that Nichia is importing, selling and offering for sale in the USA certain GaN-based laser diodes covered by one or more claims of the patent. The lawsuit seeks damages and an injunction against infringement.

Table 6.8 Cree Financial Highlights (US$ million)

	2001	2000	1999	1998	1997
Net sales	177.3	108.6	62.4	44.0	29.5
Net income	27.8	30.5	12.4	6.2	3.7
Working capital	244.2	266.0	59.9	28.3	21.1
R&D	13.0	7.1	12.1	9.9	8.0
Total assets	615.1	486.2	145.9	74.4	50.6

6.11 E2O Communications

E2O Communications Corp
26679 W Agoura Rd
Calabasas
CA 91302
USA

Privately owned E2O Communications is a provider of optical transceivers for use in high-performance, all-optical communication networks.

R&D

E2O supports 850 nm multimode fibre optic applications and is researching into 1310 nm and 1550 nm VCSEL emitters. The company claims that its strength lies in its VCSEL technology development capabilities, including its proprietary long-wavelength VCSEL device platform that is unique within the industry.

Products

The company produces small form factor pluggable (SFP) transceivers for fibre channel applications; these are only half the size of a conventional GBIC module and can double the port density in customers' equipment. The SFP-LC product family includes transceivers that operate at 1.06 and 2.12 Gb/s.

In October 2000 it introduced a low-cost, 4×2.5 Gb/s VCSEL-based, parallel transmitter/receiver link for fibre optic communication systems. Availability is ensured through volume production; for example, the links are priced at less than US$300 per transmitter/receiver pair (in volume) — only half the cost of competitive models. Similarly, the price for a 100-metre, 4-channel, bundled LC ribbon is under US$300, again less than half the cost of competing products.

6.12 Eastman Kodak

Eastman Kodak Co
343 State Street
Rochester
NY 14650
USA

Eastman Kodak was founded in 1880. The company is, through a collaborative programme, conducting research into blue laser technology, but does not as yet have any products in this field.

● Kodak reported revenues for FY 2000 of US$13 994 million, a slight reduction in FY 1999's figure of US$14 089 million and employs 78 500 people.

R&D

R&D Expenditure
FY 2000: US$784 million
FY 1999: US$817 million

Kodak's research in the area of blue light source generation is focused on sources for archival disk storage. The company's primary focus in the near term is aimed at obtaining blue laser light by second harmonic generation with diode laser pumping at longer wavelengths. Kodak has selected a quasi-phase matching approach. Materials under consideration include lithium tantalate and potassium titanyl phosphate.

The Electronics Systems Division of Kodak Research Laboratories is investigating the development of compact blue coherent light sources. This programme is funded in part by the National Institute of Standards and Technology (NIST) as part of the ATP programme. Other partners in this consortium funded under the ATP activity include IBM, JDS Uniphase, the University of Arizona and Carnegie Mellon University. SDL (part of JDS Uniphase) supplies lasers for this project.

Blue lasers will be used at Kodak for other applications beyond optical discs. Digital imaging requires output devices. For colour, conventional photographic papers must be exposed with red, green and blue sources; diode lasers are good

Figure 6.16 Eastman Kodak Five Year Sales (US$ million)

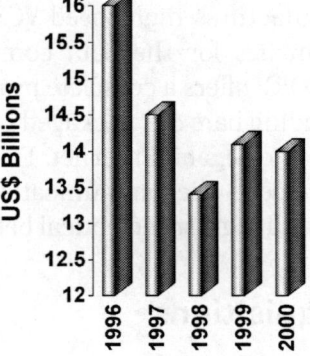

candidates for these sources. Cost reduction in the light source is a consideration for the 14-in disk application, but will be a far more crucial factor down the road for use in photo-CD (or other consumer) applications; otherwise, Kodak's products in this area will be restricted to low-volume, high-performance markets.

Products

Kodak manufactures products containing red laser diodes, such as its DryView laser imaging film for medical applications. DryView is a high-resolution, IR-sensitive photothermographic film specifically formulated for Kodak DryView laser imagers. This film delivers diagnostic-quality images, but requires no wet chemistry, or wet film processors. DryView laser imagers employ a laser diode to expose the film. The diode never makes contact with the film (and does not wear out), ensuring precise, consistent imaging.

Table 6.9 Eastman Kodak Financial Highlights (US$ billion)

	2000	1999	1998	1997	1996
Net sales	14.0	14.1	13.4	14.5	16.0
Net income	1.4	1.4	1.4	0.05	1.3
Working capital	1.5	0.8	0.9	0.9	2.1
R&D	0.8	0.8	0.8	1.0	1.0
Total assets	14.2	14.4	14.7	13.1	14.4

6.13 EMCORE

EMCORE Corp
394 Elizabeth Avenue
Somerset
NJ
USA

EMCORE Corp is a vertically integrated North American company which specializes in the design and production of MOCVD reactors and associated process technology and devices. EMCORE has sales offices, affiliate companies and distributors in all the major regions worldwide.

- The company reported net sales for FY 2000 of US$104.5 million, as opposed to US$58.3 million in FY1999.

EMCORE's MicroOptical Devices (MODE) subsidiary designs, develops and manufactures high-speed VCSELs and PIN photodiode components and subassemblies for the data communications and telecommunications markets. EMCORE offers a complete product line of VCSEL and PIN photodiode solutions, including bare die, packaged components and optical subassemblies for integration into Gigabit Ethernet, Fibre Channel, Infiniband, WDM, ATM systems and high-speed telecommunications applications, including very short reach OC-192 and high-speed optical backplanes.

Acquisitions

Uniroyal Technology Corp completed its purchase of EMCORE's share of their high-brightness LED joint-venture, Uniroyal Optoelectronics, in August 2001, in exchange for almost two million shares of Uniroyal common stock.

Figure 6.17 EMCORE Five Year Sales (US$ million)

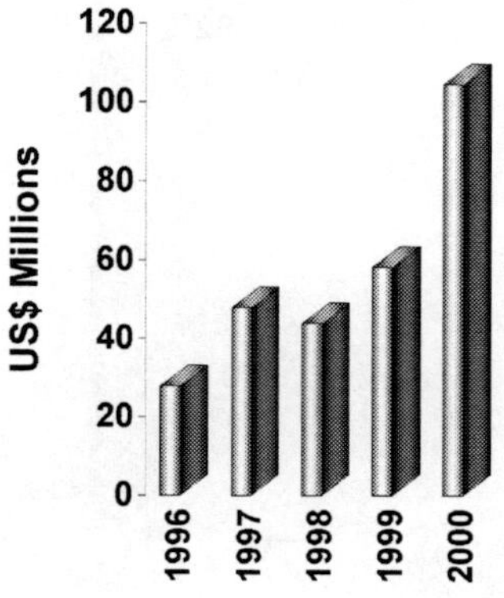

R&D

R&D Expenditure
FY 2000: US$4.4 million
FY 1999: US$4.4 million

EMCORE's published research works include:

- WDM VCSELs fabricated by selective area MOCVD growth.
- First all-monolithic VCSELs on InP: +55°C pulse lasing at 1.56 mm with GaInAlAs/InP system.
- Photoluminescence and Raman properties of MOCVD-grown $In_{0.5}(Ga_{1-x}Al_x)_{0.5}P$ layers under different growth conditions.

Facilities

MODE has a Class 1000 cleanroom. In 1998 its manufacturing capacity was tripled by the addition of 20 000 sq ft of cleanroom, reliability and test facilities. It has ISO 9001 quality certification.

Products

The company is one of the industry leaders in the provision of MOCVD systems for the R&D and mass-production of advanced optoelectronic devices such as blue emitting GaN and SiC LEDs. EMCORE is one of the world's largest producers of epitaxial wafers, grown on its proprietary *Turbo*Disc reactors. It also provides foundry services.

The company supplies MOCVD R&D and production tools. It has recently expanded its production line with the addition of the E300LDM reactor, which is the largest yield producing laser diode tool on the market. The E300LDM fulfils the needs of data communications and telecommunications module manufacturers for the development of long-wavelength, IR and visible lasers and VCSELs.

EMCORE offers a complete product line of VCSEL solutions:

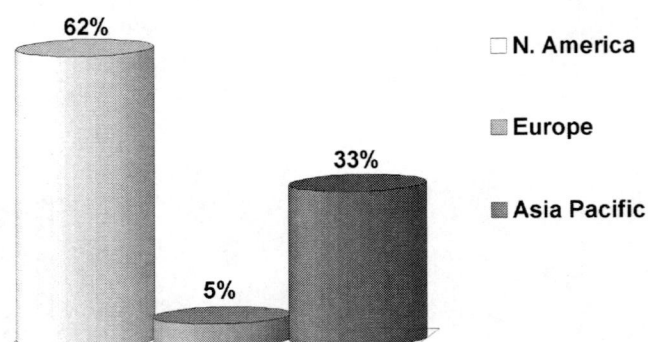

Figure 6.18 EMCORE FY 2000 Geographic Sales (%)

- In March 2001 it launched its 850 nm 10 Gb/s oxide VCSEL, the first 10 Gb/s VCSEL available commercially, is designed to work in existing optical components and is optimized for high-speed data communications over multimode fibre. The VCSEL is suitable for optical interconnect applications under 300 m, including Gigabit Ethernet and Fibre Channel.
- In May 2001 the company announced the availability of a family of CWDM VCSELs for high-speed data communications: four oxide VCSELs with wavelengths of 778, 800, 825 and 850 nm and has been selected by Blaze Network Products (which is using the VCSELs to drive the development of CWDM modules for short reach (10 Gb/s OC-192) and extended reach (1 Gb/s) data communication applications); and Intel subsidiary, Cognet MircoSystems, for use in their CWDM modules. Each VCSEL operates at 3.125 Gb/s for an aggregate bandwidth in excess of 10 Gb/s.
- The 1 × 4 VCSEL array consists of high-power near IR 850 nm VCSELs for high-speed data and telecommunications applications such as Gigabit Ethernet and Fibre Channel. This is available in die form only.
- In July 2001 EMCORE introduced its new 850 nm, 10 Gb/s GaAs photo-detector for multimode fibre optic applications. By working in conjunction with its 10 Gb/s VCSEL, the new device has been designed for Very Short Reach applications, which include serial links, LANs for Gigabit Ethernet and Fibre Channel, Infiniband and OC-192.

Alliances

EMCORE and Agilent have a three-year supply agreement (lasting until the end of 2002), whereby EMCORE manufactures Gigarray VCSEL arrays for use in parallel optical transceivers.

AMP Inc and EMCORE have a long-term strategic alliance to develop and produce VCSELs for the AMP line of fibre optic communication products.

EMCORE and JDS Uniphase have a joint development manufacturing and marketing agreement which dates from June 2000. The two companies develop, manufacture and market a family of fibre optic array transceivers (based on EMCORE's laser technology), for fibre optic communications products. EMCORE manufactures VCSEL arrays and designs gigabit speed control circuits, photodetectors, optical links and other components. JDS Uniphase handles all marketing, worldwide sales, application support, customer service and distribution functions and assists EMCORE with technical support for optical packaging and testing for the products.

Table 6.10 EMCORE Financial Highlights (US$ million)

	2000	1999	1998	1997	1996
Net sales	104.5	58.3	43.8	47.8	27.8
Net income	(25.5)	(22.7)	(36.4)	(5.6)	(3.2)
Working capital	111.6	20.7	(2.0)	12.2	1.2
R&D	4.4	4.4	3.6	–	–
Total assets	243.9	99.6	73.2	39.5	20.4

6.14 Epichem

Epichem Ltd
Power Road
Bromborough
Wirral
Merseyside L62 3QF
UK

Formed in 1983, Epichem Ltd is a private company. It is a leading manufacturer of a range of high-purity chemicals and gases used mainly in the electronics and glass coatings industries. It also has a US sales and manufacturing operation, Epichem Inc in Haverhill, MA (a metalorganics facility that was opened in 1997) and another sales office in PA.

In September 2000 Epichem officially opened a new metalorganics facility in Tsukuba, Japan. This is a joint venture between Epichem and Daido Air Products. The new facility will serve as a transfill, analytical and customer support centre initially, with planned future production capability for selected materials.

Acquisitions

During FY 2000 Epichem acquired a controlling stake in Inorgtech Ltd, a manufacturer of metal alkoxide precursors used to produce dielectric DRAM layers and non-volatile memories (ferroelectric RAMs). Inorgtech has also targeted transition metal and metal nitride precursor products whose prime application is CVD of diffusion barriers.

R&D

Epichem has been associated with many research projects with national and foreign universities and industry partners. Epichem also has a dedicated precursor research facility at the University of Salford, UK.

The development aspect focuses on transferring the research results into the production environment.

* Thermal stability studies have been performed to ensure that safe handling techniques are employed both during production and also during use by customers. Furthermore, source characterization has been investigated on a simple MOCVD kit and more recent studies relating to precursor transport relationships with temperature and flow rate are on-going.

Epichem has reported III-nitride and related work done in collaboration with academic research groups, in particular with the University of Sheffield, UK, with whom it is a partner in the Rainbow Scheme for research into precursors and dopants, such as Me_3AlNH_3, Me_3GaNH_3 and Me_2GaNH_2. Internally, it is developing lower oxygen precursors such as TMI, TMA and TMG.

The company has reported work on the use of ammonia substitutes such as hydrazine for deposition of GaN and AlN at much lower growth temperatures. Dimethylhydrazine (DMH) has also been used to grow cubic GaN on GaAs at lower temperatures ($\sim600°C$) and III/V ratios (160:1) than with ammonia. However, both hydrazine and DMH are not only toxic but also unstable. The less toxic, more stable phenylhydrazine has been tried but this has too low a vapour pressure for optimal use with MOVPE.

Epichem's work on alternative Group III precursors includes trimethylamine and good AlGaN epilayers on sapphire have been grown. Furthermore, it is working on so-called single-source precursors that combine aluminium and nitrogen, e.g. Me_3AlNH_3. Results have been mixed but GaN films have been grown at lower temperatures as this work continues.

Epichem has also collaborated with the former Defence Research Agency (DERA), Malvern, UK, in the development of low oxygen precursor materials. DERA is now known as Qinetiq. Specifically, Epichem has established a link between oxygen contamination in chemical beam epitaxy (CBE) grown AlGaAs and traces of oxygen-based solvents in the precursor's manufacture. This has enabled development of an alternative manufacturing process using amine-based solvents and improved product.

Epichem belongs to various research projects including SICOIN, RAINBOW, Widegap CPV, Admiral and CONFORM.

Facilities

During FY 2000 Epichem constructed a multi-ton trimethyl gallium plant and doubled the manufacturing capacity of its boron trichloride plant to 200 tons per annum. It also expanded its Silane capacity by 50%.

Epichem has a 20 000 sq ft manufacturing facility at Bromborough. Production facilities encompass tonnage plants for trimethyl gallium and trimethyl indium, together with smaller plants for a full complementary range of metal-organics. The company also produces boron trichloride, with a capacity of 200 tons per annum and has a silane transfilling capability of 50 tons per annum.

Epichem's Haverhill manufacturing facility was opened at the end of 1997. Since then it has seen growth of 400% and has plans to construct a further US$ 5 million plant.

Both the UK and USA production facilities hold ISO 9001 and ISO 14001 quality certification.

Products

The company's metalorganics are used in the growth of thin films of compound semiconductors such as GaAs and InP. A wide range of electronic and optoelectronic devices are made from these materials including lasers, LEDs, detectors and solar cells. It also manufactures chemicals used to grow thin films of diamond and SiC, high-temperature superconductors and optomagnetic materials.

Its main product offering is a line of ultrahigh-purity metalorganic chemicals. Owing to proprietary adduct purification techniques, Epichem's facilities yield volatile compounds of Al, Ga, In, As, P, Zn, Fe, Mg, Sb, N, Te and Cd with purities unmatched in the industry. The metalorganics are used in the growth of thin films of compound semiconductors such as GaAs and InP. A wide range of electronic and optoelectronic devices are made from these materials including LEDs, lasers, detectors and solar cells.

Epichem is a world leader in the supply of Group III alkyl compounds which are used, for example, in combination with ammonia for III-nitride growth. Epichem provides ultrahigh-purity grades of chemicals for nitride growth, such as tertiary butylamine and doping, such as bis(cyclopentadienyl) magnesium.

- Epichem recently announced EpiFill, an automated metalorganic bubbler refill system which is designed to automatically dispense chemicals such as TMG from a remote bulk container as needed by the bubblers located in as many as eight separate MOVPE systems. The epitaxial growth of compound semiconductors in modern cleanrooms is demanding increased throughput and decreased downtime. Frequent change-out of chemical source bubblers (a major cause of downtime) is eliminated by use of the EpiFill distribution system.

It is developing alternative nitrogen precursor source materials for MOVPE of III-nitrides. These are intended to replace ammonia gas popularly used but having major shortcomings. Epichem is working to develop single source precursor materials to achieve significantly lower growth temperatures and III/V ratios. This will also have other benefits, not only for doping and substrate options, but also environment benefits from more efficient reaction processes.

6.15 ExceLight

ExceLight
Communications
4021 Stirrup Creek Drive,
Suite 200
Durham
NC 27703
USA

Formerly the Electro-Optic Products Group of Sumitomo Electric Lightwave (from which it was spun-off in August 2000), ExceLight Communications is a wholly owned subsidiary of Japanese parent company, Sumitomo Electric Industries (SEI). ExceLight supplies high-performance active and passive optical components for private and public networks.

- SEI reported FY 2001 net sales of ¥1478.7 billion, an increase of 13% over its FY 2000 figure of ¥1308.6 billion.

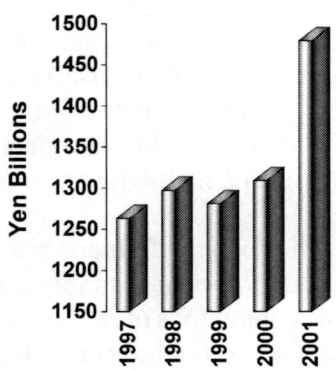

Figure 6.19 SEI Five Year Sales (¥ billion)

R&D

R&D Expenditure
FY 2001: ¥45 billion
FY 2000: ¥42 billion

The company conducts research into SMQW laser diodes using MOVPE.

- In February 2000 SEI announced the successful development of a 2-in single-crystal GaN substrate, large enough for practical use. Large-sized GaN substrates not only lower costs and fundamentally improve performance, but also accelerate development and mass production of blue lasers for next-generation DVD applications. In partnership with the Faculty of Engineering of Tokyo University, SEI expanded upon previously cultivated technology to successfully grow 2-in crystal substrates for epitaxy.
- The company was awarded US Patent No 6,273,620 in August 2001: 'semiconductor light emitting module'. A fibre grating laser module comprises a semiconductor optical amplifier and a grating fibre in which a Bragg grating is formed in the core. The semiconductor optical amplifier provides a waveguide in which the light is generated and amplified by the carrier injection, the light emitting facet and the light reflecting facet. The Bragg grating in the grating fibre and the light reflecting facet of the semiconductor amplifier forms an optical resonator. The subject of the invention is that the reflectance of the Bragg grating at the Bragg diffraction wavelength is greater than 60%.

Facilities

During 2000 ExceLight moved its operations to a new 14 000 sq ft facility in the Research Triangle Park, NC, USA.

Products

SEI has invested US$35 million in ExceLight Communications, in order to strengthen its portfolio of active, passive and optical data link components for the rapidly expanding WDM market. SEI plans to boost production capacity for optical components, high-performance WDM fibre and other related optical technologies from two to five times by 2002.

- SEI is targeting key areas of WDM technology for new products, including new, high-power 1480 nm pump lasers for Raman amplifiers with integrated wavelength-stabilizing fibre Bragg gratings, OC-48 and OC-192 optical transponders and optical sub-systems.

ExceLight produces LDs for applications such as DWDM, LAN, WAN, CATV, fibre amplifiers and SONET/SDH. It manufactures InGaP/InP, MQW laser diodes (Fabry–Perot and DFB) for low-threshold current (typically 5–8 mA). For DWDM applications, it also produces 100 GHz and 50 GHz DWDM laser sources, cooled high-power DFB laser modules for DWDM and high-power CW lasers.

- The SLT4210 Series of 1.3 μm InGaAsP/InP MQW-DFB LD modules are designed for fibre optic long and intermediate reach 2.5 Gb/s transmission applications. A laser diode is mounted into a coaxial package integrated with an InGaAs monitor PD and a single mode fibre pigtail. The SLT4260 series includes an integrated single-stage optical isolator.
- The SLV4270 series are 1.3 μm InGaAsP/InP MQW-DFB laser diode modules designed for fibre optic CATV return path applications. These modules are ideally suitable for high-capacity transmission including several video channels. A laser diode is mounted into a coaxial package integrated with a single mode fibre pigtail, a double stage isolator and an InGaAs monitor photodiode.
- The SLT5411 and SLT5413 series are 1.5 μm InGaAsP/InP MQW-DFB laser diode modules designed for a CW optical source for WDM applications used with an external modulator. The laser diode chip is mounted on a 14 pin butterfly package integrated with an optical isolator, InGaAs monitor PD, thermoelectric cooler and single-mode PM fibre pigtail.

Alliances

ExceLight is a member of the Agere Systems, Alcatel Optronics and OpNext MSA for 10 Gb/s transponder modules used in optical networking systems.

Legal Disputes

In February 2001 SEI filed a patent infringement lawsuit in the USA at the Delaware Federal Court against Furukawa Electric Co Ltd and two of its US subsidiaries: Furukawa Electric North America Inc and FITEL Technologies Inc. SEI's complaint is that Furukawa's 'Rainbow Pump' semiconductor pump laser module products infringe SEI's US Patent No 5,845,030, entitled 'Semiconductor Laser Module and Optical Amplifier'. This patent describes a critical advance in the area of semiconductor pump laser with fibre Bragg grating stabilization.

Table 6.11 SEI Financial Highlights (¥ billion)

	2001	2000	1999	1998	1997
Net sales	1479	1309	1281	1297	1263
Net income	40	24	20	33	32
Working capital	306	329	309	307	266
R&D	45	42	44	46	43
Total assets	1813	1495	1495	1429	1369

6.16 Freiberger Compound Materials

Freiberger Compound
Materials GmbH
Am Junger Löwe
Schacht 5
D-09599 Freiberg
Germany

Freiberger Compound Materials (FCM) is a private company that is a leader in GaAs technology for the manufacture of diode lasers and other substrate devices. It is Europe's largest GaAs substrate manufacturer with a greater than 20% share of the world GaAs market. Producing both semiconducting and semi-insulating GaAs wafers, FCM employs approximately 230 people. Its current largest shareholder (holding 87.5%) is Israeli company Federmann Enterprises, with the remainder belonging to Infineon AG.

The company was founded in 1957 as VEB Spurenmetalle Freiberg, a German government-owned establishment for the development and manufacture of high-purity semiconducting materials. It began growing GaAs crystals in 1980. In 1990 it was acquired by Freiberger Elektronikwerkstoffe GmbH (FEW), who, in 1991, also acquired the GaAs business of Wacker. In 1995 FEW restructured, with the GaAs division becoming FCM.

FCM's US subsidiary, Freiberger Compound Materials USA Inc in Doylestown, PA, has an agreement with French InP wafer company InPact, whereby FCM USA markets InPact's InP wafers in the USA and Canada. These substrates are crucially important for several types of diode laser device.

R&D

Currently employing 30 people in research and development, the company began research into 6-in wafers in 1994. These are primarily for the electronic device market but the technology will in due course be applicable to diode laser materials.

The company has research agreements with several German universities, including Fraunhofer-Gesellschart, Freiberg; University of Erlangen, University of Münster, University of Frankfurt-Main, ZFW Göttingen and Bergakademie Freiberg.

Freiberger started development of low-temperature gradient VGF crystal growth process for 100 mm semi-insulating GaAs in 1998.

- FCM has demonstrated successfully the LEC growth of 200 mm Si GaAs.
- The company currently has 150 mm VCZ wafers under development.

Facilities

Since the company was privatized in 1995, it has invested more than US$45 million in its new production and administration facility. It holds both ISO 9000 and ISO 14001 certifications.

Freiberger has constructed a new facility, Fab 2, for 150 mm LEC and VGF wafers. Ramp-up is scheduled for completion by the end of 2001.

Freiberger has reported results for 6-in wafers comparable to and partially better than those for its 4-in wafers. FCM anticipates it will more than double its output for 6-in wafers by 2002 (3.20m sq in/year in 2000, rising to 8.45 m sq in/year by 2002).

Products

FCM is one of the top suppliers of volume, high-quality 6-in GaAs substrates, with extensive experience in preparing epi-ready substrates for both MBE and MOCVD applications.

FCM's semiconducting wafers are offered in 3 in, 100 mm and 150 mm LEC-grown GaAs:Te and 100 mm and 150 mm VGF-grown GaAs:Si. It manufactures 3 in, 100 mm and 150 mm SI GaAs wafers.

Freiberger produces very large GaAs boules yielding more than 100 high-quality 6-in wafers. The crystals are grown by a standard LEC-process, expanding the company's expertise in 4-in growth technology. Typically, 9-in pBN crucibles and either pre-synthesized ingots or the direct synthesis technique with charges of up to 20 kg are used as starting material. The company uses a new generation of pullers equipped with a fully computerized process and diameter control system.

- In summer 2000 Freiberger announced a production-ready process for VGF (vertically gradient freeze) substrates. Substrates with guaranteed state-of-the-art structural and electrical properties coupled with epitaxy-ready surface are now in full production. Freiberger also has a growth process for 150 mm diameter ingots in development.

By the use of computer-modelling techniques, Freiberger developed a low temperature gradient VGF furnace specifically for the growth of low etch pit density (EPD) GaAs crystals. The furnace is designed to accommodate crucibles up to 360 mm long and with a maximum length to width ratio of 2, thereby allowing growth of either 100 mm or 150 mm diameter ingots.

6.17 Fujitsu

Fujitsu Ltd
6–1 Marunouchi
1-chome
Chiyoda-ku
Tokyo 100
Japan

Founded in 1935, Fujitsu Ltd is a world-leading manufacturer of goods for the computer, communications and microelectronics markets. Pertinent to this report is its line of long-wavelength laser diodes with single-mode operation and distributed feedback (DFB) lasers for fibre optic applications.

- For the year ended March 2001 the company reported net sales of ¥5484.4 billion, an increase of 4% over FY 2000's figure of ¥5255.1 billion. It employs 187 400 people worldwide.

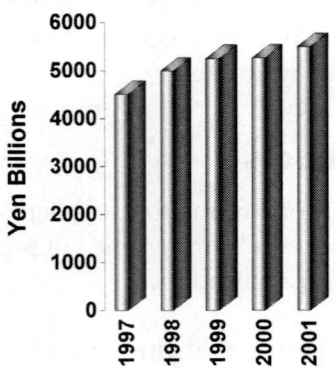

Figure 6.20 Fujitsu Five Year Sales (¥ billion)

R&D

R&D Expenditure
FY 2001: ¥403.4 billion
FY 2000: ¥401 billion

Fujitsu Laboratories Limited was founded in 1968 and is a wholly owned subsidiary of Fujitsu Ltd.

- In 1997 Fujitsu Laboratories announced it had succeeded in developing the world's first technology suitable for mass producing blue laser diodes, by achieving stable pulsed operation of a GaN-based blue laser on a SiC substrate at room temperature using MOVPE.

The Fujitsu Quantum Devices Ltd (FQD), facility in Yamanashi Prefecture, Japan, transfers microwave and optoelectronic device technology from R&D to production.

Current research in optoelectronics includes DFB laser diode modules with wavelength lockers for DWDM and 660 nm lasers for optical storage disc devices.

- Fujitsu Network Communications demonstrated the world's first 44-channel tunable laser technology during the SUPERCOMM 2001 exhibition in June 2001. This laser technology is configured with DFB laser array integration. Each laser cavity covers four wavelengths and includes temperature control and wavelength-locking circuitry. A built-in silicon optical amplifier enables large laser output power requirements for 40 Gb/s transport and is capable of being programmed down to lower settings for OC-192 or OC-48 using the same laser.

Facilities

As well as manufacturing in Japan, Fujitsu has overseas semiconductor manufacturing facilities in the USA, Ireland and Germany.

Figure 6.21 Fujitsu FY 2001 Geographic Sales (%)

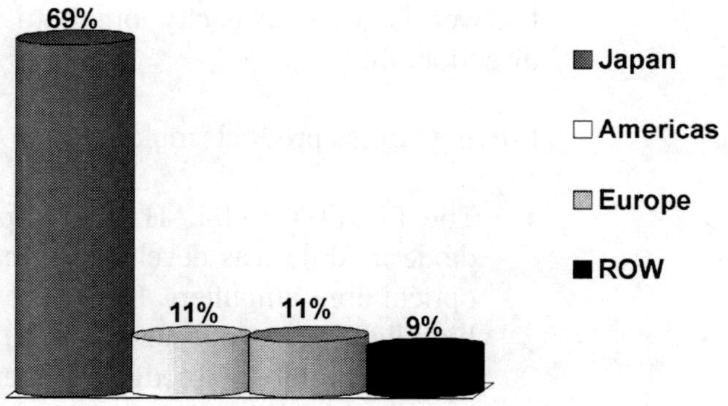

Subsidiary company FQD specializes in the manufacture of compound semiconductors. It has one of the largest GaAs wafer fabs in Japan at Yamanashi Prefecture.

- Fujitsu Compound Semiconductor Inc (FCSI) is the San Jose, CA, USA-based sales, marketing and design centre (R&D) arm of FQD. This serves as a distributor of light wave and microwave GaAs-based semiconductor products.

Products

Fujitsu is a major manufacturer in the laser diode, photodetector, CATV and telecommunications fibre optic markets. Its products include low-cost Fabry–Perot lasers, pump lasers and electro-absorption modulator integrated, directly modulated, CW source DFBs. It also provides super-high-reliability laser diodes for submarine cable systems.

FCSI provides Fujitsu products for fibre optic digital telecommunications applications in the 622 Mb/s–40 Gb/s range (laser diodes, receiver modules, GaAs ICs and LN modulators) and analogue lasers for CATV applications. FCSI also offers lasers at ITU grid wavelengths for DWDM systems.

In August 2001 FCSI announced two new 10 Gb/s (OC-192) laser drivers. Both products offer 3 V peak to peak output voltage for use with modulator integrated (MI) lasers and differential input modulators. The peak output current, output duty ratio and output offset are adjustable for both devices and both have internal 50 Ω terminations at the high-speed differential input and output connections.

- The FMM3109PG is in a 24-pin hermetic package while the FMM3109ZH is in a lower cost 6 mm × 6 mm 32-pin, plastic moulded package with heat spreader.
- August 2001 FCSI introduced two new 2.5 Gb/s (OC-48) laser drivers. Both devices have a D-FF built in and the re-timing and bypass mode is selectable. Both are ECL compatible, have duty rate adjustment and output shutdown switch. The FMM3174VI is a MI laser driver and the FMM3175VI is a direct modulated laser driver. Both units are housed in a small size 16-pin package (SSOP-16).

Fujitsu Network Communications Inc anticipates that by 4Q 2001 it will have in production 22-channel tunable laser capabilities for its FLASHWAVE OADX, the world's largest capacity long-haul system capable of carrying 1.76 Tb/s bi-directionally.

Current Fujitsu products include:

- The FLD148G3NL-E/-H/-J, 1.48 mm bandwidth, Fabry–Perot-type laser diode module was developed primarily as a light source for excitation of optical fibre amplifiers. Use of the MQW structure in active layers enables high light-power output (Pf = 140 mW or more). The device also features a built-in monitor photodiode cooler thermistor (the FLD148G3NL-L model includes a built-in isolator).
- The FLD3F8CZ is a 1310 nm MQW DFB CATV laser — a middle power laser capable of carrying 78 channels. It features a built-in thermoelectric cooler, thermistor and monitor photodiode. Transmission spans of 15 km are possible without amplification.
- The FLD5F6CX-H is a 1550 nm MQW DFB CW laser featuring a built-in thermoelectric cooler, thermistor and monitor photodiode. For 2.5 Gb/s DWDM systems, it features 10 mW output power.
- The FLD5F10NP is a 1550 nm MQW DFB modulator integrated laser. It features CW laser operation and is suitable for 10 Gb/s long-haul transmission.
- The FMM3109PG is a 10 Gb/s modulator integrated laser diode modulator driver, featuring 6.8 V power supply.

Table 6.12 Fujitsu Financial Highlights (¥ billion)

	2001	2000	1999	1998	1997
Net sales	5484	5255	5243	4985	4504
Net income	9	44	(14)	6	46
Working capital	294	523	643	532	415
R&D	403	401	395	387	353
Total assets	5200	5135	5112	5123	4728

6.18 Furukawa

Furukawa Electric Ltd
Compound
Semiconductor
Department
6–1 Marunouchi
2-Chome
Chiyoda-Ku
Tokyo 100
Japan

Founded in 1884 and currently employing more than 8300 people, Furukawa Electric is a large, vertically integrated supplier of a diverse range of microelectronic and optoelectronic materials and components. These products include diode lasers as well as GaAs epitaxial wafers and substrates. Furukawa is a major supplier of pump lasers and claims to have a 70% share of the 1480 nm pumping laser module market. It sells its optical devices under the brand name of its subsidiary company FITEL.

- For the year ended March 2001, Furukawa reported net sales of ¥827 billion, an increase of 18.7% over its FY 2000 sales of ¥696.6 billion.

Figure 6.22 Furukawa Electric Five Year Sales (¥ billion)

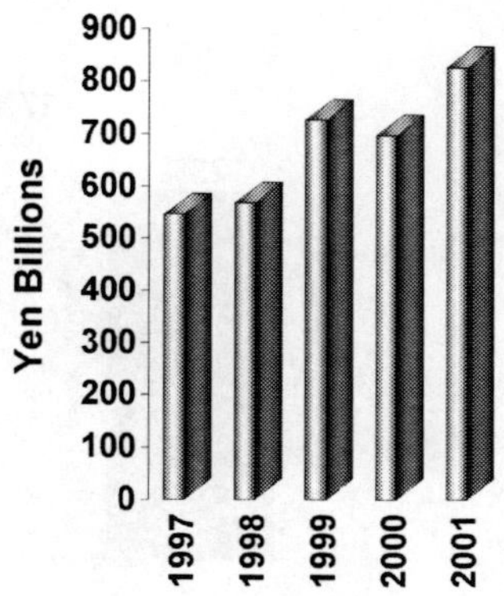

Acquisitions

In May 2000 Furukawa Electric invested ¥2 billion to gain a 60% share in Optigain Inc. Founded in July 1992, Optigain is based in Peace Dale, RI, USA.

During July 2001 Furukawa announced that it had agreed to acquire Lucent Technologies' Optical Fiber Solution (OFS), for US$2.75 billion. Furukawa will now have the second largest share of the world optical fibre market.

In January 1999 Furukawa's US subsidiary, FITEL, merged with Uniphase Corp (now JDS-Uniphase), for US$6.1 billion. Furukawa was the largest shareholder in the new company.

R&D

The company has developed MBE growth capabilities. For example, it has published work in the application of gas-source MBE to long-wavelength semiconductor diode lasers.

It is a member of the SiC Hard Electronics programme for microelectronic devices based on silicon carbide.

Furukawa Electric has been promoting the development and marketing of semiconductor lasers since the late 1980s and has made efforts to improve the output power performance of the 1480 nm and 980 nm semiconductor lasers in particular, concentrating the company's extensive technologies including the compound semiconductor, optical power coupling, heat dissipation and metal material technologies.

- Furukawa announced in March 2001 that it had developed a semiconductor laser pumping module for the 1480 nm band that has a high output power of 400 mW. Volume production is scheduled for the end of 2001.

Figure 6.23 Furukawa Electric FY 2001 Geographic Sales (%)

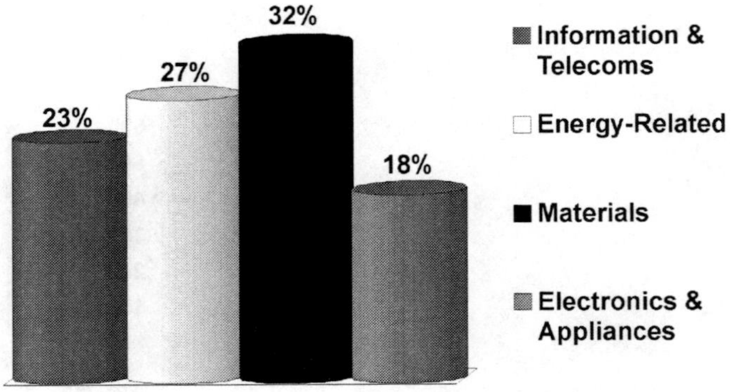

Facilities

In February 2001 Furukawa announced it would invest ¥10.2 billion to improve its productions systems for WDM optical components. Included in this investment was the founding of a volume production plant in Thailand for source lasers and fibre Bragg gratings (FBGs). FBGs are used for wavelength stabilization of 1480 nm pump and Rainbow pump lasers.

The company holds ISO 9001 certification.

The company also announced it would set up a new plant in the Chiba works for research and development as well as for production of high-value added products, such as WDM products.

Furukawa has its own SI LEC GaAs production facility. This is now used primarily as a captive source of substrates for its in-house epitaxy operations. Currently it has little presence in the merchant market for substrates.

Furukawa has set up its own manufacturing capability for GaAs MOVPE epiwafers. Applications for these products include optoelectronics and microelectronics. The technology for these facilities was largely developed in-house; MOVPE reactors in use by Furukawa are largely designed and built in-house rather than off-the-shelf equipment from merchant vendors. This, says the company, enables it to optimize the growth conditions and yield of epiwafers. However, in the present business climate it may be somewhat lagging behind the state of the art available to its competitors.

During April 2001 Furukawa Electric opened a 12 000 sq ft development centre in Warwick, RI, USA, for the EDFA and Raman markets. The centre will work in conjunction with Fitel Technologies and Optigain Inc.

Products

The company is a leading manufacturer of pump lasers. Its product range includes standard-, high- and ultrahigh-power 1480 nm pump lasers, Rainbow pump FBG lasers, standard, high-power and wavelength stabilized 980 nm pumps, YAG-welded tight coupling PD modules and un-cooled signal source FP and DFB lasers.

Pump laser products include:

- The FOL1402Pxy series of pump laser modules cover standard power ranges up to 200 mW by MQW chip design. It features maximum storage temperature of 85 °C, 2 V LD reverse voltage and 1000 mA forward current.
- The FOL1402Mx is a low-cost, un-cooled 1480 nm pump laser for single wavelength amplification applications. It features a maximum LD operating current of 180 mA (at 25 °C) and 270 mA (at 70 °C). Typical LD forward voltage is 1.1 V.

The company also manufactures epitaxial GaAs wafers for compound semiconductors. The GaAs operations of Furukawa Electric include manufacturing of materials for internal use but the company also markets and sells these through its international network of offices and distributors.

Alliances

In March 2001 Furukawa Electric and Mitsui Chemicals launched a joint development programme for 980 nm pump laser chips for optical amplification applications, such as WDM systems. The two partners will integrate their technology and intend to produce high output modules.

Furukawa has a licence agreement with Tyco Electronics for Tyco's AMP LIGHTRAY MPX optical interconnects.

The company is a member of a measurement testing round-robin formed by Japanese companies involved in GaAs production.

Table 6.13 Furukawa Financial Highlights (¥ billion)

	2001	2000	1999	1998	1997
Net sales	827.0	696.6	726.7	568.4	545.7
Net income	167.4	35.2	2.8	5.8	5.2
Working capital	181.8	69.2	108.9	105.9	69.5
Total assets	1504.2	972.3	880.3	636.0	623.5

6.19 Hamamatsu

Hamamatsu Photonics KK
325–6 Sunayama-cho
Hamamatsu City
Shizuoka Prefecture
430-8587
Japan

Founded in 1953, Hamamatsu Photonics KK is a manufacturer of optoelectronic devices.

- Hamamatsu reported sales for the year ended September 2000 of ¥51.6 billion, an increase of 26.5% over its previous year's sales of ¥40.8 billion.

Figure 6.24 Hamamatsu Two Year Sales (¥ billion)

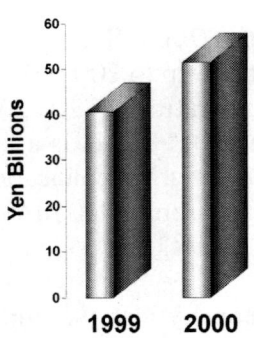

R&D

R&D Expenditure
FY 2001: ¥6.3 billion
FY 2000: ¥6 billion

The company conducts research into high-power diode lasers. It spends approximately 13% of sales on R&D per annum.

Products

Optoelectronic sales for FY 2000 were ¥40.3 billion (77% of total net sales), an increase of 25.8% over the previous year. Hamamatsu's range of optoelectronic devices includes pulsed laser diodes, LEDs, photodiodes and IR detectors.

The range of pulsed laser diodes are suitable for applications such as range finders, in optical fibre equipment and as light-triggered control systems:

- The L6690, L7055-04, L6720 and L7060-02 feature 870 nm peak wavelength, 0.5 ns rise time and are available in 9 mm diameter packaging.

The S7858 is a photo IC for laser beam synchronous detection, suitable for print timing detection for laser printers and digital PPC. It features a switchable gain function, high sensitivity and stable output during laser power and temperature variations.

Table 6.14 Hamamatsu Financial Highlights (¥ million)

	2000	1999
Net sales	51 559	40 773
Net income	1569	613
Working capital	21 901	18 582
R&D	6261	5959
Total assets	89 519	71 953

6.20 Hitachi

Hitachi Ltd
6 Kana-Surugadai
4-chome
Chiyoda-ku
Tokyo 101
Japan

Hitachi Ltd is one of the world's largest electronic and electrical equipment manufacturers, employing 341 000 people. Founded in 1910, today it produces computers, semiconductors, household appliances, power generating equipment and industrial machinery.

- Hitachi reported net sales for FY 2001 of ¥8417 billion an increase of 5% over FY 2000's figure of ¥8001 billion.
- Hitachi has five operating divisions: Information Systems and Electronics, Power and Industrial Systems, Consumer Products, Materials, Services and Other Products.

R&D

R&D Expenditure
FY 2001: ¥435.6 billion
FY 2000: ¥432.3 billion

Apart from Japan, Hitachi has R&D facilities in Germany, Ireland, Italy, the USA and the UK.

Facilities

Hitachi has facilities in Japan at Chiyoda-ku, Musashi, Takasaki, Kofu, Komoro, Tokyo, Saitama, Hokkaido and Yamagata. Overseas facilities include Kedah Darulaman and Penang, Malaysia; Singapore; Irving, TX, USA; Landshut, Germany; St Petersburg, Russia; and Ballinasloe, Ireland. It also has a US$1.5 million regional semiconductor design centre in Singapore.

Products

The Information Systems and Electronics division accounted for 42% of FY 2001 net sales. Optical laser products include lasers and modules for optical communication and visible lasers for pointers and scanners. SONET products

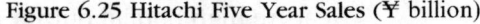

Figure 6.25 Hitachi Five Year Sales (¥ billion)

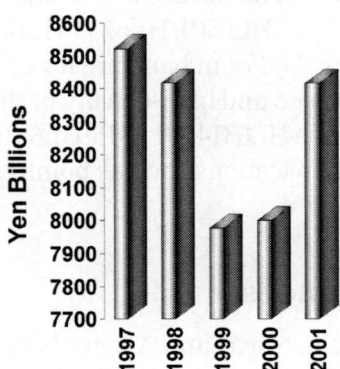

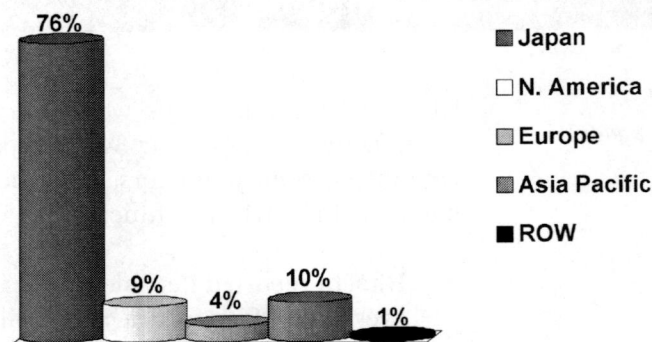

Figure 6.26 Hitachi FY 2001 Geographic Sales (%)

operate at speeds from 156 Mb/s to 10 Gb/s; MQW, DFB, Fabry–Perot and 1/4 phase shifted DFB lasers for fibre communication with wavelengths from 1310 nm to 1550 nm and visible lasers with wavelengths as short as 635 nm.

- In June 2001 Hitachi announced an industry first, the Maru Beam series, a new line of red laser diodes in the 635 nm band offering a nearly circular beam with an aspect ratio of 1.2. Hitachi claims it is now able to achieve operation current of approximately 40% lower than its previous diodes. By producing a more uniform light pattern, the high-efficiency 635 nm LDs make it easier to design laser markers, optical fibre test equipment and other light wave products.
- The first Maru Beam products are 5 mW laser sources with the HL6335G series (9 mm package) and the HL6340MG series (5.6 mm package). The diodes in both series are highly efficient with an operating current for a 5 mW output at only 25 mA, which in portable products enables extended battery life through the increased efficiency.

Hitachi's range of laser diodes include:

- The HL1569AF is a 1.55 μm InGaAsP DFB LD (with integrated EA modulator), having an MQW structure. Suitable as a light source for high-bit-rate, long-haul fibre optic communication systems (such as 2.5 Gb/s external modulation systems for up to 600 km), it features long-wavelength output of 1550 nm.
- The HL6312-13/HL6319-22/HL6712-14/HL6722 industrial red laser diodes feature high optical output of 15 mW (HL6321~22G), suitable for levellers, bar code readers and similar systems that require high performance.
- The HL6501/HL6733/HL6738 series of visible laser diodes in the 650/685 nm band are for optical storage laser diodes. These meet the high-speed and high-density performance of optical disc drives.
- The HL6314/HL6315/HL6316/HL6720/HL6724 series are LDs for consumer applications such as pointers, game machines and optical detection equipment.

Alliances

Hitachi, Nortel and Agilent have an agreement on the packaging of diode lasers for telecommunications, concluding in a common size and pin-out structure for LDs of 13.2 × 7.6 mm and pin-compatibility with 14-pin DIL packages.

Table 6.15 Hitachi Financial Highlights (¥ billion)

	2001	2000	1999	1998	1997
Net sales	8417	8001	7977	8417	8523
Net income	104	17	(328)	12	90
R&D	436	432	497	511	504
Total assets	11 247	9983	9848	10 292	10 341

6.21 Honeywell

Honeywell International
Inc
101 Columbia Road
Morristown
NJ
USA

Formed in 1999 and employing 120 000 people, Honeywell International is the consequence of a merger between Honeywell Inc and AlliedSignal Inc. Honeywell became a wholly owned subsidiary of AlliedSignal, which then changed its name to Honeywell International Inc.

- For FY 2000, Honeywell International reported net sales of US$25 billion, an increase of 5% over FY 1999's figure of US$23.7 billion.
- Honeywell is the world leader in the production and application of VCSEL technology.

R&D

R&D Expenditure
FY 2000: US$818 million
FY 1999: US$909 million

Honeywell's Technology Center (HTC) in Bloomington, MN, USA, conducts research into military and aerospace areas, as well as AlGaN-based materials and UV detectors. R&D projects include the development of other wavelengths, even lower power devices, 10–1000 element arrays and the integration of VCSELs with lenses or silicon electronics.

Honeywell and GE together developed VCSEL-based optical backplanes using GE's multichip module technology.

Figure 6.27 Honeywell Five Year Sales (US$ billion)

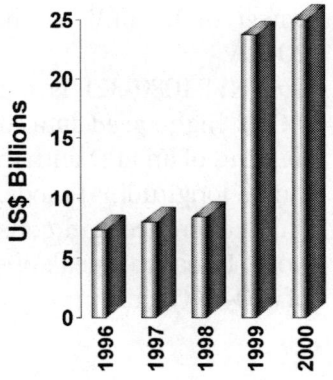

Figure 6.28 Honeywell FY 2000 Geographic Sales (%)

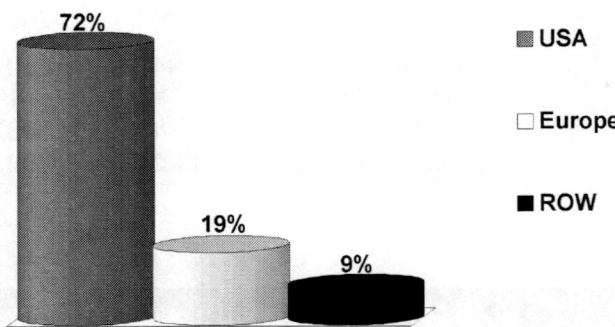

- Honeywell's VCSEL product won the award for best new electronic product from Design News in 1996 and an R&D 100 Award from R&D Magazine in 1997.

Facilities

The Optoelectronics Group is part of the MicroSwitch Division of Honeywell based in Richardson, TX. This group was originally founded in 1968 as Spectronics Inc and was acquired by Honeywell in 1979.

Products

The company is the world leader in VCSEL technology. VCSELs provide the performance advantages of lasers for the same price as LEDs; like LEDs, they emit light vertically from the wafer surface which means their fabrication and testing is fully compatible with standard IC procedures and equipment and also means that arrays of 10, or even 1000, are feasible. However, VCSELs are much faster, use lower power and produce a smaller divergence beam than LEDs.

VCSEL arrays can improve the speed and resolution of laser printers and enable low-cost, high-quality colour printers. For sensor applications, VCSELs can increase the range of a proximity sensor used on a manufacturing line, or provide high-resolution encoding. VCSEL arrays can also increase the capacity and access speeds of optical storage and enable two-dimensional bar code scanners.

- The HFE4080-321 high radiance VCSEL is a 1 GHz, high-speed device designed for drive currents between 5 mA and 15 mA. It features output power of 1.8 mW, 15° beam divergence and peak operating current of 20 mA.
- The HFE4080-321 is a high-performance 850 nm VCSEL packaged for 1 GHz, high-speed data communications. It combines many of the desirable features of an LED with the desirable features of a laser diode operating in a single longitudinal mode, but with multiple transverse modes, thus reducing coherence and consequent modal noise in multimode fibre applications. It features peak operating current of 20 mA and output wavelength of 830–860 nm.

Table 6.16 Honeywell Financial Highlights (US$ million)

	2000	1999	1998	1997	1996
Net sales	25 023	23 735	8427	8028	7312
Net income	1659	1541	572	471	403
Working capital	3447	2150	1169	939	914
R&D	818	909	482	447	353
Total assets	25 175	23 527	22 738	20 118	18 322

6.22 Infineon

Infineon Technologies AG
PO Box 800949
D-81541 Munich
Germany

Infineon Technologies AG, the former semiconductor operations of Siemens AG, was spun-off from its parent in April 1999 and renamed Infineon. Siemens remains its largest shareholder.

- Infineon reported sales of €7.3 billion for FY 2000, as opposed to €4.2 billion in FY 1999. Captive sales to Siemens accounted for 10.5% of FY 2000 sales.
- In July 2001 Infineon announced that it would proceed with a public offering of up to 60 million newly issued shares, including approximately 7.8 million shares included in the underwriters' over-allotment option. The shares will be issued by Infineon from its authorized capital in a capital increase without pre-emptive rights.

Acquisitions

Vishay Intertechnology announced in June 2001 that, subject to government approvals, it had acquired the IR components business of Infineon Technologies for approximately US$120 million. The deal includes facilities in San Jose, CA, USA, and Malaysia.

In April 2001 Infineon Technologies entered into a definitive agreement to acquire Catamaran Communications Inc, San Jose, CA, USA, for US$250 million

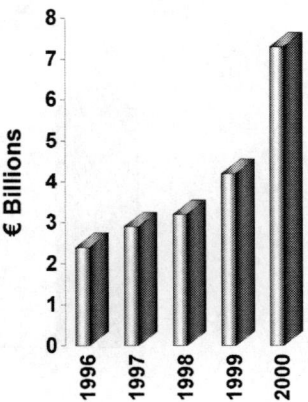

Figure 6.29 Infineon Five Year Sales (€ billion)

in common stock. Catamaran specializes in ICs for next-generation 40 Gb/s and 10 Gb/s segments of the optical networking market.

It was announced on 14 August 2001 that Osram GmbH had acquired all Infineon Technologies AG shares (49%) in their joint venture, Osram Opto Semiconductors GmbH & Co OHG, at a cost of €565 million. With this acquisition, Osram will strengthen its market position in the field of optoelectronic semiconductors, while Infineon will focus its resources more strongly on its core business activities (such as LAN, WLAN and network access within the wireline communications group). The transaction took immediate effect.

R&D

R&D Expenditure
FY 2000: €1025 million
FY 1999: €739 million

Infineon participates in national research programmes, including the NOVA-LAS programme for high-power diode lasers.

- The company was awarded US Patent No 6,271,049 in August 2001: 'method for producing an optoelectronic component'. A laser chip is secured to a semiconductor wafer that is provided with metal structures. Thereafter, for each laser chip, one lens-coupling optical element that deflects the beam path of the laser chip is positioned on the semiconductor wafer. The laser chip is driven and the space between the laser chip and the lens-coupling optical element is varied and set such that a predetermined beam condition with respect to the location of the optical image plane is met.

Facilities

The company has optoelectronic facilities in Regensburg, Germany (for which it uses EMCORE reactors); and assembly and test in Penang, Malaysia. Infineon holds ISO 9001 quality certification.

Figure 6.30 Infineon FY 2000 Merchant Sales (%)

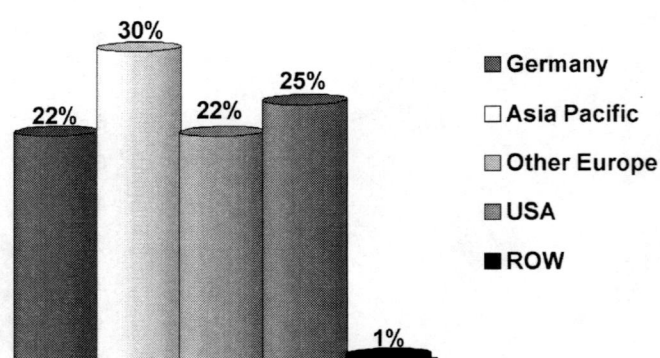

- Germany
- Asia Pacific
- Other Europe
- USA
- ROW

Infineon was the first in the world to have a GaAs manufacturing facility exclusively for 6-in high-volume production in Munich.

In November 2000 Infineon opened a new development centre in Echirolles, near Grenoble in France. The centre will develop chips for high-speed data transmission using SiGe and CMOS technologies, targeting the fibre optic market.

Products

Infineon manufactures high-performance laser diodes (some in conjuction with its opto joint venture, OSRAM Opto Semiconductors) and VCSELs.

The company produces passively cooled laser bars for applications such as end pumping of rods and fibres, soldering, direct material processing and medical. The LD bars feature a collimated radiation source for CW operation in a sealed housing, very low fast and slow axes divergence angle and a reliable strained layer InGa(Al)As/GaAs material.

The Infineon multimode family offers high performance transceivers with VCSEL diodes for 5 V and 3.3 V power supplies. The modules are designed for a broad range of private telecommunications systems, e.g. LAN, WAN and Gigabit Ethernet applications.

The single-mode family offers serial transceivers with high-reliability 1300 nm laser diodes. Single-mode transceivers can be used in a broad range of private data communications and public telecommunications systems with transmission distances of up to 40 km. The transceiver is a single unit comprised of a transmitter, a receiver and an SC or ST duplex receptacle. The next-generation Small Form Factor LC modules are only half as wide as current 1-in duplex SC transceivers.

- In April 2001 Infineon announced that its Corporate Research Centre had achieved a breakthrough in 1300 nm VCSEL technology. Previously, 1300 nm lasers were available only in edge-emitting laser technology. 1300 nm VCSELs extend the potential range of fibre optic systems, providing inherent cost, reliability and power advantages compared to edge-emitting laser technology. The new 1300nm lasers can be modulated up to 10 GHz, providing the output power required for fibre optic transmission systems operating at OC-192 data rates (10 Gb/s). Infineon utilized the company's long-standing expertise in MBE technology and production of 850 nm VCSEL components for developing the 1300 nm technology. The vertically emitting laser is grown on a GaAs substrate and has an active region consisting of multiple quantum wells of InGaAsN.

The Infineon fibre channel multimode transceiver is part of the company's Small Form Factor transceiver family. The device is a single unit comprised of a transmitter, a receiver and an LC receptacle. This transceiver supports the LC 'connectorization' concept. It is compatible with RJ-45 style backpanels for high-end data communications and telecommunications applications while providing the advantages of fibre optic technology. The module is designed for low-cost SAN, LAN, WAN and fibre channel applications. It can be used as the network end device interface in mainframes, workstations, servers and storage devices and in a broad range of network devices such as bridges, routers, hubs and local and wide area switches. This transceiver operates at 2.125 Gbit/s from a single power supply (+3.3 V).

Examples of Infineon's optoelectronic products include:

- The SPL MB81-E and SPL MB94-E passively cooled laser bars feature collimated radiation source for CW operation, strained layer InGa(Al)As/GaAs material. For end pumping of rods and fibres and medical applications.
- The SPL 2Yxx laser diode has a InGa(Al)As strained layer quantum-well structure and an efficient radiation source for CW and pulsed operation. Applications include pumping of solid-state lasers and medical uses.
- Similarly, the Infineon SPL 2Fxx is a laser diode in an TO-220 package with FC connector and 1.5 W CW.
- The Infineon fibre channel multimode transceiver is a single unit comprised of a transmitter, a receiver and an LC receptacle. This transceiver supports the LC 'connectorization' concept. It is compatible with RJ-45 style backpanels for high end data communications and telecommunications applications. The transmitter contains a laser driver circuit that drives the modulation and bias current of the laser diode. The currents are controlled by a power-control circuit to guarantee constant output power of the laser over temperature and ageing. The power control uses the output of the monitor PIN diode (mechanically built into the laser coupling unit) as a controlling signal, to prevent the laser power from exceeding operating limits.

Table 6.17 Infineon Financial Highlights (€ million)

	2000	1999	1998	1997	1996
Net sales	7283	4237	3175	2885	2350
Net income	1126	61	(775)	(95)	117
Working capital	870	444	887	560	945
R&D	1025	739	637	457	370
Total assets	8853	6445	4760	4595	3562

6.23 IQE

IQE plc
Pascal Close
Cypress Drive
St Mellons
Cardiff CF 3 0EG
UK

International Quantum Epitaxy (IQE) is a provider of semiconductor materials for the optoelectronics and high-frequency microelectronics markets. It manufactures the wafers from which optoelectronic devices are produced.

IQE was formed in 1999 after the merger of two companies, the Cardiff-based Epitaxial Products International (EPI) and the Bethlehem, PA, USA-based Quantum Epitaxial Designs (QED). In June 2000 EPI became IQE plc, whilst QED changed its name to IQE Inc.

- IQE reported sales for the year ended December 2000 of £30.1 million, an increase of 58% over FY 1999's figure of £19 million.

Figure 6.31 IQE Three Year Sales (£ million)

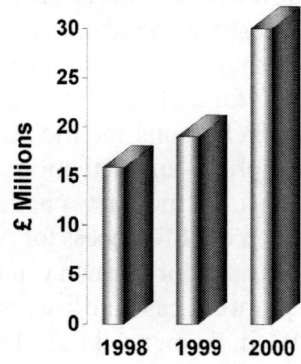

- It followed this with half-year 2001 sales of 326.1 million, almost double its half-yearly sales for the corresponding period in FY 2000.

Acquisitions

Towards the end of FY 2000, IQE acquired Wafer Technology Ltd of Milton Keynes, UK, a global supplier of III-V compound semiconductor substrates and high-purity polycrystalline materials. Wafer Technology is now a 100% subsidiary of IQE, although it still trades under its former name.

In November 2000, IQE formed IQE Silicon Compounds Ltd, a wholly owned subsidiary that provides an outsource service for silicon-based epitaxial structures.

R&D

R&D Expenditure
FY 2000: £1.9 million
FY 1999: £1.3 million

In September 2001 IQE plc announced that it had been involved in the development, with Motorola, of a new semiconductor materials technology. Motorola invented and patented the process; IQE's involvement was to make this technology into a production-worthy process.

- They have demonstrated the world's first 8- and 12-in GaAs on strontium titanate (STO) wafers.

This technology successfully combines the best properties of silicon technology with the speed and optical capabilities of high-performance III-V materials, particularly for optical communications. Other potential markets include data storage, lasers for such consumer products as DVD players, medical equipment, radar, automotive electronics, lighting and photovoltaics. Until this discovery, there was no way to combine light-emitting semiconductors with silicon ICs on a single chip and the need to use discrete components had compromised the cost, size, speed and efficiency of high-speed communications equipment and devices.

IQE has been involved in many projects researching into laser technology. These include:

- With the US Air Force Research Laboratories Sensors Directorate: to develop 850 nm GaAs/AlGaAs VCSEL technology for high-speed and low-threshold applications.
- With a major industry partner: to develop a low-cost, high-yield MBE process for high-quality 980 nm laser diode wafers.
- With Lassons Technology and NASA: to develop GaAs-based MQW devices for photo-emf detection.
- With an industry partner and Penn State University: to develop a solid source MBE process for Al-free 850 nm and 905 nm lasers.
- With major industry partners: to develop a MBE process for InP-based quantum cascade laser applications.
- With Siemens, LETI, University of Strathclyde, IMEC and Epichem: to develop OPTical Interconnects for VCSELs using As/N compounds (OPTI-VAN) MOVPE. InGaAs/GaAsP strain balanced structures for long-wavelength emission and InGaAsN epitaxial layers for 1.2/1.3 μm VCSEL applications.
- With Cardiff University: to develop 650/635 nm visible laser development for DVD and laser pointers.

Facilities

IQE Cardiff has approximately 40 000 sq ft of manufacturing space with the capacity to house up to 16 MOVPE production reactors.

IQE claims to have the world's highest-volume 6-in p-HEMT and HBT epiwafer manufacturing capacity; it was the world's first merchant epiwafer supplier with multiple 6-in MOVPE and MBE epiwafer production capabilities.

Both of IQE's production facilities (Cardiff and Bethlehem) hold ISO 9002 certification.

Products

IQE Cardiff manufactures 3-, 4- and 6-in epitaxial wafers grown by MOVPE, whilst IQE Bethlehem manufactures MBE wafers of the same dimensions for GaAs, AlGaAs and InGaAs, on GaAs and InP substrates. Applications for these wafers include visible laser, VCSELs and DFB, DH and QW lasers for fibre optics.

A major portion of IQE's business is the manufacture of InP-based wafer for lasers for fibre optic telecommunications. It produces InP HBT, MBE grown epiwafers.

The company is a leading merchant supplier of VCSEL epiwafers grown by MOVPE. These are characterized by high-quality QW uniformity, reflectance uniformity of <1%, low dopant diffusion and low defect levels. IQE offers VCSEL epitaxial wafers on 2-, 3- and 4-in GaAs and 2- and 3-in InP substrates for telecommunications, data communications and data storage applications. Available wavelengths of VCSEL products include: visible (<700 nm), 780 nm, 850 nm, 980 nm, 1300 nm and 1550 nm.

IQE is a leading supplier of GaAs HBT epitaxial wafers, offering both GaInP and AlGaAs emitter layer HBT epitaxial wafers on GaAs substrates for wireless and telecommunications applications.

Alliances

AIXTRON and IQE Cardiff have a long term purchase agreement for AIXTRON reactors.

In August 2001 IQE and QinetiQ (formerly the Defence Evaluation and Research Agency (DERA)), signed a Heads of Agreement for a research joint venture in advanced semiconductor materials. The new company will undertake R&D and provide materials assessment services on behalf of its two parents and their customers. It will initially operate from the existing QinetiQ facility at Malvern, UK, although it is planned to have a separate facility in Malvern fully operational within three years.

In August 2001 IQE Silicon Compounds and European Semiconductor Manufacturing Limited (ESM), announced a programme to qualify a volume SiGe BiCMOS process for use in ESM's foundry.

Table 6.18 IQE Financial Highlights (£ million)

	2000	1999	1998
Net sales	30.1	19.0	15.9
Net income	1.8	0.8	0.6
Working capital	124.7	25.4	5.8
R&D	1.9	1.3	1.5

6.24 Japan Energy

Japan Energy Corp
10–1 Toranomon
2-chome
Minato-ku
Tokyo 105-8407
Japan

Founded in 1905, Japan Energy Corp is a vertically integrated company and the world's leading producer of non-ferrous metals. The Compound Semiconductor division is a leading producer of InP substrates and high-purity metals such as indium, cadmium and tellurium. It is also a supplier of GaAs epiwafers as well as optoelectronic devices such as detectors. The company also markets its products through its well-known brand name ACROTEC.

- For the year ended March 2001 Japan Energy reported sales of ¥2197.6 billion, an increase of 13% over its FY 2000 sales figure of ¥1941.6 billion.
- Geographically, more than 90% of the company's net sales for FY 2001 were attributed to Japan, with very little elsewhere.
- In FY 2000 Japan Energy announced that it would exit the GaAs substrate market, although it still manufactures epitaxial wafers.

Facilities

Japan Energy has a manufacturing plant in Isohara, Japan, as well as another facility at its North American subsidiary, NIMTEC Inc, in Chandler, AZ. Technology is transferred to the Chandler plant from Isohara.

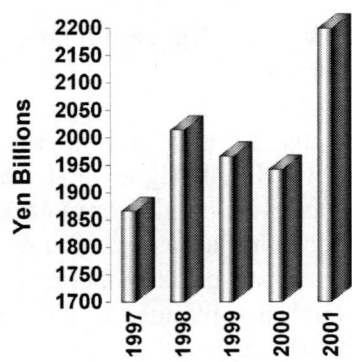

Figure 6.32 Japan Energy Five Year Sales (¥ billion)

Products

The company is a manufacturer and global supplier of high-purity sputter targets for semiconductor processing.

Japan Energy supplies both MBE and MOVPE GaAs epiwafers. Other materials manufactured by the company include:

- Single crystals: InP (dopant: none, Sn, S, Zn, Fe), CdTe and CdZnTe.
- Source materials: InP polycrystalline.
- High-purity metals: In: 7N, 6N; Cd: 7N, 6N; Te: 7N, 6N and Cu: 6N.
- It also manufactures CMT wafers and X-ray detectors.

Alliances

Japan Energy is a member of a measurement testing round-robin formed by Japanese companies involved in GaAs production.

Table 6.19 Japan Energy Financial Highlights (¥ billion)

	2001	2000	1999	1998	1997
Net sales	2197.6	1941.6	1966.1	2014.6	1866.0
Net income	49.5	(42.3)	19.0	(46.3)	20.7
Total assets	1838.6	1766.4	1642.1	1643.8	1661.8

6.25 JDS Uniphase

JDS Uniphase Corp
163 Baypointe Parkway
San Jose
CA 95134
USA

In June 1999 Uniphase Corp and JDS FITEL Inc completed a US$1.6 billion merger to create JDS Uniphase (JDSU). It manufactures and markets fibre optic telecommunications components and modules, laser subsystems and laser-based semiconductor wafer defect examination and analysis equipment.

- For the year ended June 2001, JDS Uniphase reported net sales of US$3.2 billion, as opposed to US$1.4 billion for FY 2000.

Figure 6.33 JDS Uniphase Five Year Sales (US$ billion)

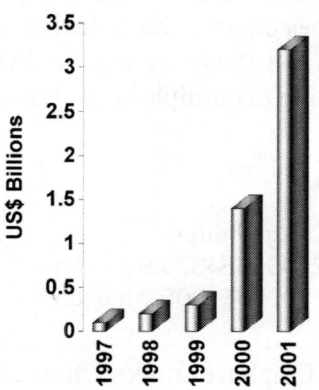

- The company employed 23 000 people at the end of FY 2000, but, owing to the industry slowdown which has affected JDSU dramatically, a realignment programme has been effected, including the reduction of approximately 30% (2 million sq ft) of its facilities and the loss of approximately 16 000 jobs by the end of 2001. Although it must be pointed out that through its numerous acquisitions, JDSU would have realigned some facilities and jobs to streamline its operations anyway.

Acquisitions

One of the reasons for JDSU's current financial troubles is its continued acquisitions. During FY 2000, JDS Uniphase acquired: AFC Technologies; Ramar Corp; EPITAXX Inc; SIFAM Limited; Oprel Technologies Inc; IOT Limited; Optical Coating Laboratories Inc (OCLI); MEMS manufacturer, Cronos Integrated Microsystems Inc; Chinese company, Fujian Casix Lasers Inc and E-TEK Dynamics Inc.

It followed these in January 2001 with a small equity investment in Avantas Networks Corp of Montreal, Quebec, Canada. Avantas is a privately owned telecommunications network testing equipment company. The terms of the investment were undisclosed.

In February 2001 JDSU acquired Optical Process Automation Inc (OPA) of Melbourne, FL, USA, for an undisclosed sum. OPA designs and manufactures automated and semi-automated systems for the manufacture of fibre optic components and modules.

However, by far its biggest (and perhaps most unfortunately ill-timed investment) was the February 2001 completion of the merger between JDSU and SDL Inc. Each outstanding share of SDL Inc common stock was exchanged for 3.8 shares of JDSU common stock and SDL became a wholly owned subsidiary of JDSU. This coincided with the industry slowdown and decreasing prices, and JDSU ended up paying a lot more for the stock than it was actually worth. As part of the merger requirements, JDSU had to sell its R&D laboratories in Zurich, Switzerland (it had purchased this from IBM in 1997 and has now sold it to Nortel Networks).

JSDU merged with SDL to take advantage of its existing complimentary business in telecommunications components. SDL is a manufacturer of laser diodes and optoelectronics ICs, which also in 1999 acquired Polaroid's fibre laser business. SDL was the first company in the world to successfully commercialize the integration of multiple lasers on a single semiconductor device.

R&D

R&D Expenditure
FY 2000: US$325.9 million
FY 1999: US$306.7 million

JDS Uniphase employs more than 1000 people in research and development and spends between 8% and 10% of net sales on R&D each year. The company has an R&D facility in Chicago for telecommunications products.

Facilities

JDS Uniphase has facilities worldwide, including the USA, Canada, UK, China and Taiwan.

The company opened its new facilities in Ewing Township, Mercer County, NY, USA, in December 2000. The West Trenton Division (formerly Epitaxx) and three other NJ facilities have necessitated the expansion of an additional 250 000 sq ft of space.

In January 2001 JDS Uniphase opened a 320 000 sq ft manufacturing facility in Shenzhen, China. The facility contains 150 000 sq ft of Class 10 000 cleanroom, is ISO 9002 certified and manufactures a variety of passive optical components for DWDM. Future plans include the expansion into active components and modules.

During June 2000 the company purchased a semiconductor facility in Research Triangle Park, NC, USA. This 177 000 sq ft facility expanded the Cronos MEMS manufacturing capabilities, with production beginning in early 2001.

- In February 2001 the company became the first TL 9000-Hardware/ISO 9001:1994 certified producer of MEMS with the certification of its MEMS division in Research Triangle Park.

Products

JDSU is a manufacturer of high-power laser diodes and diode-pumped solid-state lasers. It also produces passive fibre optic products such as WDM, optical switches and isolators. JDS Uniphase Netherlands BV in Eindhoven, was the former Philips Optoelectronics BV which the company acquired in 1998. JDSU Netherlands markets and produces semiconductor lasers, photodiodes and amplifiers for the telecommunications, cable television, multimedia and printing markets.

JDSU semiconductor lasers are comprised of 10 to 150 thin layers grown by epitaxial techniques.

- The SDL-2300 series of LDs comprise partially coherent, broad-area emitters with relatively uniform emission over the emitting aperture. Operation is multi-longitudinal mode with a spectral envelope width of approximately 2 nm FWHM. The far-field beam divergence in the plane perpendicular to the p/n junction is nearly Gaussian, while the lateral beam profile exhibits the complex pattern typical of broad-area emitters. Emitting apertures for SDL-2300 models range from 50 to 500 µm, providing CW power output of up to 4 W with excellent reliability.
- The SDL-6300 4 W CW InGaAs laser diodes in the 910–980 nm wavelength region are required for optical pumps for Er+ and Yb+ doped solid-state lasers, such as double-clad fibre lasers.
- The SDL-5800 series of laser diodes are capable of transmitting data at 1.25 Gb/s and 125°C. Based on the company's 980 nm pump laser, the SDL-5800 is a strained-layer InGaAs single-mode structure, fabricated using an MOVPE-grown QW active region. Applications include avionic LANs, satellite data links and harsh environment remote sensing.
- The 6380 series of laser diodes provide unprecedented brightness and reliability at 4 W, for next-generation applications. It is fabricated using an MOVPE-grown, strained quantum well active region. The 6380 features an entirely new epitaxial design that simultaneously optimizes power, brightness, efficiency and reliability. The product's inherently high efficiency, together with JDSU's unique low thermal resistance assembly, minimizes junction temperature rise, thereby increasing device operating lifetime.

In July 2001 JDSU introduced its CW and DML source lasers for fibre optic systems operating in the L-band. The new L-band source lasers complement JDSU's current line of C-band lasers and provide for a variety of applications in the long-haul, metro, access and CATV markets:

- The CQF 935 CW DFB laser is designed for WDM systems and is available in the 1530–1610 nm range in both 0.8 nm (100 GHz) and 0.4 nm (50 GHz) spacing. It includes a TEC, optical isolator and polarization-maintaining fibre and is available in 10 mW and 20 mW optical output powers. It is suitable for external modulation at 2.5, 10 and 40 Gb/s.
- The CQF 975 CW is designed for DWDM systems and includes an integrated wavelength reference module.

Figure 6.34 JDS Uniphase External Sales by Segment (%)

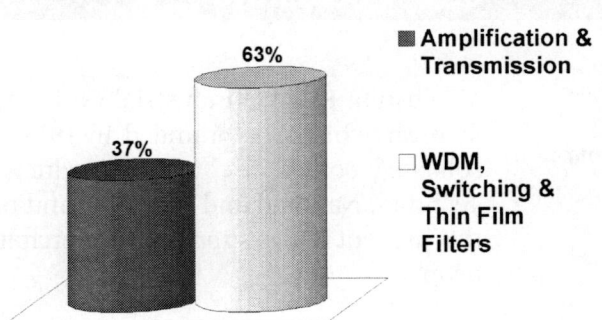

- The CQF 915 DML is a 1550 nm laser designed for 2.5 Gb/s WDM systems. It features a TEC and built-in cooled isolator. It provides 10 mW optical output power and is available with either 1800 or 3240 ps/nm dispersion.

JDSU's PL980S-5000 Series of butterfly-packaged 980 nm pump laser modules are used in EDFA-based telecommunications systems. Both fibre Bragg grating wavelength-stabilized and conventional versions are available. The laser diode is a JDSU Generation 5 device coupled to provide up to 200 mW of fibre-coupled power. The fibre pigtail comprises 250 µm Corning PureMode 1060 specialty optical fibre. The PL980S-5000 series combines the industry's leading 980 nm pump laser chip and package technologies, assuring optimum reliability.

Alliances

SDL, San Jose, CA, USA, has an agreement until the end of 2002 to provide high volumes of multi-mode pump lasers to IPG Photonics of Sturbridge, MA, for cladding-pumped fibre products. Multi-mode pump lasers employ a broad emitting aperture to allow operation at power levels as high as several watts and are used as cooled or un-cooled optical pump sources for active, cladding-pumped fibres. SDL's 6380 multi-mode pump laser and other multi-mode pump lasers will be employed in IPG Photonics' industrial fibre lasers, Raman fibre lasers and various optical amplifier products. These pumps will also be used in IPG Photonics' new generation of low-cost metro amplifiers for use in the emerging metropolitan access networks.

JDSU is a member of the Agere Systems, Alcatel Optronics and OpNext MSA for 10 Gb/s transponder modules used in optical networking systems. Other members include Agilent, Ericsson Microelectronics, Mitsubishi Electric, NEC and ExceLight.

Table 6.20 JDS Uniphase Financial Highlights (US$ million)

	2001	2000	1999	1998	1997
Net sales	3232.8	1430.4	282.8	185.2	113.2
Net income	(50.6)	(904.7)	(171.1)	(19.6)	(17.8)
Working capital	2264.3	1325.7	314.8	121.4	110.2
R&D	325.9	306.7	210.4	40.3	33.3
Total assets	17 572.2	26 389.1	4096.1	332.9	180.7

6.26 Matsushita

Matsushita Electric
Industrial Co Ltd
1006 Oaza Kadoma
Kadoma City
Osaka
Japan

Matsushita Electric Industrial Co Ltd (MEI) is a worldwide electronics manufacturer which was founded in 1918. The company is made up of 321 consolidated companies, whose products are marketed under the Panasonic, Technics, National and Quasar brand names. Much of its products are for captive use, but it also supplies the merchant market. Matsushita employs 293 000 people.

Figure 6.35 Matsushita Five Year Sales (¥ billion)

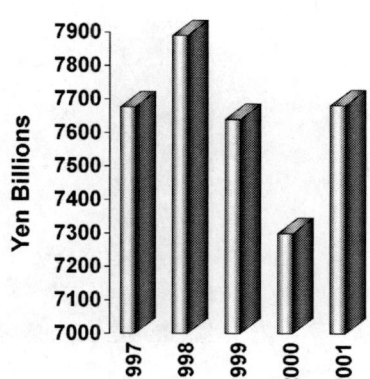

- The company reported net sales for the year ended March 2001 of ¥7681.6 billion (US$61.5 billion), an increase of 5% on its FY 2000 figure of ¥7299.4 billion.
- The Components segment reported sales for FY 2001 of ¥1617.8 billion (US$12.94 billion), an increase of 6% over ¥1530.5 billion in FY 2000.

R&D

R&D Expenditure
FY 2001: ¥543.8 billion
FY 2000: ¥525.6 billion

The Central Research Laboratories are situated at Matsushita Research Park, Kansai Science City, on the outskirts of Kyoto. Matsushita has two R&D divisions (Corporate Research and Corporate Product Development), seven laboratories and two development centres. The Electronics Research Laboratory is located at Takatsuki, Osaka.

On 1 October 2000 MEI absorbed Matsushita Research Institute Tokyo Inc (MRI), a wholly owned subsidiary. This is part of MEI's restructuring programme to integrate advanced technology research operation with the Group. MRI will be integrated into the Advanced Technology Research Laboratories. MEI is strengthening its research operations in the Tokyo region, concentrating on accelerating R&D technologies for growth areas such as the Internet.

The Opto-Electro Mechanics Research Laboratory conducts research into the development of lasers and related devices such as the creation of short-duration laser pulses and the improvement of laser-based micro-machining techniques.

- In 1999 Matsushita developed the industry's first low-noise, two-wavelength monolithic lasers for optical pickup applications.
- Matsushita has also conducted research into violet GaN LDs, blue-green semiconductor lasers (focusing on ZnMgSSe/GaAs strained MQW, grown by MBE) and optical disc technology.

Figure 6.36 Matsushita FY 2001 Geographic Sales (%)

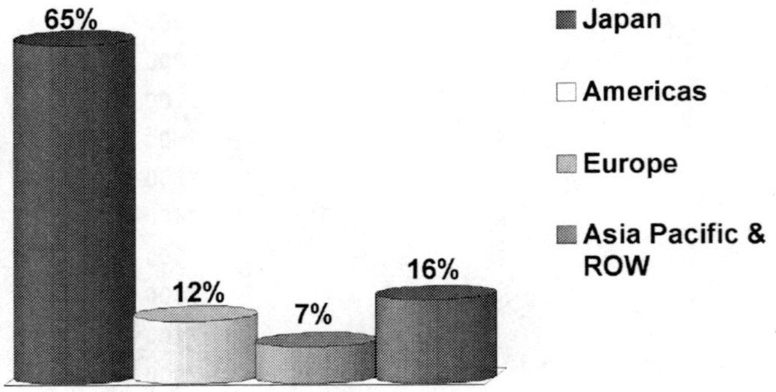

■ Japan
□ Americas
▨ Europe
■ Asia Pacific & ROW

Matsushita has recently received two US patents:

- US Patent No 6,274,518 'method for producing a group III nitride compound semiconductor substrate'. A method for producing a group III nitride compound semiconductor substrate including (a) forming a first semiconductor film over a substrate, the first semiconductor film made of a first group III nitride compound semiconductor and provided with a step; (b) forming a second semiconductor film made of a second group III nitride compound semiconductor having a different thermal expansion coefficient from that of the first group III nitride compound semiconductor on the first semiconductor film; and (c) cooling the substrate and separating the second semiconductor film from the first semiconductor film. Thus, a large-area group III nitride compound semiconductor substrate can be produced in high yields and with high reproducibility.

- US Patent No 6,273,950 'SiC device and method for manufacturing the same'. A method for manufacturing a device of silicon carbide and a single-crystal thin film, which are wide band gap semiconductor materials and can be applied to semiconductor devices such as high-power devices, high temperature devices and environmentally resistant devices, is provided by heating a silicon carbide crystal in an oxygen atmosphere to form a silicon (di)oxide thin film on a silicon carbide crystal surface and etching the silicon (di)oxide thin film formed on the silicon carbide crystal surface to prepare a clean SiC surface. The above SiC device comprises a clean surface having patterned steps and terraces, has a surface defect density of 10^8 per sq cm or less, or has at least a layered structure in which an n-type silicon carbide crystal is formed on an n-type Si substrate surface.

Facilities

Matsushita has plants in Japan at Nagaoka, Arai, Uozu, Tonami and at Okayama. Overseas plants are located in Singapore; Shanghai, China; Karawang, Indonesia; and Puyallup, WA, USA. Lasers are produced at the Okayama factory, whilst hologram units are manufactured at Nagaoka.

Products

Matsushita's Components segment was responsible for 21% of net sales for FY 2001. The segment produces optical lasers, integrated circuits, discrete devices, charge-coupled devices, Schottky barrier diodes, cathode ray tubes, image pickup tubes, tuners, resistors, capacitors, ceramic components, speakers, magnetic recording heads, liquid crystal display devices, plasma display panels, electric motors and electric lamps.

The company's laser product range includes 3 μm band InGaAsP semiconductor lasers, GaAlAs semiconductor lasers, red laser diodes, IR LDs, long-wavelength LDs and LD modules, distributed feedback (DFB) laser modules for analogue communication, long-wavelength detector modules, hologram units, photo ICs, photodetectors and laser driver ICs. It manufactures high-power, high-performance semiconductor lasers with lower power consumption for optical discs — such as for use in high-speed writing CDR drives.

- In June 2000 Matsushita announced the industry's first two-wavelength laser array that contains two semiconductor lasers: a 100 mW high-power IR laser for CD-R recording at 8 × speed and a 10 mW low operating current red laser for DVD playback. Using this technology enables integration of the optical pick-up used for high-speed CD-R recording/playback and DVD playback, thereby reducing both size and unit cost. The semiconductor laser array design technologies for the high-power IR laser and the low operating current red laser were devised by using a two-dimensional laser simulator which was developed in collaboration with Stanford University.
- In September 2000 the company developed the world's highest power (180 mW, approximately double that of previous devices), DFB semiconductor laser with an oscillation wavelength of 1.55 μm. It is suited for long-haul, large-capacity optical fibre communications. The laser was manufactured using Matsushita's proprietary mass-transport grating (MTG) technology.

Late last year Matsushita announced that it had developed the industry's thinnest hologram unit (3.5 mm thick, an improvement of 67%), that supports CD-R/re-writable (RW) recording at 8 × speed. All the light emitting and detecting elements used for optical pickups for CD-R/RW drives are integrated onto this unit. Size reduction was attained by reducing the number of optical parts from six to one. High reliability has been achieved by use of a laser with an emission optical power of 100 mW (pulse).

- HUL7202 is a hologram unit for optical information processing applications such as car CD players. It features 0.3 mW laser beam output.

Launched in summer 2001, the Panasonic DVD-LA95 is the world's first portable DVD player that can read DVD-RAM discs. DVD-RAM is not only compatible with both audio/video and PC applications, but its optical format also boasts superior rewritability, allowing roughly 100 000 rewrites per disc. The LA95 can not only play back a DVD-RAM disc created from either a PC or DVD-RAM recorder, but it also has the ability to accommodate multiple digital entertainment software, including DVD-Video, DVD-Audio, DVD-RAM, DVD-R, CDs, CD-R and CD-RW formats. The DVD-La95 also incorporates the industry's largest (9-in) LCD screen.

Alliances

Matsushita has an alliance with Stanford University and uses a two-dimensional simulator (developed in collaboration with Stanford) to enable the company to produce laser designs that feature low power consumption, solving the problem of heat dissipation.

Matsushita controls the technology behind DVD's Content Scrambling System (CSS) and has issued licences to chip manufacturers enabling them to decode DVD data and develop DVD-equipped PCs. The hardware licences are necessary because the film industry must ensure that DVD data is not copied illegally inside a computer.

SDL (now JDS Uniphase) and Matsushita Graphic Communication Systems have an agreement for the supply of fibre-coupled laser diodes for use in digital pre-press equipment.

The company is a member of IrDA (Infrared Data Association), which promotes interoperable, low-cost IR data interconnection standards.

As mentioned previously, Matsushita collaborates with Stanford University to produce high-performance, lower power consumption lasers.

Table 6.21 Matsushita Financial Highlights (¥ billion)

	2001	2000	1999	1998	1997
Net sales	7681.6	7299.4	7640.1	7890.7	7675.9
Net income	41.5	99.7	13.5	93.6	137.9
Working capital	1625.1	1793.3	1818.4	1894.4	2074.5
R&D	543.8	525.6	500.0	480.5	434.9
Total assets	8156.3	7955.1	3252.7	3356.2	3585.3

6.27 Mitsubishi

Mitsubishi Electric Corp
Mitsubishi Denki Build-ing
2–3 Marunouchi
2-chome
Chiyoda-ku
Tokyo 100
Japan

Founded in 1921, Mitsubishi Electric Corp is a leading international electrical and electronic equipment manufacturer, comprised of 139 consolidated affiliated companies.

- Mitsubishi reported total sales for the year ended March 2001 of ¥4129.4 billion (US$32.7 billion), an increase of 9% on FY 2000's figure of ¥3774.2 billion.
- Sales for the Electronic Devices segment amounted to ¥714.3 billion, or 16% of total sales in FY 2001.

R&D

R&D Expenditure
FY 2001: ¥196.4 billion
FY 2000: ¥166.7 billion

Mitsubishi employs 4000 people at its research and development centres in Japan, the USA and Europe.

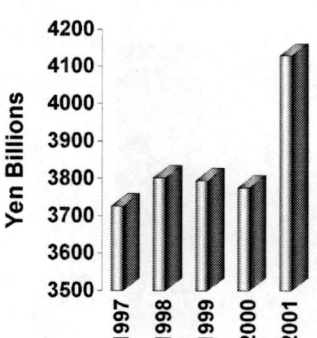

Figure 6.37 Mitsubishi Five Year Sales (¥ billion)

The Optoelectronic and Microwave Devices Laboratory of the Semiconductor Group is responsible for the development of III-V compound semiconductor devices such as GaAs ICs and laser diodes.

A world first for Mitsubishi Electric was the 1998 launch of the FU-436SDF-4M1B and 4M1C type of DFB laser diodes for CATV network applications. These did not require assisted cooling but offered high reliability and performance equivalent to that of other designs using cooling.

- Mitsubishi announced in March 2001 that it had attained high reliability results with its 40 Gb/s electroabsorption (EA) modulator and 10 Gb/s 1.3 mm Fabry–Perot (FP) laser which are under development. The 40 Gb/s EA modulator was developed on a semi-insulating substrate to achieve high bandwidth performance because the substrate can reduce parasitic capacitance. The 1.3 mm FP laser achieved a low power penalty when operating at 10 Gb/s and 85°C and demonstrated a lifetime of more than 200 000 hours.
- Mitsubishi developed a wavelength monitor integrated LD module that is scheduled for production in late 2001. This module provides wavelength-stabilized output with large wavelength tuning characteristics corresponding to DWDM systems. The wavelength monitor consists of a lens collimating output beam from an LD, a wavelength filter translating wavelength information into intensity and a pair of photo diodes translating optical intensity into an electrical signal. It features an ultra wide-band discrimination range of 500 GHz and a newly devised wavelength monitor allows the co-packaging LD to tune 10-channels in 50 GHz spaced DWDM systems. The integration of a wavelength monitor in the module provides compactness for the DWDM transmitting system.

Facilities

As well as semiconductor facilities in Japan, Mitsubishi also has facilities in Germany, USA and China. Optoelectronic modules are produced in Japan at Kamakura, whilst other optoelectronic devices and related GaAs devices are manufactured at Kita-Itame.

Mitsubishi spent ¥50 billion during 2000 to upgrade its semiconductor manufacturing facilities, principally the Saijo factory.

Figure 6.38 Mitsubishi FY 2001 Geographic Sales (%)

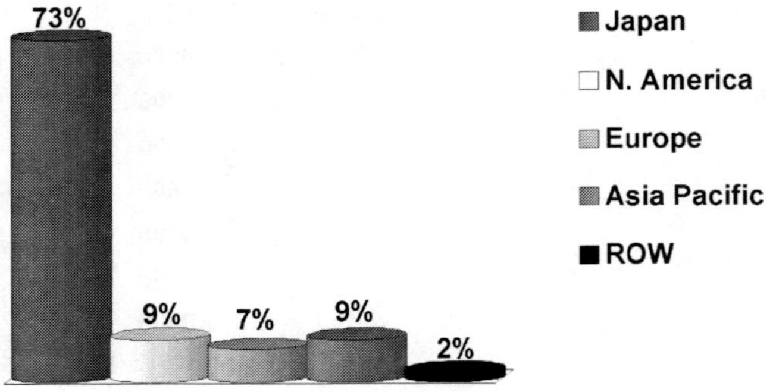

- Japan
- N. America
- Europe
- Asia Pacific
- ROW

Products

Mitsubishi manufactures short- and long-wavelength FP and DFB lasers, electroabsorptive (EA) modulated lasers and pump lasers.

In autumn 2000 the company began volume production of its ML8XX2, 980 nm pump laser diode for DWDM optical fibre amplifier applications. The device features high power output of 250 mW at 70°C.

In March 2001 it introduced three new DFB LD modules and a new DFB-LD discrete component that offer a 2.5 Gb/s data transmission rate. The cooled FU-672PDF and FU-675PDF wavelength monitor modules feature unparalleled stability power benefits in long-haul and metropolitan area network applications. The un-cooled FU-645SDF module reduces power consumption and module size. The un-cooled ML7XX16 discrete component's technology offers a wide temperature range and a reduced error bit rate. The products are designed to meet ITU, SDH STM-16 and SONET OC-48 standards.

- The ML9XX11 series are MQW DFB LDs emitting light at around 1550 nm. It features a typical low threshold current of 10 mA for wide temperature range digital transmission system applications.
- The ML1XX2 series of high power AlGaInP laser diodes feature a wavelength of 658 nm and standard CW light output of 30 mW (50 mW pulse). It is suitable for optical disc drive applications.
- The ML9XX6 series of InGaAsP LDs features a wavelength of 1550 nm and is suitable for long-distance optical communications systems.
- The ML9XX8 series of InGaAsP high-power LDs features an emission wavelength of 1480 nm and standard continuous light output of 150 mW. This is suitable for optical fibre amplification.
- The ML0131 is a laser driver IC for use in 2.5 Gb/s light wave applications. It features high-speed operation and low power dissipation.
- The ML6XX16 is a high power, AlGaAs, MOCVD laser for optical disc drive applications. It provides stable, single transverse mode oscillation with emission wavelengths of 785 nm and standard light output of 30 mW (40 mW pulse).
- The ML7XX11 series of InGaAsP MQW DFB laser diodes emit around 1310 nm. They feature a low threshold current of 10 mA and high-speed pulse response (typical rise/fall time of 0.2 ns) and are suitable for use as a light source in long-distance digital transmission systems.

- The ML9XX15 is a series of InGaAsP, 1550 nm DFB LDs with EA modulator. It features a high bit rate of 2.5 Gb/s and is suitable for long-distance digital transmission systems.

Alliances

Mitsubishi is a member of the MSA for 10 Gb/s transponder modules used in optical networking systems. This was started by Agere Systems, Alcatel Optronics and OpNext.

Table 6.22 Mitsubishi Financial Highlights (¥ billion)

	2001	2000	1999	1998	1997
Net sales	4129	3774	3794	3801	3725
Net income	125	25	(45)	(106)	9
Working capital	174	371	381	359	443
R&D	196	174	182	198	190
Total assets	4182	4003	4189	4238	4106

6.28 NEC

NEC Corp
7–1 Shiba 5-chome
Minato-ku
Tokyo 108-01
Japan

NEC Corp, which was founded in 1899, is a leading international supplier of communications systems and equipment, computers, industrial electronic systems and electronic devices.

- For the financial year ended March 2001, NEC recorded net sales of ¥5409.7 billion (US$42.9 billion), an increase of 8.4% on FY 2000's figure of ¥4991.4 billion. It employs 150 000 people worldwide.

NEC announced that with effect from 1 October 2001 it would spin-off its Compound Semiconductor Device Division into a new company, called NEC Compound Semiconductor Devices Ltd. It designs and manufactures a wide range of optical semiconductor devices including lasers, photodetectors, optocouplers and solid-state relays as well as silicon and GaAs RF and microwave semiconductor devices.

NEC Compound Semiconductor Devices Ltd is a wholly owned subsidiary of NEC, with an IPO planned within two years. It was launched with paid-in capital of US$105 million and is expected to generate first year sales of US$1 billion. NEC projects sales to grow to US$3 billion by fiscal 2005. Operating on a 'fabless' business model, it is headquartered in Kawasaki, Kanagawa Prefecture, Japan, with research and product development facilities in Kansai. Manufacturing will initially be carried out at current NEC and contractor facilities. Upon divestiture, the new company was free to pursue new manufacturing relationships.

Acquisitions

In August 2001 NEC transferred its laser printer business to Fuji Xerox (a leader in the colour laser engine market), including R&D, manufacturing, marketing

Figure 6.39 NEC Five Year Sales (¥ billion)

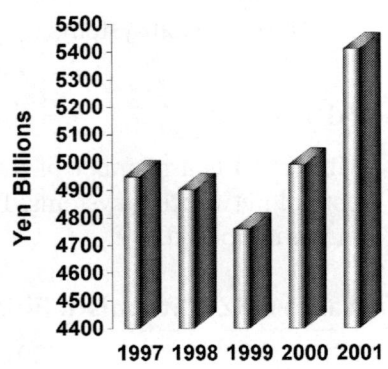

and maintenance service operations currently performed by NEC and its subsidiaries to Fuji Xerox. NEC also transferred all shares of NEC Niigata Ltd (NEC's major development and manufacturing centre for laser printers), to Fuji Xerox.

R&D

R&D Expenditure
FY 2001: ¥360 billion
FY 2000: ¥330 billion

NEC has seven R&D facilities in Japan, three in the USA and two in Germany. The NEC Optoelectronics Research Laboratories are based in Tsukuba and the NEC Kansai Electronics Research Laboratories both deal with optoelectronic device research. NEC also has a six-storey, semiconductor production technology R&D Centre at Sagamihara plant, Tokyo, which is concerned with process technologies ranging from 0.18 to 0.07 μm.

In March 2001 NEC announced that it has set a new DWDM transmission capacity world record. It successfully transmitted 10.9 Tb/s (273 channels, each with 40 Gb/s data speed) of information over 117 km. This record-setting capacity will allow simultaneous long-distance transmission of more than 1 m high-speed ADSL channels over a single line of fibre. NEC achieved this ultrahigh capacity by using three breakthrough technologies: its exclusive S-band amplification technology allowing light amplification in the 1490 nm wavelength region; optimized system design technology for the ultimate 3-band (the S-band plus the conventional C- and L-bands) transmission and ultra-dense channel multiplexing technology employing polarization mux/demux techniques.

Facilities

NEC produces optoelectronic devices in Japan at Kansai and Otsuki and at one of its subsidiaries, Kyushu Electronics Corp. It also has a production facility in Taiwan and marketing and sales offices worldwide. New technologies developed at the Optoelectronics Laboratories at Tsukuba and Electronics Research Laboratories at Kansai are transferred to the ULSI research laboratory and the Kansai plant.

Figure 6.40 NEC FY 2001 Geographic Sales (%)

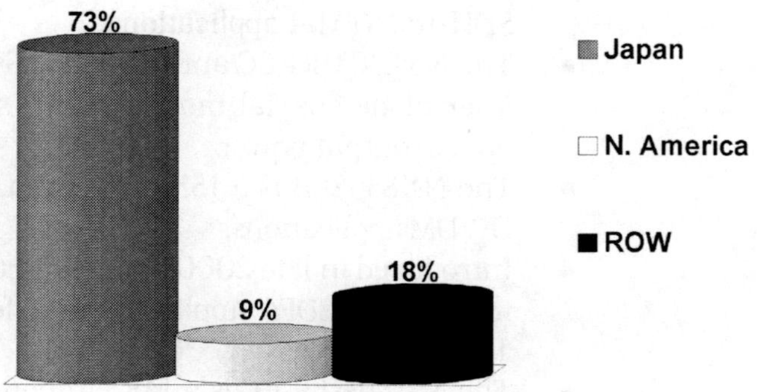

NEC manufactures optical modules at its facility in Yamanashi, producing WDM optical amplifiers, LD modules, optical waveguide modules and optical communications system equipment modules.

Products

NEC's Electron Devices Segment is comprised of the Semiconductor and Electronic Components Groups. This segment accounted for 23% of the company's net sales in FY 2001.

The company produces high output visible lasers for magneto-optical discs and visible lasers for DVD and DVD-ROM. Visible lasers are also widely used for bar code readers, laser points and measuring sensors.

NEC developed PDIC photodetectors for CDs and DVD that are key devices for the pick-ups as well as light sources. The PDIC photodetectors have high speed and sensitivity and are particularly suitable when used in conjunction with laser diodes for the design of pick-ups for DVD.

DWDM systems require one laser diode (LD) per channel. With wavelength spacing less than 1 nm, each laser diode must provide stable performance in multiwavelength, multiplexing environments. Wavelength stability and compact design are key parameters for DWDM light sources.

- NEC's new NX8570SA LD is offered in wavelengths specified in ITU-T standards between 1530 nm and 1565 nm with 0.8 nm spacing. The 1550 nm band, CW LD module features fibre output power of over 20 mW.

NEC produces the NDL3233SF/SU visible light semiconductor laser with a wavelength of 685 nm and optical power output of 30 mW for MO disk devices and the NDL3321ST/SU with a wavelength of 650 nm and optical power output of 5 mW for DVD and DVD-ROM devices. It is developing the NDL3330SF/SU with a wavelength of 658 nm and optical power output of 30 mW for DVD-RAM and high-density MO disk devices.

Examples of NEC's optoelectronic product line include:

- The NX7301BA-CC and NX7301CH-CC are 1310 nm InGaAsP MQW-FP laser diode coaxial modules for 156 Mb/s and 622 Mb/s. They are suitable for SDH and STM-4 applications.
- The NX8303BG-CC and NX8303CG-CC are 1310 nm MQW structured DFB laser diode coaxial modules with single-mode fibre. They feature 2 mW optical output power.
- The NX8562LB is a 1550 nm, InGaAsP, MQW DFB laser diode module for DWDM applications.
- Introduced in late 2000, the NX7460LE is a 1480 nm InGaAsP MQW-FP LD module for EDFA applications. It features typical high output power of 140 mW.
- The NX7462LE-CC is a 1480 nm pumping laser diode module with optical isolator for an EDFA that can expand the transmission span and compensate optical losses. The device is an InGaAsP MQW structured FP LD that features high output power (120 mW), high efficiency and a stable fundamental mode.
- The NX8564LE-CC is an EA modulator integrated, 1550 nm MQW structured DFB laser diode. The module is capable of 2.5 Gb/s applications of over 360 km ultra-long reach and available for DWDM wavelengths based on ITU-T recommendations, enabling a wide range of applications.
- The OD-J6860-0A01 is a 622 Mb/s optical transmitter for SONET/SDH/ATM systems. It features low power consumption and −15 to −8 dBm output power.
- NX8300BE-CC is a 1310 nm InGaAsP MQW DFB LD coaxial module for 2.5 Gb/s SDH/STM-16. It can achieve stable dynamic single longitudinal mode operation over a wide temperature range.
- The NX8563LB is a 1550 nm InGaAsP MQW DFB LD module for CW light source DWDM applications which features 10 mW minimum output power.

Alliances

NEC is a member of the MSA for 10 Gb/s transponder modules used in optical networking systems. Other members include Agere Systems, Alcatel Optronics and OpNext, Agilent, Ericsson Microelectronics, Mitsubishi Electric, JDS Uniphase and ExceLight.

The company is a member of IrDA (Infrared Data Association), which promotes interoperable, low-cost IR data interconnection standards.

NEC and Samsung SDI formed a joint venture company, Samsung NEC Mobile Display Co Ltd, to develop, manufacture and market displays based on organic electroluminescent (OEL) technology.

Table 6.23 NEC Financial Highlights (¥ billion)

	2001	2000	1999	1998	1997
Net sales	5410	4991	4759	4901	4948
Net income	57	10	(158)	41	92
Working capital	268	197	434	213	274
R&D	360	330	346	381	349
Total assets	4824	4609	5046	5075	4941

6.29 Nichia

Nichia Corp
491 Oka, Kaminaka-Cho
Anan-Shi
Tokushima-ken 774
Japan

Nichia Corp was founded in 1956 as Nichia Chemical Industries. The company is privately owned and is the world leader in blue light-emitting semiconductor optoelectronic components. Nichia is the world's only commercial supplier of InGaN laser diodes.

- Nichia Corp does not release financial details.

R&D

Nichia has focused on short-wavelength light-emitting optoelectronic devices. The Tokushima facility in Japan is the centre for R&D and production of blue LEDs and other optoelectronic materials and devices. This 10 000 sq m facility is equipped with several MOVPE reactors, at least one of which is for R&D.

Nichia's III-nitride research began in 1989. In December 1995 the company announced the successful fabrication of the world's first III-V nitride-based laser diode. In December 1996 Nichia announced it had achieved a room temperature CW diode with a life of 35 hours. The new laser structure was an evolutionary development of this, but emitting CW rather than via pulsed-wave operation.

- In 1997 Nichia published results on a CW semiconductor blue laser capable of sustained operation at room temperature. Through optimization of the epitaxy procedure blue lasers with a lifetime of 10 000 hours became available in 1998. The first device was at 400 nm, with follow-on devices at other wavelengths.
- In January 1999 Nichia announced commercial sampling of the world's first blue laser diode.

Facilities

Nichia has three manufacturing plants in Japan at Aratano, Tokushima and Tatsumi and overseas sales centres in Taiwan, the USA, Malaysia and Germany.

Products

Nichia manufactures blue and violet laser diodes, although its main product area is light-emitting diodes.

The NLHV 3000 series of InGaN high power violet laser diodes moved to volume production in June 2001:

- The NLHV 3000 series have a 405 nm peak wavelength and a 30 mW maximum optical power output, delivering up to five times the data storage capacity of red lasers commonly used in current DVD players. Manufacturers can use these violet laser diodes to expand the storage capacity of next-generation DVD discs from the current 4.7 gigabytes to more than

20 gigabytes. This increased storage allows each DVD disc to hold up to ten hours of recorded video, instead of the current two-hour limit. In addition to DVD players, the violet laser diodes can replace red laser diodes currently used in video games, laser printers and scanners.

- Nichia's NLHV500A violet laser diode was introduced commercially in October 1999. Its typical characteristics include CW optical output power of 5 mW, peak wavelength, 405 nm; operating current, 50 mA; operating voltage, 5 V.
- The NLBH500X-01 high-brightness blue laser diode has a 430–445 nm peak wavelength, 5 mW power output and is priced at US$3000 each.

Alliances

Although it has collaborated informally with academic and government institutes, Nichia has made a point of not entering into alliances with other companies for GaN-based R&D or production. It has also not entered into any licensing agreements for any aspect of GaN process technology, making it very clear that the company will defend its proprietary knowledge relating to GaN-based compounds.

However, close collaboration is expected between Nichia and its blue diode laser customers, for mutual benefit and a rapider refinement diode laser technology and customers' products:

- During 1999 PicoQuant GmbH, Berlin, Germany, announced the PLS 450/500, a new diode laser based instrument which includes a miniature sub-nanosecond pulsed blue/green light source (using Nichia's InGaN blue laser), with up to 80 MHz repetition rate.
- TOPTICA Photonics AG, Munich, Germany (formerly TuiOptics) announced a tunable blue laser for application in plasma physics such as monitoring fusion experiments (using Nichia's InGaN blue laser).

Legal Disputes

Nichia defends rigorously its patented technologies. It is in the process of bringing several lawsuits in the USA, although those are mainly pertaining to its LED technology.

In September 2000 Cree Inc filed a patent infringement lawsuit against Nichia Corp and Nichia America Corp in the US District Court for the Eastern District of North Carolina. The lawsuit seeks enforcement of US Patent No 6,051,849, entitled 'Gallium Nitride Semiconductor Structures Including a Lateral Gallium Layer That Extends from an Underlying Gallium Nitride Layer', issued to North Carolina State University in April 2000 and is licensed to Cree under a June 1999 agreement pursuant to which Cree obtained rights to a number of LEO and related techniques. Cree alleges that Nichia is infringing the patent by, among other things, importing, selling and offering for sale in the US certain GaN-based laser diodes covered by one or more claims of the patent. The lawsuit seeks damages and an injunction against infringement. North Carolina State University is a co-plaintiff in the action.

In December 2000 Nichia Corp filed counterclaims in US District Court for the Eastern District of North Carolina against North Carolina State University, Cree Inc and Shuji Nakamura. The counterclaims include patent infringement and trade secret theft. The patents included are US Patents 5,306,662, 5,578,839, 5,747,832 and 5,767,581. The patents and trade secrets cover methods of manufacturing and designs for GaN-based laser diodes, LEDs and electronics.

6.30 Nortel

Nortel Networks
Optical Components
2745 Iris Street
Ottawa, ON
Canada K2C 3V5

The optoelectronic division of Nortel Networks manufactures a complete range of active and passive components and low-speed modules for optical networks including long-haul, metropolitan and access applications.

- Nortel Networks reported revenues for FY 2000 of US$30.3 billion, an increase of 42% over the previous year's figure of US$22.2 billion.
- Nortel has not fared so well during the industry slowdown, producing a loss of US$19.2 billion for the second quarter of 2001. Moreover, it has announced 30 000 redundancies worldwide.

Acquisitions

At the beginning of 2001 Nortel acquired the Zurich, Switzerland, subsidiary of JDS Uniphase (a leading design and manufacturer of 980 nm pump lasers), for 65.7 million of Nortel's common shares. This acquisition also included related assets in Poughkeepsie, NY, USA.

Other recent acquisitions include:

- Australian company Photonic Technologies, for US$35.5 million in May 2000.
- Tunable laser company CoreTek Inc, for US$1.43 billion in Nortel Networks common shares.

Figure 6.41 Nortel Five Year Sales (US$ billion)

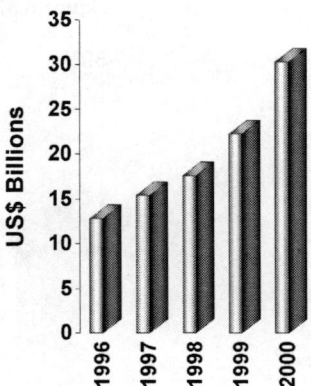

R&D

R&D Expenditure

FY 2000: US$3.8 billion
FY 1999: US$2.9 billion

Nortel Optoelectronics has research and development facilities in Ottawa, Canada, and Harlow, UK.

Facilities

Nortel has facilities in Canada, the USA, the UK, Switzerland and Australia.

Products

The Nortel Network product range includes modulator driver modules (CW lasers, fixed wavelength laser with EA modulator, fixed wavelength lasers, tunable laser with MZ modulator, directly modulated receivers, PIN preamp receivers, APD preamp receivers), tunable products (MEMS VCSEL tunable lasers, 20-channel tunable CW laser, tunable laser with MZ modulator, tunable filters), photonic line components (EDFAs, pump laser modules, pump laser chips, Raman pump unit, dynamic gain flattening filter, dynamic dispersion compensation module, optical spectrum analyser module, gain tilt monitors). It also produces passive components such as VOAs and optical add/drop multiplexers.

Nortel Networks launched 14 metropolitan optical components and modules in July 2001. These include a 2.5 Gb/s buried heterostructure transmitter, featuring maximum reach of 175 km with minimum dispersion penalty; a 10 Gb/s APD preamp receiver with best-in-class performance and very low cost; a 1310 nm 10 Gb/s transponder module with very low power dissipation; a multi-wavelength gain module, featuring a compact amplifier and integrated VOA; and 10 Gbit Ethernet switch router solutions, enabling simple, very high-speed, low-cost access to dark fibre and DWDM optical infrastructure.

The ML series of widely tunable CW laser sources are designed to support the complete range of high-performance DWDM systems at all data rates. The

Figure 6.42 Nortel FY 2000 Geographic Sales (%)

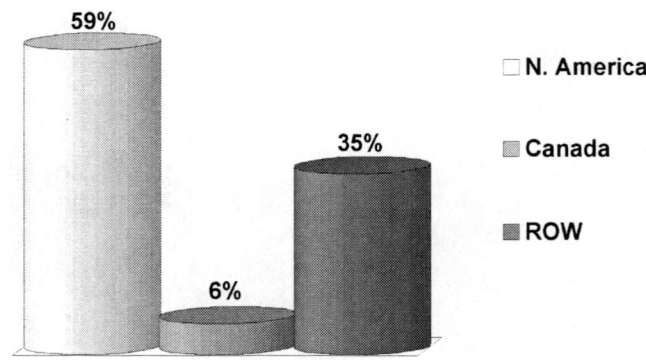

- □ N. America
- ▨ Canada
- ▨ ROW

design is based on an optically pumped VCSEL using a Nortel Networks proprietary half-symmetric cavity utilizing a MEMS structure for continuous wavelength tuning. Features include a broadband wavelength locked (locks to ITU 50 GHz grid), +13 dBm output power for use with existing external modulator, tunable across 80 channels in either the C or L bands and integral TEC.

Nortel's 20 and 40 channel AWG multiplexer/demultiplexer is designed for use in high channel count deployments at 100 GHz channel spacing. It has a reduced footprint, lower cost/port and high optical performance.

The Nortel 20-Channel CW tunable laser features a tunable InGaAsP gain coupled cascaded DFB laser, 50 GHz integrated multi-wavelength locker, high optical output power and ITU wavelengths from 1528 nm to 1604 nm.

The G0XX series 980 nm pump laser diode family features established performance and proven reliability: G03E diode option (>300 mW linear power), designed for use in submarine pump modules with up to 160 mW output power. It is suitable for very low noise wavelength locked operation with a fibre Bragg grating.

The YA07 laser driver IC is a highly integrated, low-cost laser diode driver with mean power control designed to run at data rates of up to 622 Mb/s. The YA07 has PECL/CML compatible differential inputs and requires a single +5 V power supply. Designed for SONET/SDH applications, it consists of a data input buffer, modulation switch and mean power control loop. It also features bias current limiting, laser shutdown and a means of externally monitoring the bias current, modulation current range of 5–80 mA, bias current range of 5–100 mA.

Alliances

Nortel and JDS Uniphase have a joint long-standing, high-power 980 nm pump laser supply agreement.

Nortel, Agilent and Hitachi have an agreement on the packaging of diode lasers for telecommunications, concluding in a common size and pin-out structure for diode lasers (13.2 × 7.6 mm) and pin-compatibility with 14-pin DIL packages.

Table 6.24 Nortel Networks Financial Highlights (US$ billion)

	2000	1999	1998	1997	1996
Net revenues	30.3	22.2	17.6	15.4	12.8
Net income/(loss)	(3.0)	(0.2)	(0.6)	0.8	0.6
Working capital	7.5	5.2	4.4	3.7	3.1
R&D	3.8	2.9	2.5	2.2	1.8
Total assets	42.2	22.6	19.7	12.6	10.9

6.31 Oki

*Oki Electric Industry
Co Ltd
7-12 Toranomon 1-chome
Minato-ku
Tokyo 105
Japan*

Founded in 1881, Oki Electric Industry Co is a leading supplier of electronic devices, telecommunications and information processing systems.

- Employing 25 600 people, Oki Electric reported net sales for the year ended March 2001 of ¥740.3 billion, an increase of 10.5% on FY 2000 figure of ¥669.8 billion.
- The Electronic Devices Segment accounted for 26% of net sales: ¥199.3 billion.

R&D

R&D Expenditure
FY 2001: ¥39 billion
FY 2000: ¥37.8 billion

Oki Electric's Semiconductor Technology Laboratory (STL) conducts research into surface-emitting lasers, ultra-fast optical pulse generation, wavelength conversion devices and multi-material OEICs.

The company has developed a recording medium that can be affixed with light, enabling two-colour red and black printing with a direct thermal recording method and a sol-gel solution for ferroelectric strontium–bismuth–tantalum thin films for non-volatile memories.

Facilities

Oki Electric has ISO 9001-certified manufacturing facilities in Japan at Yamanashi, Hachioji, Miyagi and Chichibu, with overseas plants in Thailand and Portland, OR, USA.

It has overseas offices in Beijing and Shanghai, China; Bangkok, Thailand; Sydney, Australia; Taipei, Taiwan; and Kuala Lumpur, Malaysia, as well as major overseas subsidiaries in the USA, Europe and Asia.

Figure 6.43 Oki FY 2001 Five Year Sales (¥ billion)

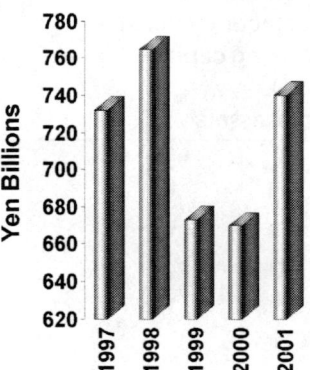

Products

The Electronic Devices Segment's optoelectronic devices include DIP and shrink-DIP pulsed laser diode modules, coaxial pulsed laser diode modules, optical communication LD modules, MINI-DIL laser diode modules, DIP and butterfly laser diode modules, coaxial pigtail laser diode modules, SLD and EELED modules, discrete LDs and LEDs, semiconductor lasers and photodiodes and discrete laser diodes.

The company produces lasers for WDM applications. DFB lasers are available in butterfly packaging for high-performance lasers and DIL packaging for lower-cost versions.

Oki's laser diode products include:

- The OL5104L-Wnnn 2.5Gbps WDM DFB laser module for DWDM and cross connect systems features an ITU grid WDM DFB laser, built-in isolator and TEC and operates in the 1530 to 1600 nm band.
- The OL5200N is a 1.55 mm laser diode DIP module with InGaAsP/InP pigtail. It features high output power and single-mode fibre. Applications include optical measuring instruments.
- In March 2001 Oki announced it had developed a semiconductor laser with an integrated EA modulator, an optical device that uses the electro-absorption effect of a semiconductor to provide high-speed modulation of a laser light, for 40 Gb/s optical communications. Oki Electric's discrete EA modulator, which uses InP/InGaAsP and owned ridge waveguide structure, has superior high-speed modulation. The Oki Electric discrete EA modulator is the first to support 40 GHz bandwidth. Sample shipments are scheduled for October 2001.
- The OL3300W surface mount type LD module is a 1.3 μm, InGaAsP/InP laser diode coupled to a single-mode fibre with a pigtail. It features maximum LD reverse voltage of 2 V, fibre output power of 7 mW and photodiode forward current of 10 mA.

Table 6.25 Oki Electric Financial Highlights (¥ billion)

	2001	2000	1999	1998	1997
Net sales	740.3	669.8	673.2	764.6	732.2
Net income	8.9	1.1	(47.4)	(13.2)	(9.3)
Working capital	81.6	155.0	183.8	190.8	147.1
Total assets	732.5	748.4	800.0	836.8	818.3

6.32 PerkinElmer

*PerkinElmer Inc
2175 Mission College
Blvd
Santa Clara
CA 95054
USA*

EG&G Inc changed its name to PerkinElmer in October 1999. The company has four business units: Optoelectronics, Life Sciences, Fluid Sciences and Analytical Instruments.

- PerkinElmer reported net sales for the year ended December 2000 of US$1.7 billion, which was an increase of 24% over FY 1999's figure of US$1.4 billion.

Acquisitions

PerkinElmer has undergone restructuring to focus on high-growth markets such as life sciences, telecommunications and semiconductor equipment.

- In May 2001 it acquired Sonoran Scanners' proprietary Laser Direct Imaging (LDI) and Computer-to-Plate (CTP) technologies. Complementing PerkinElmer's Digital Imaging and Lithography businesses, Sonoran Scanners' LDI technology uses direct CTP printing to produce a precise image.
- The company acquired Lumen Technologies Inc in December 1998 for US$253 million in cash and assumed debt.
- In January 2000 PerkinElmer made an 13% equity investment in Canadian company Bragg Photonics Inc, a manufacturer of high-speed optical components serving telecommunications optical networking applications.

R&D

R&D Expenditure
FY 2000: US$86.2 million
FY 1999: US$71.2 million

Facilities

The company has manufacturing facilities worldwide, including North America, Japan and China. PerkinElmer holds ISO 9001 certification. Optoelectronics are based in Santa Clara, CA, USA.

Figure 6.44 PerkinElmer Five Year Sales (US$ million)

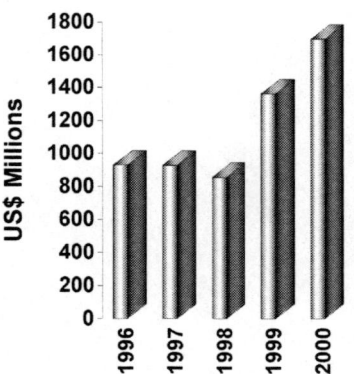

Figure 6.45 PerkinElmer FY 2000 Geographic Sales (%)

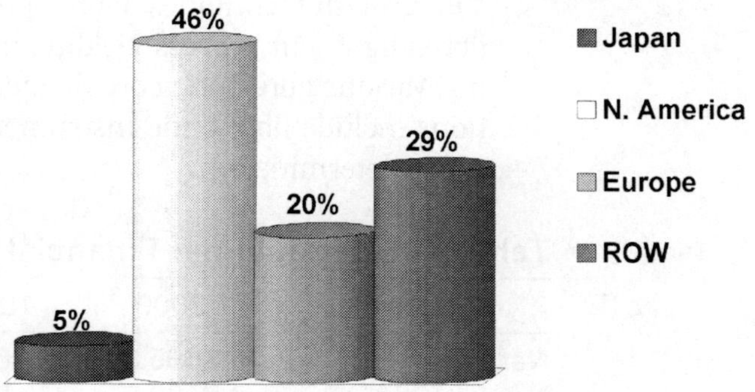

The Miamisburg, OH, USA, facility specializes in pyrotechnic and electronic devices for aircraft, defence and industrial applications such as oil well exploration; products include laser ignited devices and related products.

Products

The Optoelectronics Division manufactures products for telecommunications, sensor, lighting and imaging applications. The product line includes custom packaged laser diodes (EDFA pumps, DWDM laser diodes) as well as fibre optic and cable test equipment.

- For FY 2000, the Optoelectronics Division reported sales of US$496.9 million (or 29% of net sales), an increase of 11% over FY 1999's figure of US$447.7 million.

Examples of PerkinElmer's LD range include:

- The 805 series of devices employs elements from 75 μm wide, single sources to three stacks of 400 μm wide elements producing peak optical output power of 5.5 W to 78 W respectively. The laser diode structure is fabricated using an MOCVD epitaxial growth technique. This is a gaseous phase process that provides for very precise control of the crystal layers so that near theoretical device performance can be realized. These laser diodes are designed to provide narrow far-field emission in the plane perpendicular to the junction whilst maintaining typically 1 W/A slope efficiency. Applications include laser range finding, LIDAR, optical fusing and collision avoidance.
- The 905 nm series of laser diodes was designed specifically to address high-volume, low-cost requirements mainly for commercial applications such as LIDAR, laser range finding, intrusion alarms and collision avoidance. They are suitable for light duty factor requirements while producing high peak output power. The laser diode chips employed are modified versions of the PerkinElmer standard advanced MQW design found in the PGA series. The centre wavelength of operation (905 nm) is well matched to the peak response of Si detectors and complements the company's low-cost APD C30724 series.

- LD products in wavelengths from 850 nm to 1300 nm are produced using VPE growth techniques. Fibre optic pigtailed devices employ an advanced fibre alignment process yielding highly stable fibre to laser diode positioning. Various fibre optic core diameters can be supplied as options. Applications include fibre optic instrumentation, solid-state laser simulation and range determination.

Table 6.26 PerkinElmer Financial Highlights (US$ million)

	2000	1999	1998	1997	1996
Net sales	1695.3	1363.1	854.4	927.5	928.3
Net income	90.5	154.3	102.0	33.7	60.2
Total assets	2260.2	1714.6	1138.8	777.7	774.8

6.33 Picogiga

Picogiga
Place Marcel Rebuffat
Parc de Villejust
91971 Courtaboeuf
7 Cedex
France

Picogiga was founded in 1985 and is a leading merchant vendor of GaAs and InP MBE wafers. The company conducts a large amount of its business (69% in 1999) in the USA; to this end, in 1987, it established a US subsidiary, Picogiga Inc.

- For FY 2000, Picogiga reported turnover of €20.2 million, an increase of 82% over FY 1999's figure of €10.7 million.

Acquisitions

In FY 2000 Picogiga acquired Picopolish, a leader in the recycling of GaAs and silicon wafers.

Also in FY 2000, the company made an investment in Finnish optoelectronics company, Modulight Inc, a developer and manufacturer of high-performance optical semiconductor components for telecommunications applications, including transmission lasers, pump lasers and avalanche photodiodes. Modulight is a spin-off of the Optoelectronic Research Center at Tampere University of Technology, Finland.

Figure 6.46 Picogiga Five Year Sales (€ million)

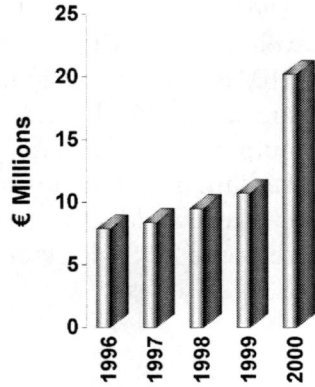

R&D

The company allocates approximately 12% of net sales to research and development. Picogiga has achieved several industry firsts:

- It was the first merchant company to use MBE for growing GaAs.
- The first to use double crystal X-ray diffraction analysis for quality control.
- The first to produce 4-in MBE wafers.
- The first to promote the use of multiwafer MBE machines for production.

Facilities

Picogiga has a 3000 sq m manufacturing facility at Courtaboeuf, France. This is the world's largest merchant GaAs MBE production facility and contains 1500 sq m of cleanrooms. Picogiga is constructing a 3000 sq m extension to the Courtaboeuf fab.

The company employs both VG Semicon and Riber reactors.

Products

The company manufactures 3-, 4- and 6-in GaAs and InP MBE wafers.

6.34 Princeton Lightwave

Princeton Lightwave Inc
2601 Route 130 South
Cranbury
NJ 08512
USA

Princeton Lightwave Inc (PLI) designs, develops and manufactures next-generation components for optical networking. It is a market leader in 1480-14xx pump lasers.

Founded in May 2000 and employing 75 people, Princeton Lightwave is a private company which was spun-off from the Sarnoff Corp. In June 2000 PLI completed its first round of financing (US$28.5 million from venture capitalist companies).

R&D

PLI's technology portfolio was established with an exclusive licence for Sarnoff's active component and telecommunications technologies, including approximately 27 patents. PLI employs 30 people in research and development.

- In April 2001 PLI announced it had achieved a significant breakthrough in optical power levels, becoming the first company to announce that it has reached the 1 W power level from a single narrow-stripe pump laser chip. This accomplishment positions PLI as a market leader in 1480-14xx pumps. The 1 W breakthrough is due to the advanced structure developed by PLI's engineering research team.

- As of July 2001 the company had a total of more than 50 patents, either filed or pending.

Facilities

PLI has a state-of-the-art, 90 000 sq ft manufacturing facility in Cranbury, including a 15 000 sq ft cleanroom.

The facility began operations for the production of components in July 2001, with ramp-up scheduled for completion by the end of 2001. The PLI manufacturing facility employs advanced manufacturing and packaging processes, featuring a fully integrated manufacturing execution system.

- The fab has a capacity of 500 4-in wafer starts per week.

Products

PLI produces high-performance Raman and EDFA pump lasers, advanced DFB lasers and Broadband GainChips, based on InP technology.

The WaveRiderBroadband GainChip (BGC), is an InP-based semiconductor chip-on-carrier. Its unique characteristics include a simplified chip design, a smaller package and tuning over S, C and L bands within the external cavity.

- The first member of the WaveRider family, the WaveRider BGC-10, is in production, with the second product, the WaveRider BGC-40, an industry first, being available in volume by the end of 2001. The WaveRider BGC-40 enables customers to reach more than 40 mW from an external cavity network source laser, while simultaneously tuning over an entire band.

Announced in March 2001, the company's WaveHarp advanced DFB laser emits high power (reaching 300 mW) in a narrow spectral band. As a wavelength-stabilized pump, the WaveHarp takes the new approach of integrating the grating on the laser chip, thus eliminating the need for an external fibre Bragg grating and reducing costs. As a source laser, it can eliminate the need for amplifiers in metropolitan DWDM networks, significantly reducing system architecture cost and complexity to enable wide-scale deployment.

In May 2001 PLI announced 340 mW of output power from a 1550 nm DFB laser coupled into a single-mode fibre, more than twice the power level of the previous record of 165 mW.

The WavePower family of pump lasers are for Raman amplifiers and EDFAs. The WavePower family is available in 14xx nm for Raman amplification and 1480 nm for EDFAs. PLI's pump design provides high power (reaching 500 mW in a module). Higher pump power translates to higher amplifier gain and more robust amplifiers. The benefits to network providers include greater bandwidth, while reducing long-haul network costs through simplified amplifier designs and all-optical transmission.

6.35 RIBER

RIBER SA
133 Boulevard National
Rueil Malmaison 92305
France

Founded in 1964, RIBER SA has more than 30 years of experience in the ultra-high-vacuum field and its applications, concentrating its activities on MBE. RIBER is one of the largest MBE/CBE manufacturers, offering a complete range of single- and multi-wafer systems for research, development and production of III-V (GaAs, InP, GaN), II-V (ZnSe, MCT) and SiGe epilayers. It supplies a complete range of effusion cells for RIBER and competing systems as well as laser ablation deposition systems.

For the year ended December 2000, RIBER reported sales of €41.2 million, an increase of 97% over FY 1999's figure of €20.9 million.

R&D

In 1978 RIBER began offering the first turnkey MBE research systems to semiconductor research institutes and it followed this during the 1980s with the commercialization of many innovations such as gas-source and multiwafer processing technologies.

Facilities

During 2000, RIBER increased production capacity and expanded its facilities in Rueil Malmaison. New assembly and test equipment is now in use, which allows RIBER to double its MBE 6000 production capacity for 2000 as compared to 1999. An additional 1000 sq m is available, which RIBER plans to use for future expansion of the existing facility to meet expected future increases in demand.

Products

RIBER designs and manufactures ultrahigh-vacuum deposition systems that perform solid-source MBE and gas-source MBE of advanced compound materials on one single 1-in substrate for R&D and up to multiple 6-in substrates for production. MBE-grown epitaxial wafers are used for the volume production of commercial micro and optoelectronic devices, such as MMICs, or pump lasers used in optical fibre networks.

Figure 6.47 RIBER Two Year Sales (€ million)

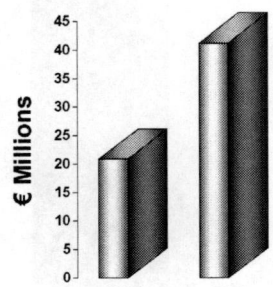

- For FY 2000 RIBER attributed 86% of sales to production machines and 6% of sales to research machines.

In addition to MBE systems, RIBER also produces a range of MBE-related components, including effusion cells up to 6000 cm^3 capacity, valved crackers, gas flow control systems and valved injectors, computerized process controllers, etc. System hardware components include source flanges, heaters, shutters and cooling panels.

RIBER's MBE systems include the MBE32 for up to 3-in wafers and the EPINEAT for both III-V and II-VI 3-in wafers. Several versions of the RIBER MBE32/CBE32, 3-in, single-wafer systems exist, adapted to the specific requirements of III-V and II-VI epitaxy. The system is also equipped with state-of-the-art gas regulation equipment developed for the III-V CBE field. The systems frequently include analytical instruments such as ellipsometers or STM apparatus.

For silicon–germanium alloys, RIBER produces the SIVA32 for 3-in wafers and the SIVA45 for 6-in wafers.

The MBE49 system is a multiple 4-in system for the production of lasers and MMICs. It can be configured for either 3 × 4-in or 4 × 4-in wafer production.

The MBE6000 is for the manufacture of 3 × 6-in III-V materials. This next-generation, high throughput, fully automated 4- and 6-in (and 9 × 4-in) multi-wafer reactor is designed for large-scale production of electronic and optoelectronic device structures such as lasers and MMICs. Its fully automated performance allows for 24 hours a day, 7 days a week operation, thus minimizing the per-wafer cost of production.

- In July 2001 Intelligent Epitaxy Technology Inc (IntelliEPI) added to its existing RIBER production systems by ordering a further three multiple 6-in MBE systems: one MBE6000 system (4 × 6 in) and two new generation MBE7000 production MBE systems (7 × 6 in).

The Compact 21 system is the new RIBER baseline MBE system for research. The reactor is a flexible and affordable system with features carefully designed to meet specifications for the research of all compound semiconductor materials

Figure 6.48 RIBER FY 2000 Geographic Sales (%)

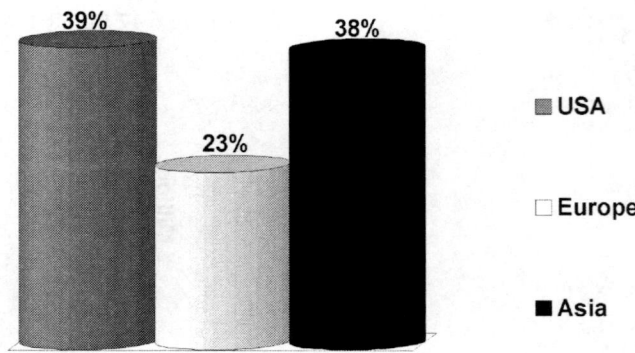

(such as III-V, II-VI, HgCdTe or GaN epi). Based on RIBER's field-proven Vertical Reactor technology, Compact 21 is a 2-in/3-in integrated system, providing, within a small footprint, all the necessary MBE facilities and characterization capabilities.

The EVA 32 system has been specially developed for the growth of metal and silicon alloys (i.e. elements sublimating at very high temperatures). The UHV epitaxy chamber is equipped with two electron beam evaporating guns, one of which can be multi-crucible and can take up to four solid source cells or gas injectors. Suitable for metal and silicon alloys epi growth, the reactor accommodates horizontal substrates up to 3 in.

In FY 2000 RIBER introduced the MBE7000 system, a highly advanced production MBE machine which can grow seven 6-in or four 8-in wafers per run. This allows a 75% increase in productivity, while keeping the same footprint as previous multiple 4-in units. Based on the MBE6000 design, the MBE7000 has a capacity of 40 8-in wafers or 70 6-in wafers per cassette load. It also features the Crystal fully automated epi process.

Table 6.27 RIBER Financial Highlights (€ million)

	2000	1999
Net sales	41.2	20.9
Net profit	7.0	2.0
Operating profit	10.3	3.4

6.36 ROHM

ROHM Co Ltd
21 Saiin Mizosaki-cho
Ukyo-ku
Kyoto 615
Japan

Founded in 1958, ROHM Co is a worldwide manufacturer of components and modules; it employs 15 000 people. ROHM has a major share of both the 650 nm laser diode for DVDs and 780 nm laser diode markets for CDs and CD-ROMs.

- ROHM reported net sales of ¥409.3 billion for FY 2001, a 14% increase on FY 2000's figure of ¥360.1 billion.

R&D

R&D Expenditure
FY 2001: ¥20.8 billion
FY 2000: ¥22.7 billion

Rohm has R&D facilities in Japan at Kyoto, Yokohama, Tokyo and Hyogo and overseas in France, Hong Kong and the USA.

Rohm has conducted research into the use of gallium nitride (GaN) for short-wavelength lasers, as well as laser diode integration with other optical devices for CD pick-ups.

Figure 6.49 Rohm Five Year Sales (¥ billion)

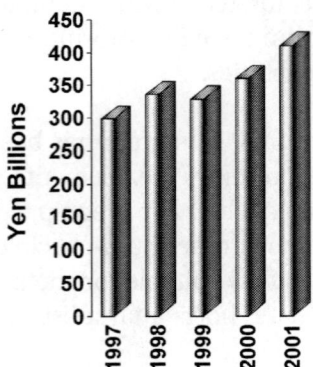

- In August 2001 Rohm was awarded US Patent No 6,274,891, 'semiconductor laser'. In a semiconductor laser which uses a semiconductor of GaN-type compound, an optimal material is used for a current blocking layer, so that it is made possible to obtain a semiconductor laser that satisfies a gain guiding structure of high light emitting efficiency or a refractive index guiding structure or both, thereby facilitating control of the noise of oscillated light (reduction of noise), control of the spread of light in lateral direction and control of the longitudinal mode.

Facilities

The ROHM group of companies has manufacturing facilities and sales offices worldwide. It has design centres in Japan as well as facilities in Kyoto, Okayama, Fukuoka and Yamanaski, Japan; Cavite, The Philippines; Patumthami, Thailand; Seoul, South Korea; Selangor, Malaysia; Austin, TX, USA; and Tianjin and Dalian, China.

- Rohm Wako in Okayama manufactures laser diodes.
- In 1997 Rohm built the industry's largest mass production line for red laser diodes for DVD applications. Initial capacity was 1 million units per month.

ROHM announced a ¥6 billion investment and the establishment of a new company, ROHM Electronics Wako (Tianjin) Co Ltd (REWT), in the Microelectronic Industrial Park of the Tianjin Economic-Technological Development Area, Tianjin, China. ROHM will also move ROHM Electronics (Tianjin) Co Ltd (RET) from its former site in the Tianjin Xiqing Economic Development Zone (RET produces glass diodes, LED lamps and transistors). ROHM has obtained 110 000 sq m of land in the industrial park to build these production facilities for the manufacture of discrete semiconductor products in order to meet increasing future demand. The new company, REWT, plans to produce semiconductor lasers, moulded diodes and chip LEDs, with production at the new facility scheduled for spring 2002.

Products

ROHM has a major share of both the 650 nm laser diode for DVDs and 780 nm laser diode markets for CDs and CD-ROMs. The company is also one of the world's largest producers of LEDs and switching diodes.

Figure 6.50 Rohm FY 2001 Geographic Sales (%)

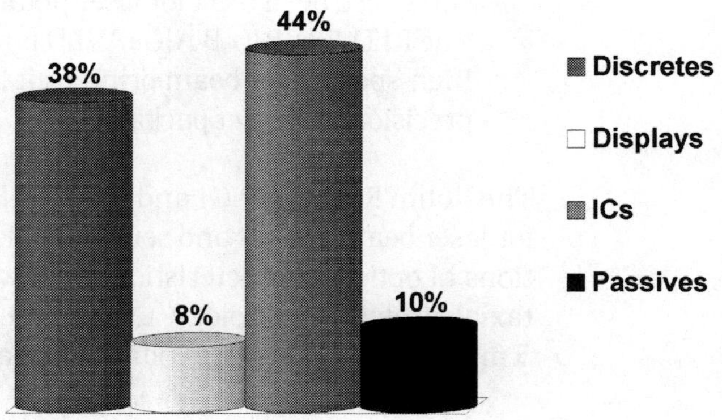

ROHM was the first in the world to mass produce semiconductor lasers by MBE. It has optimized the mass production of long-wavelength diode lasers such as AlGaAs molecular beam epitaxy (MBE) devices; it is therefore working towards extending the wavelength, hence the data storage capacity by moving to shorter-wavelength diode lasers, notably the blue diode laser.

- The RLD-78MC is the world's first mass produced LD manufactured by MBE. The characteristics are suitable for use in sensors and bar code readers. It features one-third of the dispersion compared with conventional LDs, 5 mW output and 2 V reverse voltage.

ROHM produces market-leading 780 nm laser diodes for CD-ROMs and has developed other advanced devices for DVD technology. The laser diodes use a strained multi quantum well (MQW) active layer combined with ROHM's original dual-layer heat sink, which offers a marked improvement in stability when operated at raised temperature. ROHM's RLD-65MC/PC was made from the design and production technology it has accumulated while manufacturing laser diodes for use with CDs and CD-ROMs. With the introduction of the strained MQW active layer, the operating current was reduced by 35% compared to laser diodes with conventional active layers. This improved operating characteristics during high temperatures and resulted in an increased life span with an MTTF of over 10 000 hours.

- In mid-2001 Rohm announced the RLD78PZW1, a high-power IR laser diode for high-speed CD-RW drives, ×16 and higher. This laser diode improves the reliability of pick-ups with the industry's catastrophic optical damage (COD) level by 20%. The RLD78PZW1 features 150 mW (pulse) output power, 315 mW COD level and an operating current of 120 mA.
- The RLD-78MAT1 AlGaAs series are the world's first mass produced laser diodes manufactured by MBE. They feature reduced facet reflection and one-third dispersion compared with conventional LDs. Signal-to-noise ratio is guaranteed over the entire operating temperature range. Applications for the RLD-78MAT1 series include vehicle compact disc players and navigation systems.
- The RLD-65MC and RLD-65PC are red-coloured laser diodes developed for use with DVDs. With the introduction of a strained MQW in the active layer, a low threshold current is achieved. The high operating temperature that

    reaches up to 70°C makes the LD suitable for DVD applications, although it could also been used for laser points and bar code readers.
- The RLD-78NP10-B AlGaAs LD is again manufactured by MBE. Suitable for high-speed laser beam printers, it features one-third dispersion and a high-precision, compact package.

The Rohm RLD-78PP-G1 and RLD-78NP-G1 are semiconductor lasers developed for laser beam printer and sensor applications. Rohm achieved very small variations of optical characteristics and low droop by employing its proprietary epitaxial growth technology using MBE. It features high stability wavelength, 5 mW output, 2 V reverse voltage and can be driven by a single power supply.

Alliances

Cree Inc announced an alliance with ROHM Co Ltd in December 2000. The agreement included a licence agreement under which ROHM granted Cree a five-year exclusive licence to several US patents and a non-exclusive licence under a Japanese patent for optoelectronic devices such as GaN-based blue LEDs and blue-violet semiconductor lasers based on silicon carbide. The companies signed a non-binding memorandum of understanding for the co-operative development of a packaged blue laser diode for consumer applications. In addition, Cree and ROHM have entered into an annual supply agreement for the purchase by ROHM of LED chips manufactured by Cree.

In November 1999 Pioneer Corp and ROHM Co announced a partnership agreement for the development of a GaN-based blue-violet semiconductor laser for next-generation optical disc technology. By combining each company's technological knowledge, Pioneer and ROHM will make further improvements in crystal growth technology, device production process and optimization of the device structure and will obtain a practical output level and life length. Both companies are aiming at merchandising the blue-violet semiconductor laser within two years.

The company is a member of IrDA (Infrared Data Association), which promotes interoperable, low-cost IR data interconnection standards.

ROHM and Eastman Kodak announced in April 2001 that they had formed a technical agreement for the manufacture of OEL displays. Eastman Kodak owns basic patents for low molecular OEL for OEL displays, which ROHM will utilize, together with its own semiconductor and LCD technologies to enter the OEL display market on a large scale. The company aims to complete the development of OEL displays in 2002 and then enter into the market.

Legal Disputes

Nichia Corp announced in May 2001 that ROHM Co Ltd had withdrawn its complaint against Nichia before the US International Trade Commission (ITC). This should conclude the related ITC investigation where it was alleged that certain Nichia parts infringed two US patents.

Table 6.28 Rohm Financial Highlights (¥ billion)

	2001	2000	1999	1998	1997
Net sales	409.3	360.1	328.6	335.9	297.8
Net income	86.2	66.7	52.2	61.0	45.5
Working capital	312.9	309.0	261.0	237.6	196.3
R&D	20.8	22.7	17.8	19.9	13.4
Total assets	764.5	648.3	550.4	533.8	479.1

6.37 Sanyo

Sanyo Electric Co Ltd
5–5 Keihan-Hondori
2-chome
Moriguchi City
Osaka 570
Japan

Sanyo Electric is a well-known international company producing a large range of electrical goods. The company was founded in 1947 and today consists of 156 companies employing more than 86 000 people.

● For FY 2001 Sanyo reported net sales of ¥2157.3 billion, an increase of 11% over FY 2000's figure of ¥1940.4 billion.

R&D

R&D Expenditure
FY 2001: ¥108.8 billion
FY 2000: ¥99.6 billion

The company began its research on amorphous silicon solar cells in 1975 and successfully commercialized the world's first solar cells in 1980.

Sanyo has five R&D laboratories:

● Tsukuba Research Centre.
● New Materials Research Centre.
● Microelectronics Research Centre.
● Mechatronics Research Centre.
● Hypermedia Research Centre.

Figure 6.51 Sanyo Five Year Sales (¥ billion)

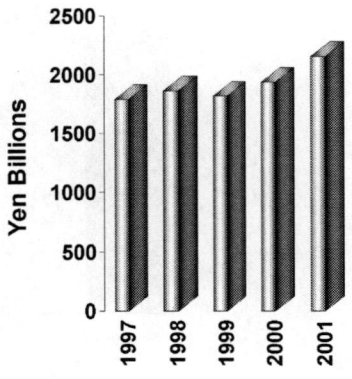

It also has eight development centres that are affiliated with each of the different business headquarters.

- The company conducts research into VCSEL applications.
- Sanyo was the first to produce 635 nm AlGaInP laser diodes.
- Sanyo developed one of the industry's smallest hologram frame lasers in 1995, which measured 4.5 × 8 × 5.8 mm.
- Sanyo has developed a thin semiconductor laser for use in compact, light-weight DVD players and DVD-ROM drives. The laser is 2.4 mm thick and weighs less than half that of cylindrical lasers. Compared to CD lasers, the optical sources used in DVD players require a higher operating current, which also creates more heat, so manufacturers must make the laser semi-conductors in cylindrical form to maximize the heat dissipation. The laser is scheduled for volume production by the end of 2001.

Facilities

Sanyo's semiconductor manufacturing facilities in Japan are located at Gifu, Gunma, Niigata and Oizumi. It has overseas plants in the USA, China, South Korea, Thailand and Taiwan.

Sanyo is in the process of constructing a clean energy solar power system at its Gifu plant. The process, termed Mega Solar is scheduled for completion in 2004. It integrates a solar battery heterojunction with intrinsic thin-layer (HIT) structure that boasts the industry's highest conversion rate and it will have an output of 3.4 MW, the highest in the world.

On completion, Mega Solar will reduce the Gifu Plant's CO_2 emissions by approximately 670 tons annually. Mega Solar will be capable of generating 3700 MW annually, equivalent to the amount of electric power consumed by 1000 homes a year.

Products

The Electronic Devices segment sales for FY 2001 were ¥447.7 billion, or 21% of net sales. Its products include semiconductor lasers, LEDs, transistors, diodes, image sensors, MOS-ICs, bipolar ICs, thick-film ICs and LCDs.

Figure 6.52 Sanyo FY 2001 Geographic Sales (%)

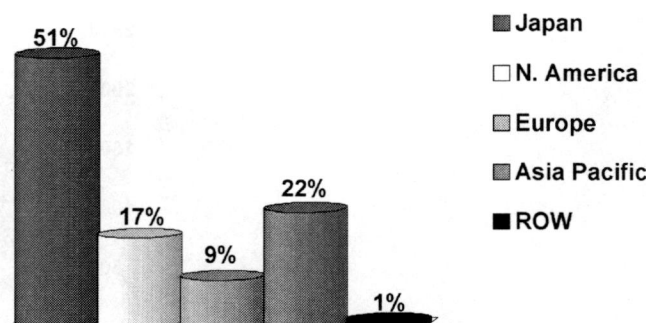

Sanyo currently markets AlGaAs IR laser diodes in the 785–830 nm wavelength range and AlGaInP-based red laser diodes in the 635–678 nm range.

- Sanyo's DL-3038-033 is an index guided, MQW 635 nm AlGaInP red laser diode with low threshold current and high operating temperature. The lasing wavelength at 635 nm is eight times brighter than that of 670 nm lasers. DL-3038-033 is suitable for applications such as bar code scanners and laser pointers. It features a low threshold current of 30 mA and low operating voltage of 2.2 V.
- The Sanyo DL-3147-141(-241) is an 645 nm AlGaInP MQW LD with low threshold current (45 mA) and high operating temperature of 5 mW at 60°C. It is suitable for applications such as bar code scanners and optical disc systems.
- The DL-3150-103 is a compact flat package-type LD, which is considerably different from conventional stem-type lasers. The new structure of the frame lead-type package enables optical systems to be lightweight and small sized. It also features a built-in PIN photodiode for light output monitoring. The DL-3150-103 is suitable for applications such as compact discs, CD-ROM systems and video disc systems.
- The DL-3147-021 is an index guided 645 nm AlGaInP LD for laser pointer applications which has a low threshold current (achieved by a strained MQW active layer). It features low threshold current of 30 mA and low operating voltage of 2.3 V.
- The DL-3147-161(-261) is an index guided 650 nm AlGaInP laser diode with low threshold current and high operating temperature (which are achieved by a strained MQW active layer). It is suitable for applications such as DVD-ROM optical disc systems.

Table 6.29 Sanyo Financial Highlights (¥ billion)

	2001	2000	1999	1998	1997
Net sales	2157.3	1940.4	1818.2	1866.4	1793.0
Net income	42.2	21.7	(25.9)	12.3	17.7
Working capital	144.3	366.5	356.6	273.5	308.4
R&D	108.8	99.6	93.7	94.7	93.6
Total assets	2945.3	2706.1	2662.5	2641.9	2518.1

6.38 Sharp

Sharp Corp
22–22 Nagaike-cho
Abeno-ku
Osaka 545
Japan

Founded in 1912, Sharp Corp is a world-leading electronics manufacturer and a leader in the optoelectronics device market, of which it claims to have a more than 15% share.

- For the year ended March 2001, Sharp reported net sales of ¥2012.9 billion (US$16.4 billion), an 8.5% increase on the previous year's figure of ¥1854.8 billion. It employs 50 000 people worldwide.

R&D

R&D Expenditure
FY 2001: ¥149.7 billion
FY 2000: ¥146.8 billion

Sharp has 18 R&D bases in Japan, including the Advanced Development and Planning Center in Nara, which conducts R&D on new materials and data-processing systems. Overseas, the company has R&D facilities in Camas, WA, USA; Oxford, UK; and Kaohsiuing, Taiwan.

Sharp's European research laboratory in Oxford, UK, conducts research in optoelectronic devices; subjects include lasers for next-generation DVD.

Sharp's laser developments include:

- In 1985 Sharp developed a 750 nm wavelength, 5 mW output, visible laser diode.
- It followed this in 1988 with the joint development (with Philips) of an industry first, a one-beam hologram laser unit.
- In 1994 Sharp developed the industries first hologram laser unit with a built-in OPIC (optical IC) for 4× speed CD-ROM.
- In 1995 Sharp achieved a long lifetime of 10 000 hours of 635 nm, 5 mW output power, red laser diode for DVD.

Figure 6.53 Sharp Five Year Sales (¥ billion)

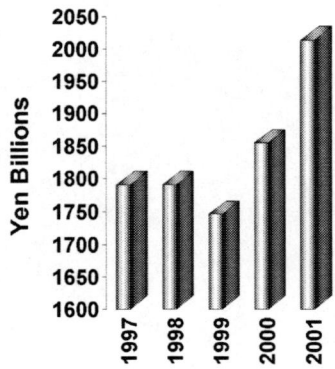

- In 1997 it developed a compact hologram laser diode for 2× speed DVD-ROM. In 1998 it developed the industry's first compact hologram laser diode for 4× to 6× speed DVD-ROM.
- In 1999 it developed the industry first compact, thin (3mm) hologram laser diode for 8× to 10× speed DVD-ROM.

Facilities

Sharp Electronic Devices has facilities in Japan at Nara (where it manufactures optoelectronic devices) and Shinjo, as well as overseas in China (Wuxi Sharp) and Indonesia.

Construction on phase I of the new Mihara plant in Hiroshima Prefectures began in June 2001. When fully operational in September 2002, the ¥18.7 billion plant will produce compound semiconductor elements including laser diodes for CD and DVD players and recorders. Initial production in phase I of the new plant will include IR and red laser diodes, with future plans including production of blue laser diodes which Sharp considers to be the light source for next-generation high-capacity data recording media. Mihara will be Sharp's main plant for the manufacture of these next-generation devices.

The Nara plant for optoelectronic devices began operations in March 2000. The plant manufactures lasers for 16× CD-R/RW drives, dual channel lasers for 16× CD-R/RW drives, dual channel lasers for DVD-ROM/CD-R drives and OPIC (optical IC) photodetectors for 16× DVD-ROM drives.

- Sharp uses Thermo VG Semicon's V100 systems for phosphide based lasers and detectors for use in CDROM and DVD systems.

In Indonesia, Sharp has a semiconductor production base, PT Sharp Semiconductor Indonesia (SSI) in Karawang Province, for ICs and optoelectronic devices. The company also has a design and development centre in Malaysia.

Products

Sharp's Electronic Components business group includes semiconductor lasers, EL and LED display systems, electronic parts such as solar cells, LSI, ICs, photosensitive converters and their applied products, photovoltaics, power devices, colour or monochrome LCD display panels and units, electronic tuners and printed circuit substrates.

Figure 6.54 Sharp FY 2001 Geographic Sales (%)

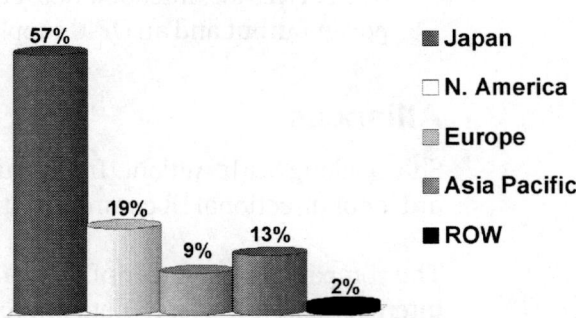

- Japan
- N. America
- Europe
- Asia Pacific
- ROW

The Electronic Components Division's sales were ¥681.5 billion (34% of net sales) in FY 2001. Sharp produces laser diodes for CD-R/CD-RW/DVD-ROM drives and holographic pick-up units for CD players. Its other optoelectronic products include CCDs used in camcorders, LEDs, photocouplers, photo-interrupters, photodiodes, OPIC devices, image sensors and IR modules.

In September 2000 Sharp introduced two new models of high-output IR semi-conductor laser (GH07812A2C) and hologram laser (GH5R412A3C). Both lasers deliver the industry's highest standard of optical output, suitable for 16× speed-write CD-R drives. These devices were in volume production in November 2000, with a total of 50 000 units per month.

- The GH07812A2C IR semiconductor laser has a maximum optical pulse output (industry's highest) of 160 mW, 780–787 nm peak oscillation wavelength and 30 mA threshold current.
- The GH5R412A3C hologram laser features a maximum optical pulse output of 144 mW, 30 mA threshold current and 773–797 nm peak oscillation wavelength.

Sharp's current laser products include:

- The GH06550B2Bis a red laser diode which features high-power output (50 mW CW) for DVD-R/RW/RAM drives. It features a typical wavelength of 654 nm.
- The GH06507B2A is a dual power supply, red laser diode for DVD-ROM drives and video DVD drives. It features maximum optical power out of 7 mW CW and typical wavelength of 654 nm.
- The GH07885C2C is a 120 mW pulse, 785 nm, high-power laser diode for 8× speed CD-R/RW drives.
- The GH07895A2C is a high-power output type compact laser diode for 14× speed CD-R applications. It features a power output of 95 mW CW (maximum) and has the ability to write with 14× speed owing to the high power output (pulse 135 mW).
- The GH17805B2AS and GH17805B2BS are insert frame structure laser diodes developed for audio CD/CD-ROM. They feature maximum optical power out of 5 mW, oscillation wavelength of 870 nm. The GH17805B2AS has a dual power supply whilst the GH17805B2BS has a single power supply.
- The GH6D407B5A is a 3 mm thickness, resin stem hologram laser for 10× speed DVD-ROM drives. It features dual power supply, power output of 6.3 mW and reverse voltage of 2 V.
- The GH7C605B5B is a 3 mm thickness resin stem hologram laser for CD-ROM drives for notebook PCs (equivalent to 40× speed). It features 4.3 mW power output and an OPIC supply voltage of 6 V.

Alliances

Sharp, along with Agilent, Intel and Microsoft, co-developed IrBus, the standard for bi-directional IR communications.

The company is a member of IrDA (Infrared Data Association), which promotes interoperable, low-cost IR data interconnection standards.

Table 6.30 Sharp Financial Highlights (¥ billion)

	2001	2000	1999	1998	1997
Net sales	2012.9	1854.8	1745.5	1790.5	1790.6
Net income	34.2	28.1	4.6	24.8	48.5
Working capital	344.1	334.0	368.2	297.6	295.3
R&D	149.7	146.8	135.1	132.3	124.7
Total assets	2003.6	1987.4	2021.9	2084.2	2048.8

6.39 Sony

Sony Corp
7–35 Kitashinagawa
6-chome
Shinagawa-ku
Tokyo 141
Japan

Founded in 1946, Sony Corp is a leading manufacturer of displays, semi-conductors, video and audio equipment, televisions, computers and information-related products, as well as in the music and image-based software markets. Sony researches and manufactures optoelectronic components for internal use as well as for sale on the merchant market.

- Sony reported net sales for the year ended March 2001 of ¥7314.8 billion (US$58.5 billion), a 9.4% increase over FY 2000's figure of ¥6686.7 billion and employed 182 000 people.

R&D

R&D Expenditure
FY 2001: ¥416.7 billion
FY 2000: ¥394.5 billion

Sony is involved in research into CD, CDR, DVD and MiniDisc technology, all of which involve lasers.

Sony has developed the SLD344YT laser diode, which achieves 6 W, the industry's highest power from a single stripe. Sony has also developed the SLD343YT 4 W laser diode and the SLD342YT 2 W laser diode, which achieve higher optical power output levels without changing the optical emission area of Sony's current line-up.

Figure 6.55 Sony Five Year Sales (¥ billion)

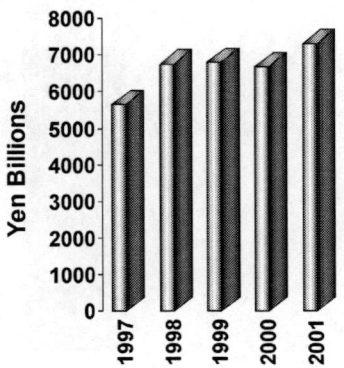

Sony accomplished CW operation of a blue GaN-based laser diode using InGaN/GaN/AlGaN in 1998; at the time, it was only the third company to do so. Researchers focused on the use of raised-pressure MOCVD to improve the crystalline quality of their material, leading to a reduction in both etch pit density (by increasing the reactor pressure from 1 to 1.6 atm) and threshold current. The resulting ridge stripe laser diodes were fabricated with a stripe width of 4 μm and a cavity length of 1 mm. The threshold current was 140 mA, operating voltage 16.8 V and emission wavelength 412 nm.

The company also produced a ZnCdSe/ZnSSe/ZnMgSSe separate confinement heterostructure laser diode which under CW operation exhibited a lifetime of more than 140 hours at 40°C and more than 400 hours at 20°C. The 514 nm laser operated with an output of 1 mW.

Facilities

Sony announced in August 2000 that it would construct a new 12-in wafer fab in Kumamoto Prefecture, Japan. The new plant will expand Sony's manufacturing capacity of small-size LCD (high-temperature polysilicon) and CCD image sensors that are currently being produced at Sony Kokubu Corp. Construction began in November 2000, with production scheduled for October 2001. Mass production of LCDs is planned to begin in 2002 at 3000 wafers per month and CCDs in 2003 at 2000 wafers per month.

Sony Shiroishi Semiconductor Inc has a processing line for the manufacture of laser couplers.

Products

The Electronic Devices Segment accounted for 17% of Sony's FY 2001 sales. Sony is the industry's leader in mass producing high-power lasers for optical disc applications. The company's optoelectronic product range includes laser diodes, laser drivers, variable capacitance diodes, CCDs, D/A and A/D converters, serial/optical communication ICs and GaAs FET devices.

- The SLD234VL is a high-power AlGaAs laser diode. It features typical wavelength of 785 nm and maximum optical output of 50 mW.

Figure 6.56 Sony FY 2001 Geographic Sales (%)

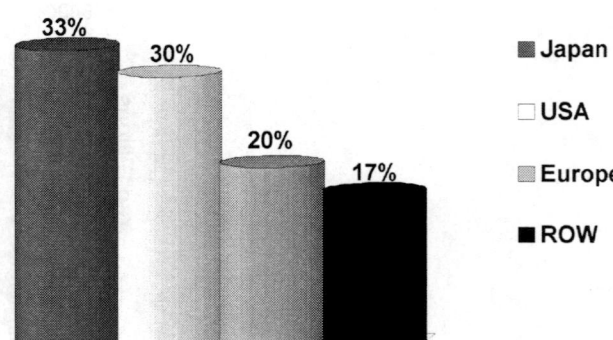

- The SLD104BUL is a high-power, low-noise laser diode developed for CD players. It features an AlGaAs double hetero structured laser diode and Pin photodiode for optical power output monitor.
- The CXB1828ER is a high-speed laser diode driver IC for SONET/SDH applications that supports bands up to 2.5 Gb/s. The device is equipped with a bias current auto power control function and a modulation output duty cycle control function.
- Sony's SLD 327 is a super-high-power laser diode, featuring 795–840 nm wavelength and 33 mW optical output power.
- The SLD1133VL-53 is an AlGaInP index-guided red laser diode designed for DVD systems or bar code scanners. It features small astigmatism (7 μm) and low operating current (60 mA). For bar code scanners, its wavelength (650 nm) is 20 nm shorter than that of the current device.
- The SLD1132VS is a AlGaInP QW red laser diode designed for laser pointers. Compared to a conventional visible laser diode (670 nm wavelength), the SLD1132Vs' typical wavelength of 635 nm is shortened by 35 nm and visibility is increased seven fold.
- The SLD322 is a super-high-power laser diode which features wavelength of 790–840 nm and 550 mW output power.
- The PBD-V30 DVD player features a Dual Discrete optical pick-up which has separate lasers optimized for CD (780 nm wavelength) and DVD (650 nm). This enables stable, accurate playback of both types of discs. It eliminates lens switching, reduces laser wear and extends laser life. It also has optical and coaxial digital outputs for digital-to-digital connections to a Dolby digital processor or receiver.
- The CXB series of transmitter ICs includes the CXB1548QY laser diode driver and Op Amp (622 MHz data rate and operating at 5 V); CXB1549Q LD driver with built-in Op Amp (with 1.25 Gbps data rate, operating at 3.3 V) and the CXB1558QY LD driver with D-FF (622 MHz and 5 V power supply).

Alliances

Sony holds a minority interest in the Corporation for Laser Optics Research, of Portsmouth, NH, USA, which holds a patent for a blue-emitting pulsed laser and uses pulsed lasers to project large-screen, full-colour video images.

Sony is a member of IrDA (Infrared Data Association), which promotes interoperable, low-cost IR data interconnection standards.

Table 6.31 Sony Financial Highlights (¥ billion)

	2001	2000	1999	1998	1997
Net sales	7314.8	6686.7	6794.6	6755.5	5663.1
Net income	16.8	121.8	179.0	222.1	139.5
Working capital	830.8	861.7	1126.8	1151.2	843.5
R&D	416.7	394.5	375.3	318.0	282.6
Total assets	7828.0	6807.2	2004.4	2036.0	1855.3

6.40 Spectra-Physics

Spectra-Physics Inc
1335 Terra Bella Avenue
MountainView
CA 94043
USA

Spectra-Physics Lasers Inc (SPSL) was founded in 1961 and today employs just over 1000 people. The company is a subsidiary of Thermo Electron Corp (which holds approximately 80% of Spectra-Physics' stock). Spectra-Physics is a leader in semiconductor-based laser and optical technologies and is one of the largest manufacturers of lasers and laser optics worldwide.

Opto Power Corp, a Spectra-Physics subsidiary, was acquired in 1992. In December 2000 Opto Power changed its name to Spectra-Physics Semiconductor Lasers (SPSL). This name change reflects the increasing importance of semiconductor lasers to the company.

- However, to coincide with the company's 40th anniversary, in June 2001 it became Spectra-Physics Inc.

This news was followed by an announcement in August 2001 that Thermo Electron had offered a cash tender (at US$20 per share) for that part of the company that it does not own. If successful, Thermo Electron intends to make Spectra-Physics a private company.

- For the year ended December 2000, Spectra-Physics reported net sales of US$186.2 million, a 32% increase on FY 1999's figure of US$141.3 million.
- Spectra-Physics is the world's largest supplier of scientific lasers.

R&D

R&D Expenditure
FY 2000: US$22.6 million
FY 1999: US$17 million

Research and development activities are divided into two main areas: telecommunications and high-power semiconductor-based lasers for the other commercial markets.

Figure 6.57 Spectra-Physics Five Year Sales (US$ million)

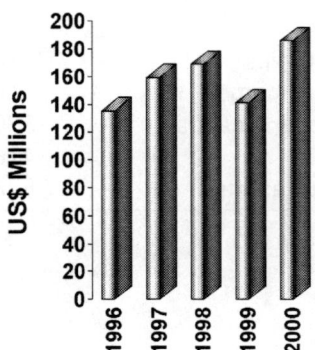

The company has close ties with the research community and has been instrumental in pioneering next-generation technology. In recent years, it has focused on the development of versatile, user-friendly, ultra-fast laser sources. These are based on the Millennia, the first high-power, all-solid-state CW visible laser.

Spectra-Physics has approximately 140 US patents and more than 20 patent applications pending. The company has a royalty-free licence to all of SDL's patents applied for or issued prior to 25 June 1993. US Patent No 6,272,159 was recently awarded to Opto Power Corp:

- 'Laser diode package with slotted lead'. A laser diode or laser diode bar mounted on a heat sink and having an insulating layer of greater thickness also mounted on the heat sink requires a lead overlying both the insulating layer and the diode to bend downward for making electrical contact. Failures have been found to occur at the bend. The provision of a pattern of slots at the bend alleviates the problem by reducing stress there. The use of such slotted leads is disclosed for diodes or diode bars individually or in a stack arrangement.

Facilities

Manufacturing facilities are located in Stahnsdorf, Germany, and at Mountain View and Oroville, CA, Tucson, AZ, USA. Spectra-Physics has Class 10 000, 1000 and 100 cleanrooms.

- The company has ten ion beam sputtering machines for the manufacture of thin-film filters.
- Spectra-Physics expanded its production capacity for passive components and launched the ZeroShift 100 GHz optical thin-film filters. The company also ramped customer shipments of ZeroShift 200 GHz filters to commercial volumes.

Spectra-Physics announced in May 2001 that it would increase production capacity, install additional equipment and accelerate new semiconductor laser development. The additional capacity will enable high-volume manufacture of the ProLite line of high-power, high-reliability semiconductor laser bars, as well as delivery of next-generation, higher power telecommunications pump chips.

Figure 6.58 Spectra-Physics FY 2000 Geographic Sales (%)

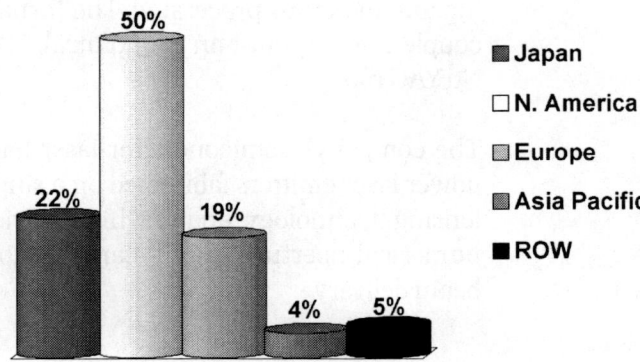

- Spectra-Physics has an industrial laser applications laboratory at its facility in Mountain View, CA, USA, to facilitate and advance the use of high-power UV lasers for high-volume micromachining tasks.

Products

Spectra-Physics manufactures both active and passive components, ranging from high-power pump lasers to thin-film WDM filters.

In January 2001 the company announced its 266 nm CW diode pumped solid-state laser, the Millennia UV, the world's first all-solid-state, deep-UV CW laser designed for use in demanding commercial and industrial applications. These include fibre Bragg grating production, wafer inspection, printed circuit board inspection and disc mastering, where the Millennia UV provides an efficient and reliable alternative to power-hungry gas lasers. This turnkey, OEM laser combines Spectra-Physics' Millennia platform with a Double DeltaConcept frequency-doubling cavity in a rugged, monolithic laser head.

In March 2001 it introduced the Inazuma diode-pumped solid-state laser that combines high output power with excellent spatial mode. Inazuma delivers 8 W of Q-switched output at 355 nm, in a near-diffraction-limited beam (M2<1.3). Inazuma is a fully sealed laser head pumped by FCbar modules, where the pump laser diodes are located in the power supply. The laser produces more than 35 W in the near IR, 20 W at 532 nm and 8 W at 355 nm. With an adjustable pulse rate from 15 kHz to 100 kHz, this laser is expected to have an impact on several materials processing applications, where the increased UV power will translate directly into higher throughput.

The 3900S high-performance, tunable, solid-state IR laser delivers CW output from 675 to 1100 nm in a low-cost package. Pumped with either an argon ion laser or a 532 nm diode-pumped solid-state laser, the 3900S solid-state Ti:sapphire laser produces up to 3.5 W of TEM00 output for the broadest range of IR applications such as fibre laser research, spectroscopy and telecommunications research.

The company also manufactures a range of industrial DPSS lasers called the Tornado series. These products were specifically designed to retain the benefits of Spectra-Physics' single-mode/lower-power products, while providing sufficient multimode output power to enable applications such as high-speed marking and materials processing. The Tornado series utilizes non-imaging optics to couple the output of an economical, 200 W diode laser stack into the end of a Nd:YAG rod.

The company's semiconductor laser bars are one-dimensional arrays of high-power laser emitters fabricated on a single, monolithic substrate. Its proprietary lensing technology enables high-efficiency coupling of bar output into low numerical aperture, small-diameter fibre bundles for the ultimate in flexible beam delivery.

- 980 nm laser bars are available for medical applications such as dermatology, plastic surgery, chiropody and wound care. The products feature a 40 W, conductively cooled, open heat sink laser bar, a 30 W fibre-coupled (850 µm diameter) laser bar and a 12 W fibre-coupled (675 µm diameter) laser bar.

Other Spectra-Physics products include:

- The Vanguard was the first high-power commercial laser to utilize saturable Bragg reflector (SBR) technology, under licence from Agere Systems. It is an all-solid-state laser that provides over 4 W quasi-CW output at 355 nm. The simple, robust mode-locking mechanism results in a pulsed, high repetition rate output, making this laser well suited to applications currently using CW gas lasers.
- The Streamline-RL is a totally integrated, turnkey Raman fibre laser system designed for the development and Telcordia reliability testing of telecommunications components. The high-power Streamline-RL consists of a rack-mounted box containing a single 35 W semiconductor laser bar pump source, the fibre laser, a cascaded Raman cavity and all associated power supply and control electronics. This product delivers over 5 W of randomly polarized output power from a SMF-28 single-mode fibre.
- Earlier in 2000 the company introduced new multi-mode pump lasers and Raman fibre lasers, active components used in the development of next-generation, high-power amplifiers for DWDM systems.

Agreements

Spectra-Physics and GSI Lumonics Advanced Manufacturing Systems in summer 2001 signed a US$5 million supply agreement for high-power, diode-pumped, Q-switched lasers based on Spectra-Physics' patented FCbar technology. The lasers will be incorporated into systems used primarily to trim components in semiconductor circuits for computer and wireless communications applications.

Legal Disputes

In June 2000 Rockwell Technologies LLC filed a complaint against both Spectra-Physics and its Opto Power subsidiary (no longer called Opto Power), in the US District Court for the District of Delaware. Rockwell alleges that both companies have infringed US Patent No 4,368,098 entitled 'Epitaxial Composite and Method of Making', by performing the patented process and by using wafers manufactured by the patented process. Spectra-Physics disputes the claim. Rockwell seeks damages, including increased damages due to the defendants' alleged wilful infringement, plus costs and attorney fees. The hearing is scheduled for February 2002.

Table 6.32 Spectra-Physics Financial Highlights (US$ million)

	2000	1999	1998	1997	1996
Net sales	186.2	141.3	169.0	159.2	135.4
Net income	1.5	(3.8)	9.6	33.9	(4.2)
Working capital	55.0	57.9	72.3	68.9	27.5
R&D	22.6	17.0	16.7	14.4	12.0
Total assets	190.8	152.3	157.0	140.5	86.8

6.41 Thales Optics

Thales Optics
Route Départementale
128
BP 46
91401 Orsay Cedex
France

The Thales Group is the new name for THOMSON-CSF, which changed its name to Thales in December 2000. Part of the defence division, subsidiary company Thales Optics is the new name for several former THOMSON-CSF optoelectronics companies:

- THOMSON-CSF Laser Diodes (a manufacturer of high-power laser diodes, laser diode bars and laser diodes for gas sensors).
- THOMSON-CSF Laser (manufacturer of laser diodes for diode pumped, lamp pumped, ultra-fast, PIV, optical and tunable lasers).
- Pilkington Optronics (based in Glasgow, UK).

Acquisitions

In May 2001 Alcatel acquired (for €795 million) Thales' 48.83% stake in their joint venture, Alcatel Space, a developer of satellite technology solutions for telecommunications, navigation, optical and radar observation, meteorology and scientific applications. Alcatel Space is now a wholly owned subsidiary of Alcatel.

R&D

Thales Group R&D Expenditure
FY 2000: €1.8 billion
FY 1999: €1.6 billion

Figure 6.59 Thales Group Five Year Sales (€ million)

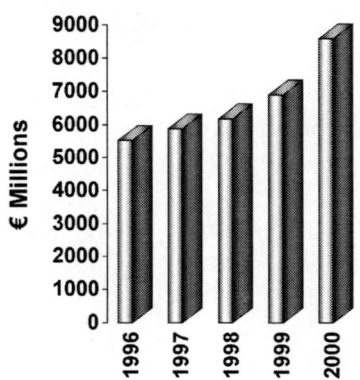

Of the €1.8 billion which the Thales Group spent on research and development in the year ended December 2000, €400 million was accounted for as company funded R&D. Current optoelectronic R&D includes research into solid-state lasers, non-linear optics and laser diode pumping technologies for next-generation equipment.

The Thales Group holds 15 000 patents, with 1800 new patents being awarded in FY 2000.

Company innovations include:

- In 1987 it developed power laser diodes.
- In 1982 its first laser diode module was commercialized.
- In 1985 it proposed the first 1300 nm diode module with full MOCVD technology.

Slab laser oscillators and amplifiers pumped by either flash lamps or laser diodes are under development. Performance levels up to 1 J at 20 Hz and 20 W at 20 kHz have been demonstrated. Wavelength shifting to the eye-safe band and to the mid-IR is achieved using optical parametric oscillators and Raman shifting. The experimental activity is supported by thermal and mechanical modelling of components and systems.

Facilities

Epitaxy, semiconductor processing, packaging, electronic design, couplings (for fibres and lenses), characterization, burn-in and selection are performed at Thales' 30 000 sq ft manufacturing facility in Orsay, France.

The company has a compact and simplified stack assembly that has been defined to produce less expensive products. Simple association with a lens duct leads to a 1 kW diode source (with an optical power density of 35 kW/cm^2 that can be easily matched to give a highly efficient end-pumping laser).

Figure 6.60 Thales Group FY 2000 Geographic Sales (%)

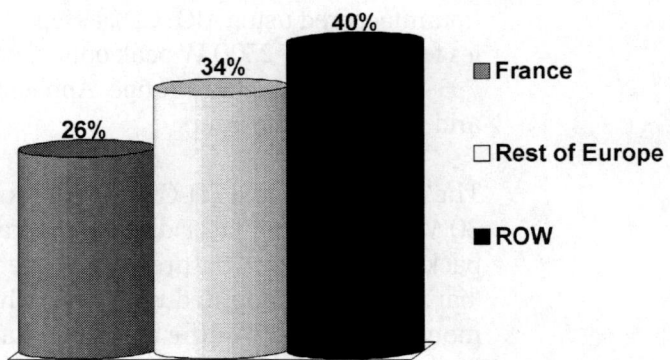

Products

Thales' major product ranges include single-mode laser diode modules for gas sensors, laser diodes for instrumentation and power laser diodes for CW or quasi-CW applications which are based on the 1 cm bar array.

Thales manufactures power laser diodes in the 795–980 nm wavelength range for solid-state laser pumping and direct use of high power. Its single diode bar or stacked array standard products are conductively or actively cooled packaged, although bare bars can also be supplied for custom applications.

The power laser diodes range includes:

- 20–30 W CW linear bar arrays with both conductively cooled and actively cooled packages.
- Fibre-coupled.
- Collimated diode bars
- Conductively cooled packaged 60–100 W QCW diode bars.
- 360–600 W QCW and 600 W to 1 kW QCW conductively cooled stacked arrays.
- 300 W to 2 kW QCW and 500 W to 2.5 kW QCW actively cooled stacked array.
- High-temperature stacked array.
- 1 kW QCW high brightness stacked array.

The TH-Q1324-D is an actively cooled LD stack specifically designed to operate at very high temperature, for applications such as pumping rods or slab solid-state lasers. Manufactured using AlGaAs QW by MOCVD, it features 1300 W peak optical power, QCW operation (200 μs/100 Hz) and up to 260 mJ energy per pulse.

The TH-Q1X01-A is a high optical power laser diode source for quasi-CW operation. These products are based upon MOCVD quantum well design to realize highly efficient 1 cm linear bar arrays manufactured using GaAlAs.

The TH-Q14xx-C series are AlGaAs QW high optical power laser diode stacks (manufactured using MOVCD), assembled on a liquid cooled heat sink. This series features 500–2500 W peak optical power, up to 500 mJ energy per pulse and 795–860nm wavelength range. Applications include pumping rod, illuminators and slab solid-state lasers.

The TH-C5520-S and TH-C5530-S products are a high performing 20 W CW and 30 W CW, 940 nm laser diode bar arrays assembled on a conductively cooled package. The LD structure is multiple emitters spaced on a monolithic 1 cm 'bar'. The bar is mounted with the active zone towards an actively cooled sub-mount. The quality of the epitaxial quantum well structure and of the process leads to high electrical to optical conversion efficiency and reliability. The compact and rugged package allows easy connection to an efficient heat exchanger. Therefore, the sub-mount temperature can be adjusted to tune the emission wavelength. With this open package the optical beam can be directly used to achieve efficient coupling.

Thales also manufactures a laser diode spectrometer system which uses one laser diode driving a power divider. The output signal of the detector goes to the control unit, which in turn optimizes the LD current for fine-tuning of the wavelength output. This systems benefits from high accuracy, fast response time, minimum maintenance and high selectivity.

Table 6.33 Thales Financial Highlights (€ million)

	2000	1999	1998	1997	1996
Net sales	8580	6890	6175	5874	5529
Net income	201	275	232	323	114
Capital expenditure	280	285	220	243	248
R&D	1800	1600	1370	1190	1190

6.42 Thermo VG Semicon

*ThermoVG Semicon
The Birches Industrial
Estate
Imberhorne Lane
East Grinstead
Sussex RH19 1XZ
UK*

Founded in 1984, Thermo VG Semicon is a division of Thermo Instruments Corp, which is part of the US$2.3 billion Thermo Electron Group. Thermo VG Semicon is a leading manufacturer of MBE equipment and holds ISO 9001 quality certification.

The V100 and V150 MBE systems are the 'industry standard' production units for the GaAs epilayer used in the fabrication of GaAs devices.

- In November 2000 the company received an order from Agilent Technologies for three V150 MBE systems for Agilent's 6-in wafer fabrication line in its Fort Collins, CO, USA, facility.
- IQE Inc of Bethlehem, PA, USA, announced in February 2000 that they would purchase up to five VG150 MBE systems to add to its existing Thermo VG Semicon machines.
- In May 2000 Thermo VG Semicon announced it had received an order in excess of US$10 million from RMFD for several VG production MBE systems.

Figure 6.61 Thermo Electron Two Year Sales (US$ million)

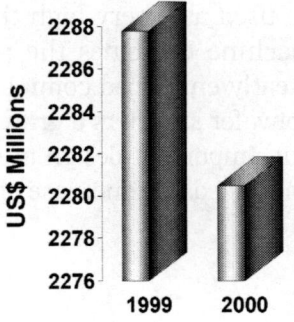

Products

In 1984, the company launched the V80H, the first MBE product specifically designed for MBE on 3-in wafers. The V80H is for pre-production and device development and has been installed at over 150 sites worldwide.

- Thermo VG Semicon is the only MBE company to offer a re-circulating hot air bakeout system. This ensures highly uniform stress-free chamber heating and enhances system reliability.

The company also produces the VG80 system, which has been installed in several universities for nitride research.

The V80H has been continuously improved since its introduction; results are evidenced in the ultrahigh-performance designs for effusion cells, including the revolutionary ThermoCell, shutters and manipulators and a positive and sure sample transfer mechanism. A fully optimized deposition configuration ensures highly uniform thin films on wafers up to 3-in in diameter. The V80H achieves world-record growth uniformity and reproducibility while still maintaining efficient usage of source materials. This reflects the optimum source geometry used in the deposition chamber.

The V90 was the world's first MBE system configured for 4-in GaAs wafers. The V90 is ideal for MBE research and pilot production of electronic devices. It is also well suited to full-scale, high-volume production of laser diodes and other optoelectronic device structures. The system can be operated manually or is available in a fully automated version.

In the late 1980s VG Semicon developed the world's first 4- and 6-in MBE systems, which was followed by development of its multiwafer production MBE system, the VG100 (for high-volume production, which was introduced in 1989).

- Sharp uses its V100 system for phosphide-based lasers and DVD pick-ups.
- Tutcore (now part of Coherent) employs its Thermo VG Semicon equipment to produce aluminium-free lasers.

The company's latest model, the V150 MBE system for GaAs, represents the state-of-the-art in multiwafer, high-throughput MBE systems and can handle multiple 6-in wafers with excellent uniformity and reproducibility. It can also be used as a very high throughput 4-in machine. The V150 production MBE machine combines the performance of the industry standard V100 with a greatly enhanced commercial yield. The V150 is designed to offer commercial epiwafer suppliers a growth system with the greatest available wafer throughput. Important design areas that have been addressed in development include enhanced up-time, ease of operation and ease of maintenance.

Table 6.34 Thermo Electron Financial Highlights (US$ million)

	2000	1999
Net sales	2280.5	2287.8
Net income	(36.1)	(174.6)
R&D	176.8	171.1

6.43 Toshiba

Toshiba Corp
1–1 Shibaura 1-chome
Minato-ku
Tokyo 105-01
Japan

Toshiba Corp was founded in 1875. It is a leading consolidated electronics company, which has both merchant and captive optoelectronic sales. Vertically integrated, Toshiba's product portfolio encompasses a wide range of electronic devices.

● Toshiba recorded total sales for the year ended March 2001 of ¥5951.4 billion, which represented an increase of 3% on FY 2000's figure of ¥5749.4 billion.

Subsidiary company Toshiba America Electronic Components Inc (TAEC) is based in Irvine, CA, USA. TAEC's product line includes laser diodes and optical transmission devices, solid-state devices, colour picture tubes, colour display tubes, liquid crystal displays, rechargeable batteries and microwave components, as well as one of the broadest IC product lines in the industry.

R&D

R&D Expenditure
FY 2001: ¥350.2 billion
FY 2000: ¥334.4 billion

Toshiba's R&D laboratories include the Information and Communications Engineering Laboratory, the Multimedia Engineering Laboratory, the Microelectronics Technology Laboratory and the Advanced Microelectronics Centre.

Figure 6.62 Toshiba Five Year Sales (¥ billion)

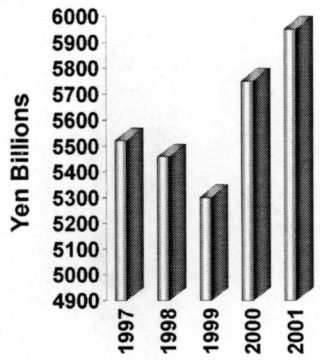

Toshiba has conducted research into:

- Room temperature pulsed operation of nitride-based MQW laser diodes with cleaved facets on conventional C-face sapphire substrates.
- Reactive ion beam etching and overgrowth process in the fabrication of InGaN inner stripe laser diodes.
- Cathodoluminescence and electron beam-induced current of two-dimensional junction laser structures on patterned (311)A GaAs substrates.

Toshiba pioneered the development of the DVD format: it announced the industry's first chipset for DVD video players in 1996.

In May 2001 Toshiba announced a prototype of a polymer organic electroluminescent (OLED) display that supports 260 000 colours.

Facilities

Toshiba has manufacturing facilities in Japan at Himeji, Tamagawa, Kitakyushu and Oita, with overseas manufacturing in Selangor, Malaysia; Pathumthani, Thailand; Braunschweig, Germany; and the USA.

Products

The Electronic Devices segment made a recovery during FY 2001 to record sales of ¥139.9 billion (US$938 million, or 22% of the company's total sales) as opposed to FY 2000 sales of ¥23.5 billion (US$190 million).

Toshiba's products include compact and lightweight visible laser diodes (VLD) featuring the shortest wavelength (635 nm) and highest power (30 mW) available and include a self-pulsating VLD developed for the DVD recording format. It also produces LDs for communication applications.

Examples of Toshiba's laser diodes include:

- The TOLD9442m MQW InGaAlP laser diode for applications such as a light source for bar code readers. It features a lasing wavelength of 650 nm and optical power output of 5 mW (7 mW CW).

Figure 6.63 Toshiba FY 2001 Geographic Sales (%)

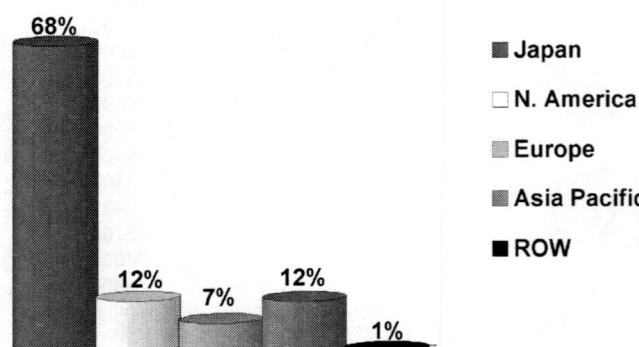

- TOLD9231m is a general purpose InGaAlP gain guided MQW laser diode. It features a lasing wavelength of 670 nm, optical power outpit of 5 mW and LD reverse voltage of 2 V.
- The TOLD9441mc is an index guided MQW InGaAlP laser diode for higher operating temperature applications. It features a lasing wavelength of 650 nm and 7 mW (CW) power output.
- TOLD9221m is a general purpose, index guided, MQW InGaAlP LD featuring 670 nm wavelength and 5 mW power output.

Alliances

The company is a member of IrDA (Infrared Data Association), which promotes interoperable, low-cost, IR data interconnection standards.

Table 6.35 Toshiba Financial Highlights (¥ billion)

	2001	2000	1999	1998	1997
Net sales	5951.4	5749.4	5300.9	5458.5	5521.9
Net income	96.2	(28.0)	(13.9)	7.3	67.1
Working capital	276.9	268.9	318.4	191.5	127.7
R&D	350.2	334.4	316.7	322.9	332.6
Total assets	5724.6	5702.2	6023.6	6062.1	5809.3

6.44 Toyoda Gosei

Toyoda Gosei Co Ltd
1 Nagahata
Ochiai
Haruhi
Nishikasugai
Aichi 452-8564
Japan

Toyoda Gosei was founded in 1949 and is 40% owned by Toyota Motors. Although the majority of its optoelectronics products are LEDs, the company is carrying out research into semiconductor lasers. Toyoda Gosei employs more than 5600 people.

- For the year ended March 1999, it reported net sales of ¥251.9 billion (US$2.1 billion), which was a 22% increase over FY 1998's figure of ¥197.7 billion.

R&D

Research and development is conducted at the Toyoda Gosei Technical Center in Inazawa, Aichi, Japan.

In April 2001 Toyoda Gosei, in partnership with Meijo University and the Japan Science and Technology Corp, announced the successful development of manufacturing technology for a bluish-purple laser diode with a short wavelength using gallium nitride. Using the blue-to-violet shorter end of the visible light spectrum, the laser increases the surface recording density of optical discs, unlike current disc readings, which use the longer, red end of the visible light spectrum. The laser has a 410 nm wavelength and an output of 3 mW. Applications for this technology will include next-generation DVDs, high-speed laser printers, full-colour projector-type laser displays and for the fine machining of

Figure 6.64 Toyoda Gosei Four Year Sales (¥ billion)

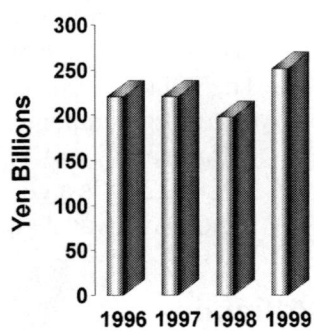

semiconductor substrates. The research took seven years at a cost of approximately ¥700 million.

- Toyoda Gosei has filed approximately 50 patent applications in connection with this research.

Toyoda Gosei has collaborated with Nagoya University and the JRDC (and separately, with Toyota) on research pertaining to GaN LEDs. The company has many patents in the area of GaN research and has published papers including the following:

- A method for manufacturing a GaN group compound semiconductor, comprising the steps of growing a first gallium nitride group compound semiconductor on a substrate.
- A light-emitting semiconductor device using gallium nitride group compound. A semiconductor device having an n-type layer of gallium nitride that is doped with silicon and has a resistively ranging from $3 \times 10^{-1}\,\Omega\,cm$ to $8 \times 10^{-3}\,\Omega\,cm$ or a carrier concentration ranging from $6 \times 10^{16}/cm^3$ to $3 \times 10^{18}/cm^3$.

Facilities

Toyoda Gosei has facilities in Japan at Nagoya, Haruhi, Morimachi, Heiwacho, Inazawa, Bisai and Nishimizoguchi with an overseas sales office in Düsseldorf, Germany.

Products

As mentioned previously, Toyoda Gosei manufactures LEDs. Its laser research has not yet been developed commercially.

Legal Disputes

Toyoda Gosei and Nichia Corp have filed counter-suits at the Tokyo District Court for patent infringements pertaining to GaN-based blue LEDs.

Table 6.36 Toyoda Gosei Financial Highlights (¥ billion)

	1999	1998	1997	1996
Net sales	251.9	197.7	220.5	220.4
Net income	4.2	2.0	2.2	2.2
Working capital	24.4	17.4	19.6	13.5
Total assets	193.7	172.5	188.0	157.0

6.45 Xerox

*Palo Alto Research Center
(PARC)
3333 Coyote Hill Road
Palo Alto
CA 94304
USA*

PARC was founded in 1970 and is one of several R&D centres which are part of the US$18.6 billion Xerox Corp.

R&D

Xerox R&D Expenditure
FY 2000: US$1.04 billion
FY 1999: US$979 million

PARC's Electronic Materials Laboratory has a long history of developing semi-conductor laser technology culminating in devices used in multiple Xerox products. In addition to expertise in near-IR and red laser technology, PARC has an active research programme that has demonstrated CW lasers that emit in the blue and near UV.

In the early 1990s PARC invented a way to fabricate lasers that emitted multiple beams from a single chip.

PARC's imager-on-a-chip technology is a way of integrating the laser and optical systems of a laser printer onto a single chip. The ultimate goal of this project is to build a full-width imager bar on a single chip, creating a scalable component for high-resolution xerographic printing.

Figure 6.65 Xerox Five Year Sales (US$ billion)

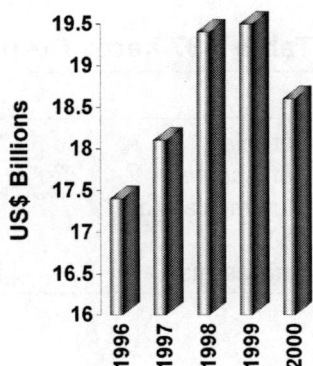

One part of the technology is a new kind of laser, one in which a thousand separate lasers are able to operate in co-ordination. The imager-on-a-chip pairs these lasers with mirrors based on MEMS. The combination of lasers and MEMS technology, not much bigger than the tip of a finger, is intended to replace the much bulkier optical systems of current printers.

Other PARC innovations include:

- In 1997 PARC demonstrated Xerox's first blue laser. Early in 1998 it became only the second organization in the world to demonstrate consistent operation of a blue laser at a continuous wavelength. (The company was a member of a consortium that in 1995 was awarded an US$4 million, two-year contract from DARPA to develop blue semiconductor lasers and LEDs.) While it may still be several years before blue lasers appear in Xerox products, the research is laying the groundwork for laser printing that approaches offset quality.
- Solid-state laser diodes optimized for printing. This kind of laser diode replaced earlier gas lasers, which were many times larger, consumed more power and were very difficult to modulate. These lasers have been at the heart of Xerox laser printers for many years now.
- The world's first multi-beam lasers. In the early 1990s PARC invented a way to fabricate lasers that emitted multiple beams from a single chip.
- In 1988 researchers demonstrated the world's first electrically injected nitride-based DFB laser — room temperature pulsed operation of an InGaN/GaN-based DFB laser with an emission wavelength of 403 nm. The DFB laser was grown by MOCVD on sapphire and consisted of an n-type AlGaN cladding layer, with GaN waveguide layers surrounding the active region of five InGaN QWs with GaN barriers. This was the first demonstration of a semiconductor-based DFB laser diode in that spectral region and in that material system.
- Xerox has a patent for its 390–430 nm VCSEL blue laser. The GaN-based laser structure is grown by selective area epitaxy and lateral mask overgrowth. By patterning of a dielectric mask on the GaN layer on a sapphire substrate, areas in a second GaN layer can have a low defect density upon which the remainder of the laser structure an be formed.

The company has a history of research in laser diodes: Spectra Diode Laboratories (SDLI, now part of JDS Uniphase) was originally a PARC spin-off.

Table 6.37 Xerox Corp Financial Highlights (US$ million)

	2000	1999	1998	1997	1996
Net sales	18632	19228	19447	18144	17378
Net income	(384)	1424	395	1452	1206
Working capital	6754	4035	3968	3074	2948
R&D	1045	979	1040	1065	1044
Total assets	29475	28814	30024	27732	26818

6.46 Zarlink Semiconductor

Zarlink Semiconductor
400 March Road
Ottawa
Canada K2K 3H4

Zarlink Semiconductor is the new name for Mitel Semiconductor, which changed its name to Zarlink in May 2001. Zarlink is a leader in optical communications device technology and employs 6000 people. The company is a wholly owned subsidiary of Mitel Corp.

- Mitel Corp announced FY 2001 net sales of C$674 million, as opposed to FY 2000's figure of C$602.8 million.
- During the first half of 2001, the company announced worldwide redundancies, with the loss of 430 jobs.

Acquisitions

Mitel has, over the years, added to its semiconductor division by acquisition:

- In 1996 it acquired Swedish company ABB Hafo AB.
- In 1998 it acquired GEC-Plessey Semiconductors.
- In 2000 it acquired fabless semiconductor company Vertex Networks Inc, whose area of expertise is WAN access markets.
- Mitel Corp spun out Optenia (formed to commercialize photonics conducted jointly by Mitel and the National Research Council of Canada). Its first products were for DWDM applications.

R&D

R&D Expenditure
FY 2001: C$125.2 million
FY 2000: C$94.4 million

Facilities

Zarlink has manufacturing facilities in Canada, the UK and Sweden. Zarlink holds ISO 9000 and ISO 14001 certification.

Figure 6.66 Zarlink Five Year Sales (C$ million)

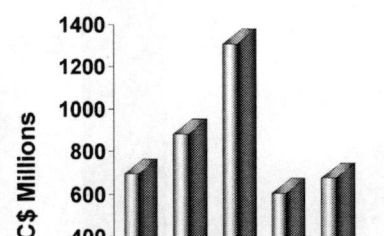

Optoelectronic products are manufactured at the Jarfalla, Sweden, facility which comprises a 400 sq m cleanroom with in-house liquid-phase epitaxial crystal growth and III-V process capabilities for VCSEL, LED and PIN diode wafer manufacturing. It also has a 600 sq m cleanroom for TO-46 diode packaging and optical receptacle assemblies.

Products

Zarlink's optoelectronic products accounted for 93% of its net sales in FY 2001. Zarlink is a high-volume partner to leading companies in the optical storage industry who use its integrated analogue and mixed-signal functions in their DVD drives and players. Zarlink's product line includes VCSELs, LEDs, PIN photodetectors and duplex devices.

In March 2000 the company introduced its 623-family of parallel fibre modules for short-reach optical interconnect in large-scale computing and communication systems. The 623-family of parallel fibre modules provide short- and medium-reach interconnects in and between network gear in the central office. The modules combine Zarlink's VCSEL and Smart OSA packaging technologies.

- The 12L485 high-speed, 840 nm, 12-channel VCSEL arrays and the 12L486 short-wavelength PIN photodiode arrays are designed for use in high-speed, multi-channel applications such as parallel Fibre Channel and parallel interconnect applications.
- The 1A440 is a 840 nm, 2 GHz VCSEL for general purpose data communication applications. It operates in multiple transverse and single longitudinal mode, ensuring stable coupling of power and low noise.
- The 2B481 and 2B482 VCSELs for high-speed data communication applications feature low power consumption, efficient coupling and low threshold current.
- The MF440 VCSEL is designed for Fibre Channel, Gigabit Ethernet, ATM and general applications. It operates in multiple transverse and single longitudinal mode, ensuring stable coupling of power and low noise. The MF440 features peak wavelength of 840 nm, 1.7 mW optical power and 6 mA threshold current.

Figure 6.67 Zarlink FY 2001 Geographic Sales (%)

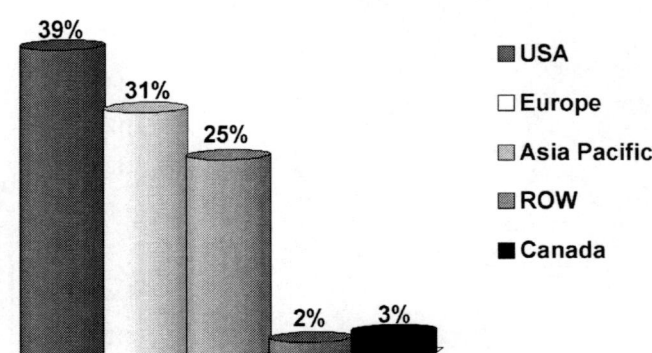

Alliances

AMP and Zarlink Semiconductor have an agreement (dating back to 1999) whereby AMP LIGHTRAY MPX products are incorporated into Zarlink's Smart OSA parallel fibre modules. The design combines Zarlink's VCSEL array and patented packaging architecture with AMP optical interconnection systems to deliver highly advanced performance to customized high-density terabit/petabit class networking, large-scale computing and communications systems.

Table 6.38 Zarlink Financial Highlights (C$ million)

	2001	2000	1999	1998	1997
Net sales	674	603	1310	881	696
Net income	(317)	56	26	92	38
Working capital	368	390	337	246	206
R&D	125	94	150	52	50
Total assets	836	1226	1300	1252	585

7 Geographical List of Universities and Selected Industrial Labs

7.1 Introduction

This section comprises an alphabetical listing of research centres worldwide active in the area of diode lasers based on wide bandgap compound semi-conductors. It includes not only academic research centres such as universities, but also industry players and government centres.

The principal focus is on diode lasers, but in certain instances this coverage extends to the related semiconductor materials and devices and includes reference to such other areas as have a bearing on the principal content of this report.

- Coverage includes not only semiconductor materials and applications but also such non-diode laser materials and systems.

Geographic coverage is as complete as a systematic search of published literature would allow. It would seem from this research that over twenty countries are sponsoring this R&D in some form or another.

The principal concentration of research is, of course, to be found in the regions with the longest history in semiconductor research, ie North America, Europe and Japan. The interest in research into laser materials and devices has never been higher.

- In the nitride area alone there are over 2000 workers who have published original or collaborative results over the past five years.

There has, however, been a considerable shift in emphasis with less interest being seen in the pursuit of non-diode lasers. This is in part due to the maturity and lower market need for these systems, and an overall shift towards diode lasers for all applications.

It should be noted that the nature of competition in industrial R&D usually precludes researchers from presenting or publishing the results of their work when it approaches commercial status. This is certainly the case in some areas of semiconductor R&D such as lasers, but owing to the enormous commercial potential for these devices in data storage and so on. Because this listing is based on published literature, it may necessarily fall short of being completely exhaustive in its coverage.

One thing is certain and that is R&D into lasers from academic and industrial parties will continue with even more vigour over the next five years. The field has attracted high calibre scientists and engineers which are likely to continue to astound the world with innovative new inventions. Much will, of course, depend on continuation of funding for all aspects of research from theoretical modelling through crystal growth to optics and processing.

7.2 Australia

Australian National University

Department of Electronic Materials Engineering
RS Phys SE
The ANU
Canberra
ACT 0200
Australia
Tel: +61 2 6249 0020
Fax: +61 2 6249 0511
E-mail: eme@rsphysee.anu.edu.au
Professor C Jagadish and his team are conducting research on understanding damage accummulation during ion implantation in GaN and application of ion implantation for p-type doping and device isolation. Also reactive ion etching and ohmic contacts on GaN. Efforts are underway to develop processing of GaN for blue light emitting diodes and blue lasers.

Macquarie University

Semiconductor Science and Technology Laboratories (SSTL)
School of Mathematics, Physics, Computing and Electronics
NSW 2109
Australia
Tel: +61 2 850 8912
Professor Trevor Tansley and his team conduct research into: photoluminescence and Raman characterization of GaN epilayers; fast UV detectors using short lifetime GaN layers grown by LT-MOVPE; electron transport in nitride semiconductors; piezoelectric properties of GaN; work on microwave-plasma and excimer laser stimulation in low temperature MOVPE of Group III nitrides; conventional MOVPE of narrow-gap III-V antimonide heterostructures; monochrome and bicolour devices in AlGaAs and double-barrier QWIPs.

University of Melbourne

School of Physics
Parkville 3052
Australia
Tel: +61 3 9344 7670
Fax: +61 3 9347 4783
Researching into nitride semiconductor film growth by pulsed laser deposition, and Raman spectroscopy.

University of New South Wales

Sydney 2052
Australia
Tel: 61 2 9385 1000
Professor Michael Gal's team are researching the optical characterization of GaN epitaxial layers.

University of Sydney

Optical Fibre Technology Centre
University of Sydney
Sydney
Australia
This group is concerned with the development of blue lasers for improved data storage, surgical techniques and other applications. The project is supported by IBM and the Federal Government of Australia. The University is also conducting research into electron microscopy characterization of defects in epitaxial and ion implanted GaN.

University of Technology, Sydney

PO Box 123
Broadway
Sydney
NSW 2007
Australia
Researching cathodoluminescence characterization of GaN epilayers, single crystals, ion implanted GaN layers. Also application of environmental scanning electronic microscopy for the defection of defects.

University of Western Australia

Perth
Australia
The team led by Professor Laurie Faraone are researching GaN based UV detectors, design, fabrication and testing; modelling of UV detectors based on GaN/AlGaN system for solar blind UV detectors; GaN based microcavities.

7.3 Belgium

Alcatel Microelectronics

Westerring 15
B-9700 Oudenaarde
Belgium
Large manufacturer of diode laser and related optoelectronics components and systems for telecoms. Studying SOI voltage reference sources in project with University Catholique de Louvain.

IMEC v.z.w.

Kapeldreef 75
B-3001 Leuven
Belgium
Tel: +32 16 281211
IMEC is an inter-university microelectronics centre, founded in 1984 by the Flanders Government in Belgium. With over 3.6 m^2 cleanroom, it is the largest of its kind in Europe. III-nitride work is concentrated on GaN on sapphire for blue LEDs and lasers based on MOVPE, using a Thomas Swan vertical rotating disk growth system. IMEC collaborates with other institutions including: Trinity College Dublin, the Technical University of Delft and the University of Ghent. At the associated University of Antwerp the UIA group is working on organic light emitting semiconductors.

7.4 Canada

McMaster University

Centre for Electrophotonic Materials and Devices
McMaster University
Hamilton
Canada L8S 4M1
This university is conducting research into compositional variations in InGaAsP films grown on patterned substrates. The InGaAsP family of semiconductors is widely used in optoelectronic devices because a range of bandgap wavelengths (1.1–1.7 μm) can be attained by varying the composition. The two

most commonly used wavelengths (1.3 and 1.5 µm), corresponding to minimum loss and minimum dispersion for glass fibres, lie within this range. The growth of III-V compounds on patterned substrates for device applications, (in particular the lattice-matched AlGaAs/GaAs system), is now commonplace, but the success of the process more often depends on the skill of the crystal grower rather than on a fundamental understanding of the many complex factors that influence the growth process. One of the advantages of using a patterned substrate is that two degrees of carrier confinement can be achieved in a single growth step, while interfacial defects or contamination can be minimized if an in situ growth process such as MBE is used.

Université de Montréal

Laboratory for the Integration of Sensors and Actuators
Montréal
Québec
Canada
LISA is a laboratory of the Engineering physics Departement of École Polytechnique of Montréal, which is an affiliated school of Université de Montréal in Montréal, Québec, Canada. research themes include high-speed microelectronics, gas sensors, physical sensors, micro-electro-mechanical systems and materials science. The work of the group includes fundamental physics as well as the fabrication and testing of porous silicon devices including solar cells and UV detectors.

High performance devices such as lasers, photodetectors (discrete and array), photo-diodes, photocathodes etc, have to be fabricated using a variety of epitaxial techniques: VPE, LPE, MOVPE, and MBE. The Solid Phase Epitaxy (SPE) technique is a FSF patented technology for the epitaxial growth of semiconductor crystals using chemical vapour deposition technique.

University of Victoria

Center of Advanced Materials and Related Technology and Department of Mechanical Engineering
Victoria BC
Canada V8W 3P6
This group is working on the growth of alloy GaInP crystals by compositional conversion of InP layers grown on GaP substrates in an LPE system this is a collaboration with Shizuoka University in Japan.

7.5 PR China

Chinese Academy of Sciences

Laboratory of Semiconductor Materials Science
Institute of Semiconductors
PO Box 912
Beijing 100083
PR China
Fax: +86 10 62322388

The researchers at this laboratory are interested in AlN, GaN, InN, and their alloys, AlGaN, GaInN, and AlGaInN as promising materials for fabricating blue-green light optoelectronic devices, especially LEDs, laser diodes and detectors.

The Institute of Semiconductors, Chinese Academy of Sciences, has made progress in diode technology after reducing the background electron concentration and the achievement of undoped and p- (Zn) doped GaN producing GaN blue-green LEDs with a metal-insulating GaN-n type GaN (m-i-n) structure with a forward voltage of under 5 V.

It has also published work on the growth of an AlGaN/GaN modulation-doped heterostructure by ammonia source MBE and has achieved high growth rate GaN deposition via GSMBE.

The Laboratory of Semiconductor Materials Science has collaborated with the Laboratory of Optical Electronics (also part of the Chinese Academy of Sciences), working on MOVPE growth of ZnTe and ZnSeTe for blue-green LEDs and lasers.

Also, at the Beijing Laboratory of Electron Microscopy, Center for Condensed Matter Physics at the Chinese Academy of Sciences, research into the micro-structures of GaN films grown by LP-MOVPE on (0112) sapphire substrates has been done in collaboration with the Laboratory of Mesoscopic Physics, Department of Physics, Peking University. Also using LP-MOVPE, the gallium diffusion through cubic GaN films grown on GaAs (100) substrates at high-temperature has been described by workers at the National Research Center for Optoelectronic Technology, Institute of Semiconductors, of the Chinese Academy of Sciences.

The Institute of Semiconductors also published work on GaN epilayers on sapphire substrates grown by GSMBE using NH_3 gas as the nitrogen source. Properties of gallium nitride epilayers grown under various growth conditions were investigated. The growth rate is up to 0.6 gm/h. Cathodoluminescence, photoluminescence and Hall measurements were used to characterize the films.

Peking University

Department of Physics and Mesoscopic Physics Lab
Peking University
Beijing 100871
PR China
Researchers at this establishment have published work into low temperature photoluminescence properties of InGaN films grown on (0112) Al_2O_3 and (0001) Al_2O_3 substrates by low pressure MOVPE.

Shanghai Institute of Metallurgy

State Key Laboratory of Functional Materials for Informatics
Chinese Academy of Sciences
Shanghai 200050
PR China
Growth and mosaic model of GaN grown directly on 6H—SiC by DC-PAMBE in collaboration with the Drude Institute for Solid State Electronics, Berlin, Germany.

Tsinghua University

School of Materials Science & Engineering
Beijing 100084
PR China
This school has published work on the diffusion of Al through the GaN buffer layer during LP-MOVPE growth.

7.6 Finland

Helsinki University of Technology

PO Box 1000
Fin 02015 HUT
Finland
Tel: +358 94 511
This university is a subcontractor to the University of Paderborn in the pan-European LAQUANI project on wide bandgap materials for solid state semiconductor blue-green lasers.

Tampere University of Technology (TUT)

PO Box 692
Fin 33101
Tampere
Finland
Tel: +358 3 365 3477
In 1995, this university produced a blue-green laser diode which lasted for one whole week, becoming the first European, ZnSe-based, blue-green laser that

operated at room temperature for more than a few minutes. TUT's laser was the world's first RT blue-green laser which had an inverted internal structure.

The university collaborates with Trinity College Dublin and University of Ghent-IMEC on II-VI research.

Tutcore Ltd, a spin-off company for the university was founded in 1991. It specializes in epitaxial growth and processing of optoelectronics devices such as laser diodes, photodetectors and solar cells. It also manufactures laser diode testing systems and produced the world's first GaInAsP production MBE reactor. Tutcore was acquired by Coherent in 1996. Another spinn-off is Modulight, which was acquired by Picogiga in 2000.

7.7 France

CEA-Grenoble

DRFMC/SP2M
17 rue des Martyrs
38054 Grenoble Cedex 9
France

The CEA-Grenoble research group is involved in the MOVPE growth of c-GaN and AlN on silicon and sapphire and has published work on the surface morphologies and crystallographic structures of GaN layers. For example, the thermal treatment under propane at 1300–1400°C — to produce a very thin cubic SiC layer — has been used to prepare silicon (001) wafers for growth of cubic GaN and AlN by ECR plasma- assisted MBE (ECRMBE).

CRHEA-CNRS

Centre de Recherche sur l'Heteroepitaxie et ses Applications (CRHEA-CNRS)
Parc de Sophia Antipolis
Rue Bernard Gregory
Sophia Antipolis
06560 Valbonne
France

The CNRS group has published many papers on MOVPE-grown GaN: yellow band and deep levels in undoped MOVPE GaN, and alternative N precursors and Mg doped GaN. The Materials Research Laboratory is also involved in work on the growth of prismatic domains in cyclotron assisted MBE of GaN/SiC (in collaboration with the University of Illinois).

CNRS is the co-ordinating group for the pan-European LAQUANI project on wide bandgap materials for solid state semiconductor blue-green lasers. Results from this collaboration include improvement of the ELOG process. The approach omits the usual mask layer preparation and uses self-organized islands as the starting point for the GaN epilayer. The CNRS project is promising since it has demonstrated the possibility of selective lateral epitaxy without the need for the ELOG mask. Most importantly, the CNRS approach is entirely in situ in a single growth chamber. This is much more convenient and promises significant savings in time and cost. Moreover, it is possible to repeat lateral growth several times on the same substrate.

CNRS is also a leading participant in the EU ESPRIT Project on III-V Nitrides — program co-ordinator is Prof Pierre Gibart). Work includes studies of the impact of GaN buffer growth conditions on photoluminescence and background carrier concentration of MOVPE-grown bulk GaN.

In the chalcogenides, the group has reported ZnSe pseudomorphic layers grown by MBE on various GaAs buffer layers. Using a dual-chamber Riber Epineat system, migration enhanced epitaxy (MEE) at low temperature was performed and resulted in the first steps of the growth decreasing the defect density by two orders of magnitude.

Also, the CNRS Laboratoire de Physique du Solide et Energy Solaire collaborated with Riber SA on the development of the EPINEAT III-V reactor.

Other studies structural and optical characterization of InGaN layers grown by MBE using ammonia as the nitrogen precursor. MBE-grown LEDs based on InGaN/GaN QWs have been fabricated showing electroluminescence peaks at 440 nm at 300K.

Groupe d'Etude des Semiconducteurs

GES-CNRS
CC074 Université Montpellier II
Place E. Bataillon
34095 Montpellier Cedex 5
France
Researchers are involved in work on Raman determination of the phonon deformation potentials in GaN as well the growth of MOVPE GaN layers on sapphire and the correlation between surface morphologies and crystallographic structures. Other research areas include strain effects on hexagonal GaN epilayers grown on sapphire, ZnO and SiC.

The Semiconductor Group's GaN team, headed by Dr Olivier Briot, has an Aixtron AIX 200/4RF-S MOVPE reactor for the growth of nitrides, in particular AlGaN based structures for detectors and electronic applications.

Institut Laue-Langerin

BP 156X
F-38042 Grenoble Cedex 9
France
This institute is investigating CdTe Deposition and growth with desorption in MBE in a collaboration with CNRS Université Joseph Fourier.

Laboratoire de Physique et Modélisation des Milieux Condensés

CNRS Université Joseph Fourier
BP 166
F-38042 Grenoble Cedex 9
France
This institute is investigating CdTe Deposition and growth with desorption in MBE in a collaboration with Institut Laue-Langerin.

LETI CEA Technologies Advancees

DMITEC-CENG
17 rue des Martyrs
38054 Grenoble Cedex 9
France
Work studies in wide bandgap semiconductors includes structural studies of homoepitaxial grown 4H- and 6H-SiC epilayers grown on off-oriented SiC substrates by TEM and AFM.

LPM (UMR CNRS 5511)

INSA de Lyon
Bat 502
20 avenue A Einstein
69121 Villeurbanne Cedex
France
Research includes SiC pn junction diodes and other electronic devices.

LTPCM - ENSEEG

Domaine Universitaire
BP 75
Saint Martin D'Heres 38402
France
Tel: +33 4 76826532
Fax: +33 4 76826677
SiC bulk crystal growth and sublimation process modelling.

Université Pierre et Marie Curie (Paris VI)

BP 169
4 place Jussieu
F-75252 Paris
Cedex 05
France
Tel: +33 1 44 27 47 21
This group is involved in research into GaN and InN. Recently published work includes: optical absorption, transmission, luminescence and Raman spectroscopy techniques applied to n-type bulk GaN samples and MBE grown epitaxial films (in collaboration with the High Pressure Research Center of the Polish Academy of Sciences, North Carolina State University, and Boston University),

and bandgap variation at the isostructural phase transformation of wurtzite InN (with the US National Renewable Energy Lab and CNRS).

Université Paul Sabatier
Laboratoire de Physique des Solides de Toulouse

Toulouse
France
This group collaborates with GES-CNRS to produce work on Raman determination of the phonon deformation potentials in [alpha]-GaN. Raman spectroscopy is used to study the effect of the built-in biaxial stress in wurtzite GaN layers deposited by MOVPE on (0001) sapphire substrate.

Thales Group

Laboratoire Central De Recherches
Domaine de Corbeville
91404 ORSAY Cedex
France
Tel: +33 1 69 33 00 00
Fax: +33 1 69 33 08 78
This industrial research lab is equipped with an Aixtron AIX 200 RF system for the growth of nitrides on 2-in wafers. It is used for research purposes within the RAINBOW project supported by the EC. This has the aim of developing the first complete European AlGaInN materials base, culminating in a mass production technology of UHB-LEDs in various colours and in the fabrication of blue lasers.

7.8 Germany

AIXTRON AG

Kackertstr. 15–17
52072 Aachen
Germany
Tel: +49 241 8909 0
AIXTRON AG was founded in 1983. It is a world leading supplier of epitaxial reactors for R&D and mass production of optoelectronic and other devices.

AIXTRON contributes more than 20% of net sales to research and development. The company's Application Laboratory in Aachen is involved in many collaborative projects with research centres worldwide. It was the first company to demonstrate MOVPE epitaxial growth on full 2-inch GaN and InGaN wafers, as well as the first to offer GaInN multiwafer technology. It has demonstrated device quality GaN with PL uniformity better than 1 nm across a 2-inch wafer, and thickness uniformity typically better than 2%.

AIXTRON is involved in several EU and government-funded R&D projects such as: the German BMBF program for high bandgap nitride semiconductors and the BMBF project "PAR-CVD" for the development of high performance parallel calculation processes for analysis and optimization of CVD processes. Several EC BRITE/EURAM Projects: RAINBOW, for Ga-In-In-N multiwafer sources; LASBE, for layers and structures for blue emitters; LAQUANI, for laser quality III-V nitrides; MIRIAD for microwave and infrared industrial applications for diamond; as well as the EC Bluemat Project for the development of II-VI light emitters.

Paul-Drude-Institut fur Festkörperelektronik

Hausvogteiplatz 5–7
D-10117 Berlin
Germany
The Drude Institute conducts studies on the growth and modelling thereof for GaN grown on substrates such as SiC. GaN layers have been grown on Si (111) substrates by MBE. Other techniques include DCPAMBE (direct current plasma assisted molecular beam epitaxy) developed in collaboration with the State Key Laboratory of Functional Materials for Informatics, Shanghai Institute of Metallurgy, Chinese Academy of Sciences.

University of Bremen

Postfach 330 440
D–28334 Bremen
Germany
In 1999 it took delivery of a 3×2-in GaN system from Thomas Swan.

Fritz-Haber-Institut der Max-Planck-Gesellschaft

Semiconductor Surfaces Group
Faradayweg 4–6
Berlin D-14195
Germany
Tel: +49 30 8413 5640
Fax: +49 30 8413 5603
This group's interests include surface and bulk band structure determination using valence level photoemission, e.g. in cubic and wurtzite GaN.

Institut für Halbleitertechnik RWTH Aachen

Templergraben 55
D-52056 Aachen
Germany
Tel: +49 241 807745
Founded in 1961 by Prof Heinz Beneking (who was director until 1989), the Institut für Halbleitertechnik, Rheinisch-Westfalische Technische Hochschule (RWTH) Aachen, is part of the pan-European Bluemat Project for wide bandgap materials. It has a staff of 13 scientists, 12 technicians and various postgraduate students is involved in a wide range of III-V and wide bandgap semiconductor research.

The university is also active in the optimization of pre-epitaxial processes to obtain high-quality ZnSe/GaAs heterointerfaces grown in a LP-MOVPE system.

This work was a collaboration with the Gerhard-Mercator Universitat in Duisburg, Germany.

Scientists from the institute became co-founders of AIXTRON AG, with whom it collaborates: RWTH Aachen developed an in-depth characterization tool for blue GaN LEDs, the device structures were grown on an AIX 200 RF at the AIXTRON Application Laboratory. Blue LED emission at 443 nm was attained. In addition to blue LEDs and lasers, the two groups also collaborate on research into AlGaInP UHB LEDs.

The institute's work on gallium nitride includes: p- and n-doping and activation mechanisms; transport mechanisms, temperature stability of ohmic and Schottky contacts, and device process technology for LED and transistor applications.

Amongst its other research activities, RWTH is currently involved in research into the growth of: ZnSe, ZnSSe and ZnMgSSe heterostructures for blue lasers; Group III antimonides for mid-infrared optoelectronic devices, magnetic sensors and for low temperature transport measurements. It is also interested in the selective growth of Group III arsenides and phosphides.

RWTH used an MOVPE process with dc-plasma activated nitrogen to dope ZnSe, grown with novel precursors such as ditertiarybutylselenide and dimethylzinc-triethylamine.

Gerhard-Mercator-Universität Duisburg

Laboratorium für Festkörperphysik
Duisberg
Germany
This group has published work on metal contacts on [alpha]-GaN. The university has collaborated with the University of Aachen in the optimization of pre-epitaxial processes to obtain high-quality ZnSe/GaAs hetero-interfaces grown by LP-MOVPE system.

Universität Erlangen-Nürnberg

Lehrstuhl für Strömungsmechanik
Cauerstr 4
D-91058 Erlangen
Germany
Researchers at this institution are concerned with a theoretical model for analysis and optimization of MBE growth of group III-nitrides, as well as MOVPE equipment for developments in blue and green solid state lasers. There have been collaborations with other institutes in Germany and also with the MOVPE equipment manufacturing company, AIXTRON, to develop blue and green diode lasers. In particular, the group has helped with the development of the reaction chamber through computer modelling for optimization of growth parameters. Techniques in use include Navier-Stokes equations in conjunction with thermodynamics of radiative heat transfer and flow dynamics and chemistry of growth reactions especially in the vicinity of the substrates and susceptor support.

The Fraunhofer Institut für Angewandte Festkörperphysik

Tullastraße 72
D-79108 Freiburg
Germany
Tel: +49 761 5159 416

The Fraunhofer Institut für Angewandte Festkörperphysik (Fraunhofer IAF), was named after Joseph von Fraunhofer, an 18th century researcher and inventor from Munich. Founded in 1957, it employs more than 200 people and is one of 47 research institutes which make up the Fraunhofer Gesellschaft.

Fraunhofer IAF's main research is into III-V electronics, especially high speed/high frequency components and circuits. It researches into extremely fast directly modulated lasers and wide-band MSM detectors which have been monolithically integrated with electronic circuits. These OEICs are the combination of few optical components, such as lasers, detectors or modulators with complex electronic circuits.

This was organized in a national network funded by the BMBF (Photonik I & II). The Photonik II network has 16 partners with co-ordination shared between the Heinrich Hertz Institut in Berlin and Fraunhofer IAF. The other partners are: Alcatel-SEL, Philips Communications, Siemens, Daimler-Benz, Hirschmann, Integrated Optic Technology, Bosch Telecom and the Universities of Dortmund, München, Dresden, Berlin and Ulm.

Researchers have recently published work on the growth of gallium nitride on c-plane sapphire substrates. The layers were grown in a horizontal MOVPE reactor at atmospheric pressure using trimethylgallium and ammonia.

Universität Hannover

Welfengarten 18
30167 Hannover
Germany
Tel: +49 511 762 5240

The university's material research department has an alliance with Riber SA to jointly develop MBE/CBE processes for nitride compounds.

Institute für Kristallzüchtung

Berlin
Germany

IKZ Epigress AB was selected as the preferred supplier of a VP 508 Silicon Carbide CVD production system to the Institute für Kristallzüchtung in Berlin, Germany. The VP 508 system for IKZ is configured and designed as a production unit including two separate reactors for un-doped and n-doped material and p-doped material, respectively.

Magdeburg University

Otto-von-Guericke Universität
Institut für Experimentelle Physik
Abt. Festkörperphysik II
Universitätsplatz 2
39016 Magdeburg
Germany
The Magdeburg semiconductor epitaxy group which is headed by Prof Alois Krost, is strongly focussed on the growth of high quality AlGaN based structures for electronic and sensor applications. The development of these and related materials for new electronic devices is one of the main goals of research and part of the Magdeburg "Center for New Materials".

To this effect, in early 2000, the University of Magdeburg purchased an AIX 200/4RF-S MOCVD system with *in situ* reflectometry for the growth of III-Nitride compounds, in particular AlGaN-based structures for detectors and electronic applications.

Technische Universität München

Physikdepartment E16
Technische Universität München
D-85747 Garching
Germany
Researchers (in collaboration with Siemens and the Max-Planck-Institut für Festkörperforschung) have published work on the identification of a cubic phase in epitaxial layers of predominantly hexagonal GaN. Epitaxial layers of GaN on c-plane sapphire are analyzed by continuous-wave and time-resolved photoluminescence at 4K and by X-ray diffraction. At the Walter Schottky Institute collaborative research into novel single source MOVPE precursors for III-nitrides has been undertaken.

Osram Opto Semiconductors GmbH & Co

Wernerwerkstr 2
D-93049 Regensburg
Germany
Tel: +49 9131 7 21493
Fax: +49 9131 7 23046
Formerly known as Siemens Optoelectronics, Osram Optoelectronics (see Company Profile) is part of the diversified global electrical engineering and electronics company, Siemens.

Room-temperature pulsed current-injected operation of InGaN multiple quantum well laser diodes was demonstrated at OSRAM Opto Semiconductors (Regensburg, Germany). The devices were grown by MOVPE on SiC substrates. Gain guided lasers with cleaved facets — one facet with high reflectivity coating — show a threshold current density below 20 kA/cm^2 at an emission wavelength of 420 nm. The turn-on voltage was 25V.

The first of EMCORE's multiwafer systems was delivered to the SiC research group at the Corporate Technology Lab in Erlangen which is under the direction of Dr Dietrich Stephani. The Siemens laboratory is considered an international technology leader in SiC research and has previously based its SiC work on EMCORE's single wafer growth systems.

Also, in Erlangen Siemens manages one of Europe's largest programs on SiC device technology. Blue light emitting SiC diodes research includes materials, technology, characterization by Siemens Research Laboratories in Munich, Germany.

Universität-Gesamthochschule Paderborn

Fachbereich Physik
Warburgerstrasse
33098-Paderborn
Germany
This university works with the University of Helsinki as subcontractors in the LAQUANI program. It has also published work on optical detection of electron nuclear double resonance on the residual donor in GaN (in conjunction with CRHEA-CNRS). Other research includes optical detection of electron nuclear double resonance on the residual donor in GaN and luminescence of Be-doped MBE GaN layers grown on Si (111).

Universität Regensburg

Institut for Festkoperphysik
D-93040 Regensburg
Germany
This institute has published work on the MOVPE of ZnSe films on dissimilar substrates such as glass and GaAs(111). It studied the low-temperature growth and doping of polycrystalline ZnSe by MOVPE using ditertiary-butylselenide (DtBSe) and dimethylzine-triethylamine (DMZn-TEN) as precursors. With these alkyls a deposition of ZnSe at less than 400°C is possible without using toxic H_2Se. By using MOVPE-grown ZnSe as a buffer layer, Cd-free CIS-based solar cells with 11% efficiency were fabricated.

Walter Schottky Institut

Technische Universität München
D-85747 Garching
Germany
GaN on c-plane sapphire have been grown by plasma-enhanced MBE — PEMBE — at low deposition temperatures.

Universität Stuttgart

Physikalisches Institut
D-70550 Stuttgart
Germany
This group is involved in work on GaInN/GaN quantum well structures grown by LP-MOVPE by picosecond time-resolved photoluminescence spectroscopy. It

collaborates with other German institutions as well as Siemens on characterization of cubic and hexagonal GaN epilayers on c-plane sapphire. The group has also published work on optical phonons and free-carrier effects for the non-destructive optical characterization of III-nitride MOVPE layers measured by IR ellipsometry in collaboration with the Center for Microelectronic and Optical Materials R&D of Electrical Engineering of the University of Nebraska-Lincoln.

Universität Ulm

Abteilung Optoelektronik
Universität Ulm
D-89069 Ulm
Germany
This institution studies the structure of MOVPE and gas source MBE grown GaN on c-sapphire and also the MBE homoepitaxy of GaN. Recently published work includes: the improved optical activation of ion-implanted Zn acceptors in GaN by annealing under N_2 overpressure, and photoluminescence study on MBE grown GaN homoepitaxial layers (in conjunction with the Polish High Pressure Research Centre and Academy of Sciences).

In 1999, Dr Markus Kamp's GaN Group in the Deptartment of Optoelectronics received the AIXTRON HVPE equipment as a member of a German GaN consortium funded by the German Ministry for Education, Science, Research and Technology (BMBF). The aim of this project is to produce thick GaN layers suited to act as substrates for the growth of high quality laser structures. The HVPE system was added to an AIXTRON GaN MOVPE reactor that has been running for several years by the University of Ulm.

The university collaborates with other German institutes such as Erlangen Nürnberg, and Forschungszentrum Jülich, Institut fur Schicht-und Ionentechnik.

Universität Würzburg

Physikalisches Institut
Am Hubland
D-97074 Würzburg
Germany
Researchers at this institution have published work on the formation of DFB structures for GaN lasers. The gratings were etched by ECR-RIE, 200 V bias, at a rate of 50–100 nm min^{-1}. PL on 26 nm GaN lines was similar to that from large mesas. It has also published work on MBE and the growth and doping of blue-green emitting ZnSe laser diodes.

7.9 Greece

Aristotle University of Thessaloniki

Physics Department
Solid State Section
54006 Thessaloniki
Greece
This institute has published work in collaboration with LETI of France on the growth of 4H and 6H-SiC films grown on off-oriented (0001) SiC substrates.

7.10 India

Anna University (University of Madras)

Crystal Growth Centre
Chunnai 600 025
India
Fax: +91 44 2352774
Researching into crystal growth for optoelectronic devices.

Shree Javendrapari Arts and Science College

Department of Physics
South Gujarat University
Bharuch 392 002
Gujarat
India
This group is concerned with the synthesis of laminar SnSe crystals by a chemical vapour transport technique.

7.11 Ireland

National Microelectronics Research Centre

University College
Lee Maltings
Prospect Row
Cork
Ireland
Tel: +353 21 904110
Fax: +353 21 270271
The NMRC is a leading European research centre. It provides microelectronic services, training for Irish industry and European companies. Also, it

undertakes research in the fields of silicon and advanced materials characterization and technology, CAD, microsystems, components, packaging and systems integration. In wide bandgap semiconductors, R&D includes fabrication of SiC Schottky diodes.

Trinity College, Dublin

Department of Physics
Trinity College
Dublin 2
Ireland.
Fax: +353 1 679 8412
This group is involved in research into II-VI materials in collaboration with the University of Gent-IMEC and Tampere University of Technology.

7.12 Italy

Università di Bari

Centro di Studio per la Chimica dei Plasmi, CNR
Dipartimento di Chimica-Università di Bari
via Orabona 4
1-70126 Bari
Italy
Researchers at this establishment have published papers on the use of remote RF plasma sources to enhance III-V MOVPE technology.

CNR-MASPEC Institute

via Chiavari 18A
43100 Parma
Italy
This group is involved in on GaN on sapphire and has recently published work on the growth mechanisms, electro-optical properties and microstructure of heteroepitaxial GaN layers.

In collaboration with the institute IRRMA and the Ecole Polytechnique Federal in Lausanne, Switzerland, the group studies the electronic properties in compound semiconductors via ab-initio molecular dynamics (FP-MD) calculations. Research focuses on the characterization of suitable substrates and on the evaluation of absorption mechanisms for the atomic species, which can be used as precursors for Ga and N, via FP-MD simulations at finite temperature.

7.13 Japan

Chubu University

Department of Electrical Engineering
Chubu University
1200 Matsumoto-cho
Kasugai
Aichi 487
Japan
Fax: +81 568 51 1141
This group has published work investigating the deep hole trap level of nitrogen-doped ZnSe, grown by MOVPE.

Fujitsu Laboratories Ltd

10-1 Morinosato-Wakamiya
Atsugi 243-01
Japan
Fujitsu is a diversified corporation supplying electronic components for internal use and merchant sale (see Company Profile). It is one of the few commercial research laboratories to have demonstrated a working blue-green diode laser. This was achieved via a low pressure MOVPE process and obtained 12 kA/cm^2 thresholds at 15 V applied. The Fujitsu laser was grown on Shin Nippon Steel SiC substrates and was the first time that a low-pressure process had been reported to produce a nitride-based laser.

Furukawa Electric Co Ltd

Yokohama Research Labs
2-4-3 Okano
Nishi-ku
Yokohama 220
Japan
Furukawa's research includes wide bandgap semiconductors. It is an industrial partner in MITI's 5-year National Program on Hard Electronics. It has also published work on: high-quality n-type GaN grown by gas-source molecular-beam epitaxy (GSMBE). The GaN buffer layer is uniformly formed on a sapphire substrate using dimethylbydrazine, and Si-doped GaN is grown on the buffer layer using uncracked ammonia gas; and on a high-quality GaN growth by GSMBE and the high-temperature properties of a GaN Schottky barrier diode.

Hokkaido University

Research Institute for Electronic Science
Hokkaido University
Kita-12
Nishi-6
Sapporo 060
Japan
Fax: +81 11 706 4973

The Research Institute for Electronic Science collaborates with the Hokkaido University Department of Science working on the growth of II-VI superlattices for blue lasers.

University of Tokyo

Institute of Industrial Science,
7-22-1 Roppongi
Minato-ku
106-8558 Tokyo
Japan

A research group at the Institute of Industrial Science, University of Tokyo, was the first to report a blue light-emitting VCSEL. The work was the result of a collaboration between researchers in Japan, Italy and Germany. They reported the world's first optically-pumped blue — 399 nm — VCSEL in the journal Science (Vol 285, pp 1905) in October 1999.

The project team was led by T Someya from the Institute of Industrial Science, University of Tokyo, Japan. The work describes the formation of microcavities by epitaxial growth of InGaN quantum wells between nitride- and oxide-based quarter-wave reflectors. The diode lasers were fabricated as disc-shaped VCSEL structures 18 microns in diameter having a planar multilayer formed by RIE. The team has managed to overcome one of the main fabrication obstacles with crystal growth of highly reflective nitride mirrors on which InGaN active regions are grown. This difficulty arises from the large differences in thermal expansion coefficients and lattice constants between GaN and AlN.

However, whilst lasing action was achieved at room temperature it was under optical excitation. In order for this type of device to become commercially important the team must achieve coherent light output using electrical rather than optical pumping. T Someya's group was confident that all of the fundamental techniques required for electrically-pumped versions are in place and they expect to be able to report success in the next two years.

This development is important because VCSELs have a number of key characteristics which enable this relatively new family of devices to achieve impressive commercial success in many applications. VCSELs, the high performance device which can be manufactured in high volumes, very economically stand to emulate the success of their longer wavelength equivalents in taking market share from edge emitter diode lasers.

Kyoto University

Department of Electronic Science and Engineering
Kyoto University
Kyoto 606-01
Japan
Fax: +81 75 753 5898
This group is researching various wide bandgap semiconductors. For example, the characterization of p-type ZnSe grown by MOVPE. Also, the group has published work on growth characteristics of GaInN grown on (0001) sapphire via the plasma-excited MOVPE technique.

Matsushita Electronics Corporation

Electronics Research Laboratory
Saiwai-cho Takatsuki
Osaka 569-11
Japan
Matsushita is a large corporation which develops and manufactures a wide range of electronic components and systems (see Company Profile). Its researchers have projects underway involving in work researching the growth of GaN by MOVPE. In collaboration with Osaka University, the two groups have recently published work on the inheritance of zinc-blende structure from 3C-SiC/Si(001) substrate in growth of GaN by this method.

In February 1999, Matsushita announced that it had achieved continuous wave oscillation of GaN violet semiconductor laser diodes. An oscillation wavelength of 400 nm minimizes the size of its optical focussing spot to 440 nm, a technology required for next-generation DVD systems. This enables a storage capacity increase from 2 h of SDTV format video images to 7 hours of SDTV, or two hours of HDTV video images. The continuous wave oscillation was made possible with the development of a new evaluation process for the optical properties of semiconductor laser diode's active player and a new growth method for high-quality crystal. Matsushita also conducts research into blue-green semiconductor lasers, and optical disk technology.

Matsushita researchers have collaborated with a number of Japanese university departments. For example, they have published results of a collaboration with Osaka University on the growth of GaN on 3C-SiC.

Meijo University

Department of Electrical and Electronic Engineering
1-5-1 Shiogamaguchi
Tempaku-ku
Nagoya 468
Japan
Tel: +81 52 832 1151
The research group at this university is headed by Prof I Akasaki and Prof H Amano. Prof Amano has presented work on the fabrication and properties of MOVPE-grown GaN/InGaN inner stripe geometries. Work has also been

published on GaN MODFETs (with Cornell University) and into free excitons in GaN (with Linköping University). Other work being performed in collaboration with Linköping University includes the crystal growth of GaN in the area of luminescence studies such as PL measurements on GaN/AlGaN modulation doped QWs.

Also, the High-Tech Research Center at Meijo University is involved in collaboration with the Department of Electrical and Electronic Engineering having published work on the radiative recombination in InGaN/GaN MQW structures.

Meijo University has an Aixtron AIX 200 series for the growth of nitride structures.

The Meijo University group, in collaboration with Hewlett Packard Labs in Kawasaki, Japan, have reported a blue-green diode laser claiming to have fabricated the world's first ridge guided, cleaved-facet nitride laser.

Mie University

Dept. of Electrical & Electronic Engineering
1515 Kamihama
Tsu Mie 514-8507
Japan
This group has published a number of papers on the structural properties of III-nitrides. These include compositional analysis of GaInN on GaN so as to determine the compositional pulling effect within epitaxial layers.

Mitsubishi Cable Industries Ltd

Central Research Laboratory
4-3 Ikejiri
Itami
Hyogo 664
Japan
This laboratory has published work (in collaboration with Nagoya University) on novel precursors and dopants for MOVPE growth of n-type and p-type GaN films.

Nagoya University

Furo-cho
Chikusa-ku
Nagoya 464-01
Japan
This group is striving towards the commercialization of LEDs based on InGaN. It announced results of a stimulated emission at 380 nm from a InGaN/AlGaN diode at room temperature. It has also published work on novel precursors and dopants in collaboration with Mitsubishi Cable Industries.

The Nagoya Institute of Technology (in a collaboration between two departments: the Research Center for Micro-Structure Devices and the Department of Electrical and Computer Engineering) works on GaN-based wide bandgap materials grown on sapphire using MOVPE. It has published work on the first observations of stimulated emission from current injected InGaN/AlGaN double heterostructure diodes. Other work includes a collaboration with Northwestern University on the simultaneous growth of two differently oriented GaN epilayers on sapphire by MOVPE.

NEC Fundamental Research Laboratories

NEC Corporation
Tsukuba
Ibaraki 305
Japan
NEC researchers are undertaking work in the area of epitaxial growth of III-nitride structures. Published work includes descriptions of ELO-growth using laterally overgrowth using both MOVPE and HVPE and theoretical studies of the behaviour of precursors.

The Fundamental Research Laboratories collaborates with the co-located NEC Opto-Electronics Research Laboratories on, for example, microstructural studies of GaN films grown on GaAs substrates by hydride VPE.

NEC Corp

Optoelectronics and High Frequency Device Research Laboratories
34 Miyukigaoka
Tsukuba
Ibaraki 305
Japan
Fax: +81 298 50 1107
Email: shiraish@uhl.cl.nec.co.jp
Amongst considerable world-class R&D for a wide range of optoelectronics subjects, NEC (see Company Profile) is conducting research into selective-area metallorganic MBE of GaAs using metallorganic chloride gallium precursors.

Nichia Chemical Industries Ltd

Department of Research and Development
491 Oka
Kaminaka
Anan
Tokushima 774
Japan
Nichia has almost single-handedly been responsible for the renaissance of optoelectronic devices based on the III-nitride wide bandgap semiconductors. Long-lifetime violet InGaN multi-quantum-well/GaN/AlGaN separate-confinement heterostructure diode lasers have been successfully fabricated using epitaxially laterally overgrown GaN by reducing a large number of threading dislocations originating from the interface between GaN and sapphire substrate. The threading dislocations shorten the lifetime of the LDs through an increase of the threshold current density.

In 1994 Nichia developed the first blue InGaN/AlGaN DH LEDs and then developed blue-green InGaN single-quantum-well (SQW) structure LEDs in 1995. Subsequently, UV as well as amber LEDs have also been developed. UV, blue, green, amber and red InGaN-based LEDs were obtained using an InGaN active layer instead of a GaN active layer.

Nichia made the first demonstration of room temperature operation violet laser light emission in InGaN MQW/GaN/AlGaN-based heterostructures under pulsed operation. InGaN MQW/GaN/AlGaN SCH LDs were fabricated on the ELOG substrate grown by MOVPE.

Nippon Telegraph and Telephone Corp (NTT)

NTT Atsugi R&D Center
3–1 Morinosato-Wakamiya
Atsugi
Kanagawa 243-01
Japan
Nitride semiconductor research by scientists at the NTT labs includes studies of phase separation in wurtzite InGAlN quaternary system. Modelling results are compared with composition data obtained experimentally from InGaN grown by MOVPE and from this, the characteristics of the ternary alloys of InAlN, InGaN and GaAlN can be predicted.

Osaka University

Department of Electrical Engineering
Faculty of Engineering
2-1 Yamadaoka
Suita
Osaka 565
Japan
This group (in collaboration with Matsushita Electronics corporation), has published work on the inheritance of zinc-blende structure from 3C-SiC / Si(001) substrate in growth of GaN by MOVPE. GaN was grown on SiC / Si (001) at temperatures between 600 to 800°C by low pressure MOVPE, using TMGa and NH_3.

RIKEN Semiconductor Laboratories

The Institute of Physical and Chemical Research (RIKEN)
2–1 Hirosawa
Wako
Saitama 351-01
Japan
This group is interested in gallium nitride and has reported possibilities for the use of GaN quantum dots in laser structures to reduce the threshold current. It has also demonstrated reduction in defect density in GaN films using very thin AlN buffers on 6H-SiC.

Rohm Electronics

21 Saiin Mizosaki-cho
Ukyo-ku
Kyoto 615
Japan
Rohm electronics is a diversified developer and manufacturer of electronic and optoelectronic components including LEDs and diode lasers (see Company Profile). Researchers are Rohm's R&D labs are investigating increasing the wavelength range of devices including p-containing compounds grown by MOVPE. They are also researching the use of GaN for short-wavelength lasers, as well as laser diode integration with other optical devices for CD pickups.

Rohm and Pioneer in December 1999 announced a partnership agreement to develop a GaN-based blue-violet semiconductor laser for next-generation optical disc technology. This partnership agreement will enable the two companies to progress the development of GaN-based blue-violet lasers much more efficiently. By combining each company's technological knowledge, Pioneer and ROHM will make further improvement in crystal growth technology, device production process and optimization of the device structure and will obtain a practical output level and life length. Both companies are aiming at merchandising blue-violet semiconductor lasers within two years.

It is investigating increasing the wavelength range of devices into the green-blue region including p-containing compounds grown by MOCVD. The company has also undertaken III-nitride R&D for five years. This work includes light-emitting devices such as GaN layers on dissimilar substrates such as silicon.

Shizuoka University

School of Electronic Science & Engineering
3-5-1 Johuku
Hamamatsu 432-8561
Japan
This institute has been working on GaN grown on AlN buffer layers grown by atmospheric pressure MOVPE on sapphire substrates. It has collaborations with other Japanese universities and also with Sanken Electric Co Ltd.

Shizuoka University

Research Institute of Electronics
Johoku 3-5-1
Hammamatsu 432
Japan
This group is working on the growth of alloy GaInP crystals by compositional conversion of InP layers grown on GaP substrates in an LPE system as part of a collaboration with the University of Victoria, in Canada.

Showa Denko KK

Sakaida Research Department
Central Research Laboratory
1–1 Ohnodai 1-CHOME
Midori-Ku, Chiba 267-0056
Japan
Tel: +81 43 226 5240
Fax: +81 43 226 5262

An AIX 200RF MOCVD system was installed at Showa Denko Research Laboratory, Chiba, within the framework of the NEDO sponsored program, "The light for the 21st century".

Showa Denko is developing new structures for nitride applications focusing on LED development. Dr Okuyama, head of the MOCVD research group, said that he had decided on AIXTRON because of the reliable, proven technology of AIXTRON's MOCVD systems. With this technology he expects an important move into the white light market.

Sony Corporation Research Center

Ikeda Research Center
174 Fujitsuka-cho
Hodogaya-ku
Kanagawa-kan 240
Japan

Sony is a large Japan-based developer of electronic and opto components and the products based on them. Scientists at Sony's R&D labs were among the first to have published results on the operation of a blue-green laser diodes based on ZnS/Se MQW structures grown by MBE. Research into the chalcogenide wide bandgap semiconductors was undertaken at the Ikeda Research Center in Japan. This involved the optical characteristics of ZnCdSe/ZnSSe single quantum wells grown by MBE on n-type GaAs (001) substrates.

It is thought that this line of R&D has since been either cutback or has ceased altogether.

More recently Sony's scientists have reported the growth of III-nitrides by photo-assisted MOVPE.

Sumitomo Electric Industries Ltd

Basic High Technology Laboratories
1-1-3 Shimaya
Konohana-ku
Osaka 554-0024
Japan

This is a research division of Sumitomo Electric (see Excelight Company Profile) and has a fairly broad area of interest in compound semiconductor materials, devices and systems, particularly crystal growth and epitaxy for optoelectronic and related devices. The laboratory has recently successfully grown one-inch diameter single crystal ZnSe by a rotational chemical vapour transport method.

Tohuku University

Institute for Materials Research
Sendai 980-8577
Japan
This group has published results on the growth of GaN compounds on dissimilar substrates by MOVPE such as mixed-perovskites. Other workers at the Tohuku University have published studies of growth of SiC on Si using a specially-designed UHV deposition system. The Department of Electronics has also developed an inert-gas curtain and combustion method for the preparation of films of diamond and SiC/Si.

University of Tokushima

Dept of Electrical & Electronic Engineering
2−1 Minami-Jpsanjima
Tokushima 770-8506
Japan
This group is based closed by the Nichia Chemical Industries R&D labs and some degree of collaboration has been underway for some years. He department has published a number of papers on III-nitride work including a collaboration with its affiliate the "Satellite Venture Business Lab" on 2D electron gas in AlGaN/GaN heterostructures grown by MOVPE.

University of Tokyo

Research Center for Advanced Science and Technology
4-6-1 Komaba
Meguro-ku
Tokyo 153
Japan
This group recently published work on the study of A,B, and C free excitons, deriving deformation potential from the measurements of photoreflectance. It has also reported work on the conduction-band-edge formation in $GaP_{1-x}N_x$ alloys, using PLE spectroscopy (in collaboration with the Department of Physics). The Imaging Science and Engineering Laboratory at Yokohama has also published work on the growth of GaN on sapphire, using a specially designed two-flow horizontal MOVPE reactor.

Other research includes step structures on the $\{0\,0\,0\,1\}$ facet of SiC single crystals grown by the modified Lely method, which were examined through both optical microscopy and atomic force microscopy to discuss the dependence on the polytype and the polarity. The results were compared with those of the VPE method.

Tokyo University of Agriculture & Technology

Department of Applied Chemistry
Faculty of Technology
Koganei
Tokyo 184
Japan
Fax: +81 423 86 3002
This group is involved in the growth of InGaN by MOVPE.

Toshiba Advanced Semiconductor Devices Research Laboratories RD Center

Toshiba Corp
1 Komukai Toshiba-cho
Saiwai-ku
Kawasaki 210
Japan
The Toshiba R&D group has successfully demonstrated a blue-green diode laser: a threshold current of 530 mA, and a density of 10.6 kA/cm^2. P-type doping was achieved without post-growth annealing using Mg doping with an N$_2$ carrier gas. It is based on a new structure, termed the "Inner Stripe Laser Diode" where the current stripe is defined by a blocking layer underneath the p-contacts. The Toshiba group suggests that the relationship between device degradation and dislocations has two components, temperature rise and nano-pipes.

Toyoda-Gosei Co Ltd

1st New Market Devel HQ
Tech Div
1 Nagahata
Ochiai
Haruhi-cho
Nishikasugai-gun
Aichi-pref. 452
Japan
Toyoda-Gosei is a company focused primarily on the automotive sector (see Company Profile) however it has a strong interest in opto devices especially blue LEDs. It is a commercial supplier of these devices in Japan and in other countries.

Toyoda-Gosei is involved in research into blue LEDs, having published work on LED technology regarding p-n junction devices which radiate 2.5 Cd (450 nm, 3.6 mW, 5.1%) via a donor-acceptor transition.

The company has been granted a number of patents for nitride-based LEDs and also for a diode laser based on AlGaInN having a double heterojunction structure (the active layer held between layers with a greater bandgap). The diode laser is made up of mirror surfaces formed by cleaving the multi-layered coating and the sapphire substrate in directions parallel to < 0001 > (c axis) of the sapphire substrate.

In T-G's device, the intermediate ZnO layer is selectively removed by wet etching with a ZnO-selective etchant solution to form gaps between the sapphire substrate and the lowest sub-layer of the semiconductor laser element layer (which is cleaved with the aid of the gaps), the resulting planes of cleavage are used as the mirror surfaces of the laser cavity.

Toyohashi University of Technology

Dept of Electrical & Electronic Engineering
Toyohashi 441
Japan
This group is known for studies of AlN and related compounds via Raman spectroscopy, PL and CL techniques.

Toyota Central Research & Development Laboratories

41-1 Nagakute
Aichi 480-1192
Japan
Toyota is a global supplier of automotive components, systems and vehicles. It also has an interest in semiconductor materials: for example, workers at the Central R&D Labs have published work on the crystal growth of SiC by a modified Lely method which showed an absence of micropipes and other differences from other growth methods.

7.14 Korea

Chungnam National University

220 Kung-Dong
Taejon 305-764
South Korea
Research subjects at Chungnam National University have included Mg-doping and Si-doping of GaN epilayers by LP-MOVPE. The university is also a participant in the project entitled "Development of Blue Light Emitting Diode (III)", which is sponsored by the Ministry of Trade, Industry & Energy, Korea.

Dongguk University

Dept of Physics
Pildong 3–26
Chung-ku
Seoul 100-715
South Korea
Tel: +82 2 278 3429
Fax: +82 2 322 1146
Research interests include SiC and related materials — epitaxial growth, characterization, and device fabrication.

Jeonbuk University

SPRC
Jeonju 561-751
South Korea
The institute has published results of a collaboration with Kyunghee University and KIST (see below) on the characteristics of Si-doped GaN compensated with Mg.

Kyunghee University

Dept of Physics
Kyungki-do 131-701
South Korea
This institution has published results of a collaboration with Jeonbuk University and KIST (see below) on the characteristics of Si-doped GaN compensated with Mg.

Korea Research Institute of Standards and Science

New Materials Evaluation Center
Daeduk Science Town
Taejon 305-600
South Korea
Fax: +82 42 868 5032
This renowned institute is involved in a variety of semiconductor materials and devices R&D. In the wide bandgap semiconductor area it has published work on various aspects of GaN. For example, on the polytypes in GaN films grown by MOVPE on (0001) sapphire substrates. This work was in collaboration with OPTEL Semiconductor Co, of Jksan, Chunbuk, South Korea.

The institute has also published results of a collaboration with Kyunghee University and Jeonbuk University on the characteristics of Si-doped GaN compensated with Mg.

In collaboration with the University of Korea, this institute is working on mobility enhancement and yellow luminescence in Si-doped GaN growth by MOCVD for optoelectronics; and the formation mechanism of 'volcano-like' structural defects in multiple periods of InAs quantum dots on GaAs. In this respect it has collaborated with the University of Korea.

Korea University

Department of Materials Engineering
Seoul 136-701
South Korea
This institute is working on mobility enhancement and yellow luminescence in Si-doped GaN growth by MOCVD for optoelectronics in a collaboration with Korea Research Institute of Standards and Science.

Samsung Advanced Institute of Technology

Samsung Group
Taepyung-ro
Chung-gu
Seoul
South Korea
Tel: +82 2 751 2082

Samsung is a large industrial corporation with a wide range of components and system products made and marketed on a global basis. To support next-generation productions it has an ongoing R&D program in the III-nitride materials and devices such as LEDs and diode lasers (see Company Profile).

Samsung is particularly interested in the R&D of blue diode lasers for digital storage disk applications. InGaN/GaN MQW laser diodes have been grown by multiwafer MOVPE. In early 1999, Samsung Advanced Institute of Technology in Suwon, South Korea, added to its achievements in nitride semiconductor technologies with the successful development of high brightness blue and green LEDs and the pulsed operation of a blue laser diode at room temperature.

Other work published includes the investigation of nanopipes in GaN, grown by MOVPE on c-plane (0001) sapphire. Well-defined etch-pits were found to form on the surface of GaN films, with aqueous solutions of H_3PO_4 (85 mol%). Using optical microscopy, SEM, and TEM, the relationship between etch-pits and crystalline defects in GaN films was investigated.

School of Materials Science & Engineering

Seoul National University
Room 32-203
Seoul 151-742
South Korea

This group is working on asymmetric growth behaviour of selectively grown InP on vicinal surfaces by low-pressure MOCVD.

7.15 The Netherlands

Technische Universiteit Eindhoven

Semiconductor Physics Group
COBRA Interuniversity Research Institute
Eindhoven University of Technology
Physics Department
Semiconductor Physics
PO Box 513
5600 MB Eindhoven
The Netherlands
Tel: +31 40 2452277
Fax: +31 40 2461339

This institute has published work on III-V nitride semiconductors for blue light-emitting devices grown by all alkyl chemical beam epitaxy. This uses the ultra high vacuum growth chamber of MBE and the source materials of MOVPE. In general, growth temperatures in CBE are about 100°C lower than in MOVPE while source material consumption is decidedly less.

An important research line using the CBE system is to prepare double hetero-junctions in a p-i-n diode structure for short wavelength emission. The group has had to solve two major problems: the structural quality of the material and the control of n- and especially p-type doping. The Group III elements required for these devices (Ga and In) are already present and aluminium sources are also being developed. Amines (alkylated nitrogen compounds) are to be used as the Group V source instead of the ammonia as is commonly used.

The group of semiconductor physics collaborates with device and system oriented groups in the COBRA-institute. Collaboration also exists amongst others with Philips Research Laboratories, IMEC Leuven, IMEC Gent, Imperial College London, the University of Essex, the University of Nottingham, the Max-Planck-Institut Stuttgart, the Paul-Drude-Institut Berlin, the Walter-Schottky-Institut München, the Bilkent University Ankara, the University of Honolulu and the University of the Andes Merida.

University of Nijmegen

Exp. Solid State Physics III
RIM
University of Nijmegen
Toernooiveld 1
6525 ED Nijmegen
The Netherlands

This department has collaborated with universities in Poland, Germany and UK on the morphological and structural characteristics of homo-epitaxial MOVPE growth of GaN.

7.16 Poland

High Pressure Research Center

Warsaw 01 142
Sokolowska 29/37
Poland
This group is working on gallium nitride and has published papers on: temperature distribution in the chamber used for crystal growth of GaN under high pressure of nitrogen, and surface morphology of grown and annealed bulk GaN crystals. The centre subsequently became the company, Unipress.

The High Pressure Research Center collaborates with a number of other institutes around the world through the interest in the GaN single crystals that are made. Examples of these collaborations include the North Carolina State University, Pierre et Marie Curie Université and Boston University.

It has also published work on the surface morphology of both as grown and annealed bulk GaN crystals (in conjunction with Warsaw University and Silesian Technical University) and a photoluminescence study of MBE grown GaN homoepitaxial layers (with the Polish Academy of Sciences and Universität Ulm).

Silesian Technical University

Department of Chemistry
Silesian Technical University
44-100 Gliwice
Poland
This department (in a collaboration with the High Pressure Research Centre and University of Warsaw) has reported on work on the surface morphology of as grown and annealed bulk GaN crystals.

Unipress (formerly Polish Academy of Sciences, see above)

Al Lotnikow 32/46
02-668 Warsaw
Poland
This group is involved in the study of photoluminescence on GaN homoepitaxial layers grown by MBE (in conjunction with Universität Ulm and the High Pressure Research Center University of Warsaw).
GaN epitaxial layers on GaN single crystals have been grown using MBE with an ammonia source. It has also reported on the thermodynamic properties of nitrogen in pressures up to 20 kbar and temperatures up to 1600°C.

Warsaw University

Institute of Experimental Physics
Warsaw University
Hoza 69
Warsaw
Poland

This group is involved in the study of the surface morphology of both as grown and annealed bulk GaN crystals. It has published work in conjunction with the High Pressure Research Center and Silesian Technical University.

7.17 Russia

Ioffe Physical-Technical Institute

Russian Academy of Sciences
St Petersburg
Russia
Tel: +812 515 9988

Researchers at the Ioffe Physico-Technical Institute in St Petersburg, Russia are working with FTIKKS Enterprise on devices based on epitaxial films on 6H- and 4H-SiC. FTIKKS has grown good quality, structurally perfect SiC epitaxial films of n-type and p-type conductivity, and fabricated devices which are able to work at high temperatures. UV photodetectors and JFETs prepared by a sublimation technology have shown good serviceability up to 700K.

The group has an interest in a number of high temperature sectors using SiC power MOS-based devices for amplifiers and power conversion. 6H SiC films of n-type conductivity have been used for oxidation tests on MOS structures.

The Ioffe Institute is also involved in work on a theoretical model for analysis and optimization of Group III-Nitrides growth by MBE. Another project includes mesa-structure and contacts as well as GaN epitaxial growth, having published work on the growth of GaN layers grown by HVPE on P-type 6H-SiC substrates, which was done in a former collaboration with Cree Research.

The institute is also investigating the chemical processes which occur inside AIXTRON multiwafer planetary reactors. The models account for gas flow, heat transfer, and multicomponent mass transport along with gas phase and surface chemical reactions.

Technologies and Devices International Inc in collaboration with Crystal Growth Research Center, (St Petersburg, Russia) and PhysTech WBG Research Group (St. Petersburg, Russia) is developing GaN bulk crystal fabrication technique by hydride vapour phase epitaxy (HVPE) of GaN on SiC substrates and selective removal of the substrate by RIE.

MV Lomonosov Moscow State University

Department of Physics
119899 Moscow
Russia
This group of researchers have published work on the optical characterization
of superbright blue and green InGaN/AlGaN/GaN LEDs.

General Physics Institute of RAS

38 Vavilov Street
117942 Moscow
Russia
Tel: +7 095 132 8342
Fax: +7 095 135 8234
Russian university researching a wide range of semiconductors and electronics
including laser processing of metals for SiC surfaces for ohmic contacts.

Russian Research Center "Applied Chemistry"

St Petersburg 197198
Russia
Researchers at this centre are involved in MBE growth of Group III-nitrides and
have produced a theoretical model for their analysis and optimization.

Ukraine Academy of Sciences

Department of Optoelectronics
Institute of Semiconductors
252650 Kiev
Pr Nauky 41
Ukraine
This group studies light-emitting structures and photodetectors based on III-V
semiconductors. Amorphous SiC structures deposited on flexible substrates
have been developed.

7.18 Singapore

Nanyang Technological University

School of Electrical and Electronic Engineering
Nanyang Avenue
Singapore 639798
Singapore
This institute is active in a wide range of semiconductor research including the
structural identification of cubic phase in hexagonal GaN. The work involves
epitaxial growth of III-nitride epilayers on sapphire using a gas source MBE sys-
tem. The work was carried out in collaboration with the Chinese Academy of
Sciences.

7.19 South Africa

University of Port Elizabeth

Department of Physics
PO Box 1600
Port Elizabeth 6000
South Africa
Fax: +27 41 504 2573
This group is interested in wide bandgap II-V semiconductors, having recently published work on the growth of Zn_3As_2 on InP by atmospheric pressure MOVPE.

7.20 Spain

Department Ciencia de Materiales e IM y QI

Universidad de Cádiz
Puerto Real
11510 Cádiz
Spain
This group is investigating the influence of the surface morphology on the relaxation of low-strained In Ga As linear buffer structures. This is a collaboration with fellow-Spanish university Ciudad Universitaria and the University of Liverpool.

Ciudad Universitaria

Department Ingenieria Electrónica
ETSI Telecommicación
UPM
Ciudad Universitaria
28040 Madrid
Spain
This group is investigating the influence of the surface morphology on the relaxation of low-strained In Ga As linear buffer structures. This is a collaboration with fellow-Spanish University of Cadiz and the University of Liverpool.

Ciudad Universitaria has published work on the effect of III/V ratios and substrate temperature on the surface quality and properties of III-nitride epilayers grown by plasma assisted MBE on silicon wafers. This system utilizes an Oxford Research CARS25 nitrogen activation unit.

These research activities are supported not only by national funding agencies such as the CICYT Project but also by the ESPRIT LAQUANI program.

Universidad Politecnica de Madrid

Politecnica de Madrid
Ingeniera Electronica
ETSI Telecomunicacion
28040-Madrid (UPM)
Spain
This university is a LAQUANI program subcontractor. It has published work on yellow band and deep levels in undoped MOVPE GaN, in collaboration with CRHEA-CNRS and Nottingham University.

7.21 Sweden

Lund University

Lund Laser Centre (LLC)
PO Box 118
SE-221 00 Lund
Sweden
Tel: +46 46 222 0178
Fax: +46 46 222 4250
Web: www-llc.fysik.lth.se/index.htm
The Lund Laser Centre is an organization for co-ordination of research and teaching in lasers, optics and spectroscopy at Lund University. The Lund Laser Centre has the status of a European Major Research Infrastructure. Units from the Technical, Natural Sciences and Medical Faculties participate.

Units belonging to the centre include Division of Atomic Physics, including the Lund High Power Laser Facility and Medical Laser Centre.

The Lund Laser Centre is established at the Lund University, directly under the Rectorate, as an organization for laser, optics and spectroscopy research. It constitutes the largest unit in the field in the Nordic countries. The research is being performed in a unique, multi-disciplinary atmosphere and fosters close collaboration with industry. The Centre is characterized by a very strong exchange of ideas, expertise and resources between different projects, where advanced electro-optics form a common denominator. Members in the Centre are the Research Divisions of Atomic Physics, Atomic Spectroscopy, Chemical Physics and Combustion Physics, as well as the Lund University Medical Laser Centre. Further, a National High-Power Laser Facility features spear-head technology installations in the field of lasers and laser spectroscopy. Two further collaborative centres, the Combustion Centre and the Environmental Measurement Techniques Centre, are associated. The particular set up of the Lund Laser Centre, gives a unique and informal access to relevant knowledge also outside the field of lasers, optics and spectroscopy to allow well-balanced, synergetic research projects to be pursued. Further, the presence of the MAX-lab Synchrotron Radiation Facility and the IDEON Research Park in close walking distance to the LLC adds to its scientific environment.

Linköping University

Department of Physics and Measurement Technology
Linköping
S-581 83
Sweden
Tel: +46 13 281797
This university is involved in a collaboration with the Swedish Space Corporation and Epigress. Prof Janzen of Linköping University (the group leader) has published investigations on convection-free growth of SiC in the microgravity of space. Other collaborations include: with Meijo University on free excitons in GaN, and with Outokumpu/Semitronic AB in the development of SiC materials, using a VP 508 hot wall CVD system from Epigress. Other long-term collaborations include PL studies with the Department of Electrical and Electronic Engineering, Meijo University, Japan.

Linköping University has also demonstrated high temperature gas sensor CV characteristics measured to 1050°C.

UMEÅ University

The Laser Physics Group
Department Of Experimental Physics
S-901 87 UMEÅ
Sweden
Tel: +46 90 786 67 54
Fax: +46 90 786 66 73
E-mail: Ove.Axner@Physics.umu.se
Research is conducted within the following areas: development of laser-based (spectroscopic) techniques for ultra-sensitive trace-element detection, laser-induced fluorescence in graphite furnaces, and diode laser spectrometry, non-intrusive manipulation of biological objects and intra-cellular microsurgery by: optical tweezers and laser scalpel laser cooling and trapping.

7.22 Switzerland

CSEM Centre Suisse d'Electronique et de Micro-technique SA

Jaquet-Droz 1
CH-2007 Neuchatel
Switzerland
Tel: +41 32 7205 347
Fax: +41 32 7205 730
Organization specialized in the research, development, engineering and production of state-of-the-art realizations in microelectronics, microtechnology and bio-inspired systems for industry. CSEM has developed hot filament CVD diamond coatings for sensing devices like special AFM tips, UV sensors, electrochemical microelectrodes, heating elements and gas sensors.

Ecole Polytechnique Federale de Lausanne

Institut de Micro- et d'Opto-electronique
Ecole Polytechnique Federale de Lausanne (IMO-EPFL)
CH-1015 Lausanne
Switzerland
This research laboratory is a member of the LAQUANI project involved in the R&D of GaN-based wide bandgap semiconductors.

IBM Zurich Research Laboratory

Saumerstrasse 4
CH-8803 Rueschilon
Switzerland
Tel: +41 1 724 8203
The European branch of IBM's Research Division, this lab employs 200 people. One group is researching into novel light-emitting materials and structures for LED-based display technologies. IBM Zurich collaborates with various institutions including the Universities of Ulm, Boston and Neu-Technikum Buchs.

7.23 Taiwan

Arima Optoelectronic Corp

6th Floor
No 327 Sung Lung Road
Taipei
Taiwan ROC
This Taiwan-based company is one of the many recent start-ups in Taiwan which have targeted high performance epi-based semiconductor optoelectronic components. They have also chosen to compete in the market for blue-green LEDs and related devices. To this effect they have been equipping themselves with a range of advanced epitaxial growth systems, principally from European or US suppliers.

In mid-1999, AIXTRON AG supplied Arima with its first nitride reactor system but this was installed not at its Taiwan base but rather at the University of Bath, UK. This AIX 200/4RF-S system is for the growth of GaN- and SiC-based LED and LD structures. The co-operation targeted the development of processes for mass production of blue and green LEDs. However, the company ordered additional AIXTRON Planetary Reactors® of the latest G3 series, ie the AIX 2400G3 HT system, which will be used for the mass-production of GaN-based LEDs

EPISTAR Corp

Science-Based Industrial Park
Hsinchu
Taiwan ROC
In 1999, Epistar has recently placed an order for an AIXTRON multiwafer nitride Planetary reactor with a capacity of 6 × 2-in wafers for the growth of AlGaInN blue and green LEDs. Together with another recently ordered system, the AIX 2600G3 for AlInGaP, Epistar is committed to become the leading supplier of ultra high brightness LEDs of all wavelengths.

EPISTAR has several AIXTRON Planetary Reactor systems of second and third generation. These enable the production of the brightest absorbing substrate AlGaInP LEDs in the market. The purchase of a nitride system is an logical extension of EPISTAR's capacity in addition to the existing reactors.

National Central University

Semiconductor Physics Laboratory
Department of Physics
Chung-li
Taiwan ROC
This institute is investigating a range of wide bandgap semiconductors including the MOVPE growth and device fabrication via RIE.

The National Central University of Taiwan is equipped with an AIX 200 series MOVPE systems for research into III-nitride materials and devices.

The group collaborates with other departments in Taiwan and also with the University of Florida and with Bell Labs.

I-Shou University

Electronic Engineering Department
Ta-Hsu Hsiang
Kaoohsiung County
Taiwan ROC
This department has published work on substrate misorientation effects when growing ZnMgSe epilayers on GaAs by MBE.

United Epitaxy Company (UEC) Ltd

9F No 10 Li-Hsin Road
Science-Based Industrial Park
Hsinchu
Taiwan ROC
Tel: +886 3 567 8000
A manufacturer of semiconductor optoelectronics components, in 1999, UEC ordered an AIXTRON Planetary Reactor for the growth of AlGaInN with a capacity of 6 × 2-in wafers for the growth of III-Nitride optoelectronic devices.

7.24 Tunisia

University of Tunisia

Faculté des Sciences
5000-Monastir
Tunisia
The group has published work on alternative N precursors and Mg doped GaN grown by MOVPE.

7.25 UK

AEA Technology plc

F5 Culham
Abingdon
Oxfordshire
OX14 3DB
UK
Tel: +44 1235 463407
Fax: +44 1235 464253
AEA Technology is an independent science and engineering services business which solves technical, safety and environmental problems for industries and governments globally. Involved in many aspects of HT electronics, including semiconductors (diamond, silicon carbide, SOI), sensors, undertaking research, development, manufacture, and consultancy.

University of Bath

Bath
BA2 7AY
UK
Tel: +44 1225 826826
In 1999, the University of Bath took delivery of an Aixtron AG AIX 200/4RF-S MOVPE system for the growth of GaN- and SiC-based LED and LD structures. The reactor has been ordered by Arima Optoelectronic Corp, Taiwan, for a close collaboration with the University of Bath. The cooperation targets the development of processes for mass production of blue and green LEDs.

The group, led by Prof Dr WN Wang who joined in March 1999, is using the reactors for the growth of GaN- and SiC-based LED and LD structures, particularly towards the production of blue and green UHB-LEDs.

University of Cambridge

Pembroke Street
Cambridge
CB2 3QZ
UK

This group specialises in the characterization and growth of GaN and related alloys. As well as commissioning a new TEM, Thomas Swan Ltd (see Company Profile) donated a 6 × 2-in capable MOVPE GaN reactor as part of a collaborative R&D programme funded by EPSRC.

The Dept of Materials Sciences and Metallurgy has collaborated with universities in Poland, Germany and The Netherlands on the morphological and structural characteristics of homo-epitaxial MOVPE growth of GaN.

Epichem

Power Road
Bromborough
Wirral
Merseyside
L62 3QF
UK

Epichem Ltd (which was formed in 1983), manufactures a range of high purity chemicals and gases used mainly in the electronics and glass coatings industries (see Company Profile).

Epichem has reported III-nitride and related work done in collaboration with academic research groups. In particular, with the University of Sheffield, UK, with whom it is a partner in the Rainbow Scheme for research into precursors and dopants. Internally, it is developing lower oxygen precursors such as TMI, TMA and TMG.

Epichem belongs to various EC BRITE/EURAM/ESPRIT/LINK research projects including: SICOIN, RAINBOW, Widegap CPV, Admiral and CONFORM, Bluemat Joule III for Wide Gap Chalcopyrites for Advanced Photovoltaic Devices.

Heriot-Watt University

Department of Physics
Edinburgh EH14 4AS
UK
Fax: +44 131 541 3088

This group is working on II-VI materials for blue lasers and collaborates with the University of Strathclyde on ZnSe. In 1992, Heriot-Watt were the first group in Europe to demonstrate blue ZnCdSe/ZnSe MQW diode lasers. It is also working on diamond thin films and coatings.

IQE plc

Cypress Drive
St Mellons
Cardiff SC3 0ET
UK
IQE plc was formed in 1999 from the merger of Cardiff-based Epitaxial Products International Ltd, and Quantum Epitaxial Designs of Bethlehem, PA, USA.

EPI has since its formation in the late-1980s been a participant in many national and international R&D projects. IQE was a participant in the EC BRITE/EURAM project "Layers and Structures for Blue Emitters" (LASBE).

Marconi Optical Components

Caswell
Towcester
Northants NN12 8EQ
UK
Tel: +44 1327 350581
Fax: +44 1327 356775
With a new name, this establishment is a supplier of range of advanced semiconductor devices but now focused on optoelectronics devices including high performance lasers (see Company Profile). Participant in many national and European R&D programs.

North East Wales Institute of Higher Education

Advanced Materials Research Laboratory
Plas Coch
Mold Road
Wrexham LL11 2AW
UK
Tel. +44 1978 293165
Fax: +44 1978 293212
These research laboratories, completed in October 1994, complement those of the Advanced Materials Research Laboratory. Current research activities include the growth of II-VI semiconductor thin films by MOVPE for blue LEDs and lasers. The new optoelectronic materials research laboratories, under the direction of Professor Stuart Irvine are concerned with photo-assisted MOVPE of II-VI semiconductor for a range of applications, including infrared detectors, solar cells, radiation detectors and visible emitters. The new facility comprises of cleanrooms for wafer preparation and fabrication, electrical characterization, X-ray double crystal diffractometry and surface science analytical equipment. A wide range of challenging research projects have been initiated from fundamentals of growth kinetics, development of in situ optical characterization of epitaxial growth, to characterization of epitaxial layers and devices.

An Engineering and Physical Sciences Research Council (EPSRC) grant has recently started to study the photo-assisted nitrogen doping of ZnSe for blue lasers and LEDs. The group is also developing techniques for optical monitoring of industrial processes and has two postgraduate projects in this area. NEWI has close links with industry and is establishing a range of international contacts with which to collaborate on specific projects. The advanced materials research laboratory was identified in 1994 by the Welsh Development Agency (WDA) as a centre of expertise for industry within Wales which has helped to establish closer ties with local industry.

NEWI is involved in MOVPE research having published work on epitaxial MgS grown by MOVPE using bis(methylcyclopentadienyl)magnesium and tertiary-butylthiol precursors on both GaP and GaAs substrates; and the thermal decomposition of DtBSe.

The Optoelectronic Materials Research Lab of the NEWI has worked with Epichem Ltd, the Merseyside supplier of OM precursor materials. The project was supported by the ESPRC under grant reference K14490. Three possible nitrogen precursors were investigated under pyrolytic and photo-assisted MOVPE — triallylamine (TAN), trimethylsilylazide (TMSiN) and bis[di(trimethylsilyl)amido]zinc (BTM).

University of Liverpool

Department of Materials Science and Engineering
Liverpool L69 3BX
UK
This group is investigating the influence of the surface morphology on the relaxation of low-strained In Ga As linear buffer structures. This is a collaboration with Spanish universities Ciudad Universitaria and Universidad de Cádiz.

University College London

Diamond Electronics Group
Electronic Engineering
Torrington Place
London WC1E 7JE
UK
Tel: +44 171 391 1381
Fax: +44 171 388 9325
Department of Electronic & Electrical Engineering for processing, characterization and packaging of electronic devices. The department has two cleanrooms, a device characterization room and has fabricated a range of electronic and opto devices such as LEDs and lasers. The Diamond Electronics Group in the Department of Electronic and Electrical Engineering at University College London was established in 1990. Activities include diamond nucleation and growth, characterization and processing, and device fabrication and evaluation. It is involved in the design and fabrication of high performance electronic and optoelectronic devices from diamond and other wide band gap materials. Materials processed: silicon, GaAs and related compounds InP and related compounds glass, quartz and diamond.

University of Nottingham

Department of Physics
University Park
Nottingham NG7 2RD
UK

The Physics Dept. and Dept. of Electrical and Electronics Engineering are working on MBE-based III-V processes. The epitaxial system used is an in-house modified Varian Gen-II MBE system fitted with an Oxford Applied Research CARS25 plasma source for active nitrogen. The group has reported growth on a range of different substrates such as GaAs and gallium phosphide and doping using a range of different doping sources such as silicon and beryllium.

The group is partnered with CNRS in the LAQUANI program for MBE research in GaN materials and devices.

Workers at Nottingham have reported on yellow band and deep levels in undoped MOVPE GaN in a collaboration between Universidad Politecnica de Madrid and CRHEA-CNRS.

Another collaboration with the Ioffe Institute in St Petersburg concerns GaN grown directly on sapphire, Si and GaAs via a PA-MBE technique so as to understand the processes responsible for device degradation.

Roke Manor Research Ltd

Roke Manor
Romsey
Hampshire SO51 0ZN
UK
Tel: +44 1794 833206
Fax: +44 1794 833433

A contract R&D facility, wholly owned by Siemens, specialising in Communications, Surveillance and Management Systems. General interest in optoelectronics for a range of applications.

Sharp Laboratories of Europe Ltd

Edmund Halley Road
Oxford Science Park
Oxford OX4 4GA
UK
Tel: +44 1865 747711
Fax: +44 1865 747717

Established in February 1990, with a site area of 15 500 sq m (and floor area of 5 626 sq m), Sharp Corporation Laboratories of Europe has a key pioneering role in the company's global strategy for advanced devices including those based on wide bandgap semiconductors. Sharp is focussing on research themes of LCD displays, imaging technology, optoelectronics and information systems and technology to "maximize its contribution to the evolving world of multimedia and communications".

It has made a substantial investment in people and research equipment and facilities to take a full part in research activities in Europe. It has also joined a number co-operative research programmes of the UK Government and European Union.

Its location on the Oxford Science Park was designed to develop co-operation with Oxford University and with research departments in other universities in the UK and Europe.

University of Glasgow

MBE Research Group
Department of Electronics and Electrical Engineering
University of Glasgow
Glasgow G12 8QQ
UK
Tel: +44 141 330 4798
Fax: +44 141 330 4907
E-mail: crstanley@elec.gla.ac.uk
Under Professor Colin R Stanley, the group undertakes III-V semiconductor growth by molecular beam epitaxy, low temperature quantum transport measurements, and low dimensional device fabrication. Fundamental investigations into the mechanisms which control the deposition and doping of III-Vs.

Involved in the Scottish Collaborative Initiative in Optoelectronic Sciences (SCIOS) Research Programme, providing both structures and processed devices for evaluation in the Department of Physics at Heriot-Watt University as part of a drive to develop systems capable of all-optical signal processing. A 16×16 SSEED array fabricated in Glasgow from a multilayer InGaAs-GaAs strain-balanced structure grown by MBE has been successfully demonstrated at 1064 nm.

The MBE Group also undertakes joint research with the University of Lancaster, University of Exeter, Brooklyn Polytechnic and the Ioffe Institute in Moscow, amongst others.

The Optoelectronics Research Group consists currently of around 50 members of academic staff, research fellows and technologists, postdoctoral research assistants and research students - with comprehensive technician support. The group has a multi-million pound portfolio of collaborative research grants and contracts to support its activities. Our principal source of funding is the Engineering and Physical Sciences Research Council (EPSRC).

The Optoelectronics Research Group in the Department of Electronics and Electrical Engineering at The University of Glasgow has been established for over 25 years. It is closely integrated with other research programmes in nanoelectronics, dry-etching, semiconductor growth and sensors. Much of the research work is carried out in collaboration with industrial partners and with other research groups throughout the U.K. and the world.

Also in the remit of research are high power visible lasers under Prof JH Marsh; there are extensive nanofabrication facilities available at Glasgow.

University of Sheffield

Dept. of Electrical & Electronic Engineering,
Mappin Street
Sheffield S1 3JD
UK

This is the home of the EPSRC (Engineering and Physical Sciences Research Council) III-V Semiconductor Facility. The facility includes a dedicated III-nitride reactor from Thomas Swan Ltd (Cambridge, UK).

South Bank University

School of Electrical & Electronic Engineering
103 Borough Road
London
SE1 0AA
UK
Tel: +44 171 815 7513
Fax: +44 171 815 7599

Research activities include SiC-on-Si growth using ECR PACVD, dry etching; hetero-junction devices and contacts as well as materials and device characterization.

University of St Andrews

Department of Chemistry
University of St Andrews
St. Andrews
Fife KY16
UK
Fax: +44 1978 290008

This group is working on MOVPE of wide bandgap II-VI semiconductors such as ZnS/Se. It collaborates with the North East Wales Institute; they have jointly published work on the thermal decomposition of DtBSe, both alone and in the presence of DMZn.

University of Strathclyde

Department of Physics and Applied Physics
Strathclyde
Glasgow G4 0NG
UK

This university collaborates with University of Nottingham and has published work on the morphology and cathodoluminescence of GaN thin films. It also collaborates with Heriot Watt University on ZnSe research. Other studies include GaN epilayers grown on mis-oriented sapphire substrates.

Dr S Nakamura (now Prof S Nakamura of UCSB) is an Honorary Professor in Prof O'Donel's group.

University of Surrey

Optoelectronic Devices and Materials
Department of Physics
University of Surrey
Guildford
Surrey
UK
Tel: +44 1483 259400
This group is part of the European RAINBOW project.

Thomas Swan Scientific Equipment Ltd (now part of AIXTRON AG)

Buckingway Business Park
Cambridge CB4 5UG
UK
Tel: +44 1223 519444
Thomas Swan is an MOVPE equipment supplier which has achieved many sales worldwide particularly for its GaN reactors. In 1999 it delivered a 3×2-in GaN system to the University of Bremen in Germany and a similar system in the UK to the EPSRC III-V Semiconductor Facility at the University of Sheffield.

Thomas Swan also donated one of its 6×2-in GaN reactors to the University of Cambridge, UK, as part of a collaborative R&D programme also funded by EPSRC. In Asia, a Taiwanese company purchased three of the company's standard CCS GaN 6×2-in reactors for delivery in February 2000.

Thomas Swan has traditionally worked very closely with a number of universities, but particularly the University of Gent in Belgium, to develop new products. In the future it plans to maintain these relationships but has created a its own in-house R&D unit to permit it to focus on developing systems for mainstream manufacturing. However, the announcement in 1999 of Thomas Swan's moving into the AIXTRON group meant that the company would have access to that company's extensive R&D and sales/support organization worldwide.

University of Wales

Department of Physics and Astronomy
PO Box 913
Cardiff CF2 3YB
UK
This group has published results of studies of shallow acceptor levels in III-nitride layers grown by MBE.

University of Wales, Bangor

School of Informatics
Dean Street
Bangor
Gwynedd LL57 1UT
UK
Tel: +44 1248 382686
Fax: +44 1248 361429
Email: informatics@informatics.bangor.ac.uk

Amongst the research conducted here is an experimental demonstration of chaos-based secure optical communications under an EPSRC project in collaboration with NORTEL Networks, Harlow, UK. The aim of this project is to experimentally demonstrate that chaotic external cavity semiconductor lasers can be used to provide a high data rate secure communication channel. It studies the security of a GHz data transmission system that uses the chaotic output of a laser diode to mask the data signal. The project will provide proof-of-principle experimental results, and will quantify the degree of security this encryption format can provide. This work will be accomplished by using commercially available laser diodes supplied by NORTEL Networks.

Professor K Alan Shore leads a group with research interests in optoelectronics, optical communications and nonlinear dynamics. Research projects are directed at advancing the design and applications of optoelectronic devices and, in particular laser diodes and optical waveguides. Areas of interest range from basic device operating principles through to matters of engineering design. Complementary interest is in the nonlinear dynamics of optical devices and systems.

Much of the work funded by the EPSRC, the Royal Society, and the British Council is being undertaken within national and international collaborative projects involving both universities and industry. Current international collaborations involve groups in Argentina, Australia, France, Ireland, Spain and the US. Collaborations have also been undertaken with groups in Belarus, Denmark, Germany, Italy, the Netherlands and Japan.

Recent Research Projects Include Quantum Well structures for intersubband population inversion, Packaging and hybrid integration of semiconductor lasers in collaboration with University of Bath and Nortel. For example, 'optical chaos cryptography using laser-diode transmitters' (OCCULT) in collaboration with Department of Physics, University de les Illes Balears, Palma de Mallorca, Spain. Plus the non-linear dynamics in laser diodes, a Royal Society Collaborative Project with the Institute of Physics, Minsk, Belarus.

7.26 USA

Air Force Research Laboratory

Sensors Directorate
Electromagnetic Materials/SNHX
USA
This government laboratory has published results from the synthesis and growth of GaN by a chemical vapour reaction process (CVRP) in collaboration with Solid State Scientific Corp. These studies include synthesis and bulk crystal growth of GaN single crystal c-plane platelets by sublimation of solid ammonium chloride in a carrier gas, which passes over gallium at a temperature of approximately 900°C at near atmospheric pressures.

Arizona State University

MBE-Optoelectronics Research Group
Engineering Research Center
Tempe
AZ 85287-1704
USA
This group is interested in the MBE growth of GaN on sapphire and has collaborated with other universities such as Brookhaven National Laboratory and the Univ. of Illinois, Urbana-Champaign. Another group has collaborated with Motorola Phoenix Research Centre on organic electroluminescent devices. The Arizona research has also included the characterization of ZnMnSSe quaternary alloys for visible light emitting devices.

The Centre for Solid State Science at the university is also active in wide bandgap semiconductors. For example, it has collaborated with other US institutes such as Oak Ridge National Laboratory in the low temperature CVD of 3C-SiC films grown on Si (100) using special silane mixtures to reduce dislocation density and improve film quality.

Astralux Inc

2500 Central Avenue
Boulder
CO 80301
USA
This small company is led by Prof Jacques Pankove. Researchers have published work on GaN emitters with high-power high-temperature hetero-bipolar transistors in collaboration with the University of Ulm, Universität Erlangen-Nürnberg and the Forschungszentrum Julich, Inst. für Schicht-und Ionentechnik.

ATMI Epitaxial Services

21002 North 19th Ave
Suite 5
Phoenix
AZ 85027-2726
USA
Tel: +1 602 581 3663
Fax: +1 602 581 3415

Former ATMI subsidiary Epitronics, has been renamed ATMI Epitaxial Services (see Company Profile). It is a developer and supplier of a wide range of epiwafer products and has developed a number of key wide bandgap semiconductor materials including the world's only independent source of ELOG gallium nitride substrates.

It is a participant in a number of US R&D programs including the GaN Microwave Power Amplifier program. ATMI was a member of a consortium that in 1995 was awarded an US$4 million, 2-year contract from DARPA to develop blue semiconductor lasers and LEDs. Other members of the consortium were SDL (now part of JSD Uniphase), Xerox American Crystal Technologies, Boston University and the University of Texas at Austin. The consortium members provided additional funds of more than US$4 million. Under the programme, the consortium aimed to develop advanced GaN epiwafers for use in the fabrication of blue lasers and LEDs.

AT&T Bell Laboratories

600 Mountain Ave
Murray Hill
NJ 07974
USA

This research establishment has published work on oxide substrates for GaN epitaxy such as ScAlMgO4 and ZnO. Techniques in use include plasma-based activation of reactants in conjunction with MBE. Optical Physics Research Department has published work on organic light emitting devices.

AXT

4311 Solar Way
Fremont
CA 94538
USA
Tel: +1 510 683 5900

American Xtal Technology (AXT) is a developer and supplier of a wide range of semiconductor wafer products (see Company Profile). Its portfolio now includes silicon carbide and indium phosphide and it is also researching into gallium nitride. It has a number of research programs underway in wide bandgap semiconductor crystal wafer development.

As well as internally funded programs, AXT has two US Government contracts to develop GaN substrate growth technology. To this end, AXT has an alliance with the Polish High Pressure Research Center (HPRC) to develop a bulk GaN synthesis technique, evaluate the crystal quality and realize good device performance on these GaN substrates. GaN single crystals with diameters up to 5 mm have been produced. AXT hopes to be able to eventually manufacture GaN substrates using this method on a commercial basis.

Boston University

The Center for Photonics Research
143 Bay State Road
Boston
MA 02215
USA.
Tel: +1 617 353 8899
Current research projects in the MBE laboratory address both material and device issues of the family of III-V nitrides grown by MBE and hydride VPE methods. These studies cover both bulk thin films as well as heterojunctions and multiquantum wells. Material issues include growth and doping of III-V nitrides by plasma-assisted MBE. Particular emphasis is placed on developing the InGaN alloys for optical devices operating in the visible spectrum. Issues related to miscibility of the InN-GaN pseudo-binary system are being addressed. The AlGaN alloy system is also being optimized for applications in the ultraviolet part of the spectrum.

The device program includes the fabrication of blue-green LEDs and lasers and solar-bind ultraviolet detectors. These programs are funded by ARPA in various industrial consortia. The group also has published work on SiC/GaN transistors for high-temperature applications.

The Electrical, Computer, and Systems Engineering department is investigating the development of blue LEDs and lasers in GaN wide bandgap material, with the intention to develop devices. The University successfully developed a GaN UV-blue LED to construct a UV-blue laser in 1995. It published work on growth and doping of AlGaN alloys by ECR-assisted MBE, and collaborates with many universities including: Polish Academy of Sciences, North Carolina State University and Pierre et Marie Curie Université. R&D includes GaN growth with alternative nitride sources and computer-controlled growth of GaN.

The university was also a member of a consortium that in 1995 was awarded an US$4 million, two-year contract from DARPA to develop blue semiconductor lasers and LEDs. Other members of the consortium were SDL (now JDS Uniphase), Xerox American Crystal Technologies, Advanced Technology Materials Inc and the University of Texas at Austin. The consortium members provided additional funds of more than US$4 million. Under the programme, the consortium aimed to develop advanced GaN epitaxial wafers for use in the fabrication of blue lasers and LEDs.

Brown University

Division of Engineering and Department of Physics
Providence
RI 02912
USA
Fax: +1 401 863 1387
This group is involved in II-VI materials, dealing with QW heterostructure emitters for blue-green lasers. It collaborates with Purdue University, Indiana.

California Institute of Technology

Pasadena
CA 91125
USA
Research includes W and Re contacts to GaN and other electronic devices.

University of California, Berkeley

National Lab
1 Cyclotron Road Mailstop 66
Berkeley
CA 94720
USA
Tel: +1 510 486 6303
Lawrence Berkeley Lab, a US Department of Energy facility managed by the University of California, has a strong involvement with compound semiconductors. Recent achievements include uncovering the secrets behind nitrogen's influence on GaInAs and using an electron microscope to resolve nitrogen atoms in the presence of gallium atoms in GaN.

Also at Berkeley researchers have been using the one-Angstrom Microscope at the National Center for Electron Microscopy (NCEM) to resolve N atoms in the presence of more massive gallium atoms in GaN in columns spaced only 1.13 A apart. Moreover, they have made unprecedented images of columns of carbon atoms in a diamond lattice, only 0.89 A apart.

Researchers from Berkeley Lab's Materials Sciences Division are developing preparative methods for thick GaN crystals, using the Center for X-Ray Optics fluorescence microprobe to test the quality of the crystals made. Having a clear view of what is going on inside the material will help researchers to optimize the growth process to make high-quality GaN. This group has published work on the defect spectroscopy of wide bandgap GaN. Other related research topics include ZnSe epilayers grown on dissimilar substrates such as GaAs.

University of California, Santa Barbara

Electrical and Computer Engineering and Materials Departments
Santa Barbara
CA 93106
USA
Tel: +1 805 893 8511
Fax: +1 805 893 4500
At UCSB, specific III-nitride research interests include growth of wide-bandgap semiconductors, such as GaN, and their application to blue LEDs and lasers and high power electronic devices.

In early 2000, Dr S. Nakamura the pioneer researcher and developed of III-nitride LEDs and diode lasers at Nichia Chemical Industries, became a Professor at the UCSB Materials Department.

Research at UCSB led to the first US university demonstration of a blue GaN laser diode. UCSB graduate students Michael Mack and Amber Abare having succeeded in growing, fabricating, and testing an InGaN/GaN laser diode.

Prof Steven P. DenBaars of the Materials Department, UCSB, is the lead investigator of the ARPA-funded Multi-university Nitride Consortium which will develop and transfer GaN technology to industry.

The Optoelectronics Technology Center (OTC) concentrates on the formation of advanced vertical-cavity laser and photodetector arrays as well as their combination with integrated circuits using new heterogeneous integration technologies. This technology should provide new device and materials capabilities for the next generation of parallel computer interconnects and data communications. OTC is the lead member of a multi-campus university research consortium – The Heterogeneous Optoelectronics Technology Center (HOTC) – which was established in 1997 after a national competition by the Defense Advanced Research Projects Agency. HOTC is composed of investigators from UCSB, Cornell, UCSD, UCLA, USC and UT-Austin who are teamed to provide advanced capabilities for real-time information access systems. A main aspect of the Center's charter is to encourage collaboration between university and industry leaders in the area of optical interconnects and memory.

Also, a significant commercial GaN-based company was spun out of UCSB. Originally called Widegap Technology LLC (WiTech) its name was changed to Nitres Inc in late 1999. Nitres continues WiTech's strategies in blue, green, white and UV light emitter products. It also has R&D underway in HEMTs, HBTs and UV detectors and blue diode lasers. This includes several significant government contract awards such as a $1.5 million award for the development of white LEDs from the US Navy NSWC, Dahlgren, VA. The white LEDs are GaN-based working in conjunction with a polymer-based down-conversion medium.

The group is investigating the structural and optical properties of GaN laterally overgrown by MOVPE on Si using an AlN buffer layer. This work was a collaboration with H-P research labs and results were similar to the structural properties of LEO GaN grown on GaN/sapphire.

University of California, Los Angeles

UCLA Electrical Engineering Department
Optoelectronics Group
Los Angeles
CA 90095-1406
USA

Under Prof E Yablonovitch, researchers at UCLA have investigated wide band-gap gallium nitride LEDs. The work focused on characterization of the quantum efficiency of GaN and its alloys. Blue LEDs have been fabricated using an AlGaN/InGaN double heterojunction.

The researchers are also involved in the optimization of OMVPE growth of GaN on sapphire, working on GaN buffer layers where partial coherence can be achieved during the low temperature growth stage.

Researchers in the Department of Chemistry and Biochemistry at UCLA have developed a process to produce bulk quantities of high purity, polycrystalline gallium nitride. GaN is traditionally prepared by heating materials at high temperatures and pressures over extended periods of time. Using this new process, GaN is prepared using a solid-state metathesis (exchange) reaction which occurs rapidly (less than 1 second) under moderate pressure. A modification of this new process can enable the reaction to proceed without applying any external pressure. The technique builds on earlier work previously patented by the University. Under the UCLA Technology Transfer Program this process has been made available for commercial licensing.

University of California, San Diego

Ultrafast and Nanoscale Optics Group
Department of Electrical and Computer Engineering
9500 Gilman Drive
Mail Code 0407
University of California
San Diego
La Jolla
CA 92093-0407
USA
Web: kfir.ucsd.edu/Facilities/facility.shtml

Many research projects are carried out in close collaboration with industrial partners for successful technology transfer. Along with well funded research programs and the strong interest of eleven faculty members, the photonics/opto-electronics program enjoys state-of-the-art facilities for opto-electronic materials growth (such as MBE and MOUVD) and characterization, (such as STM, X-ray, PL) device micro-fabrication, (such as E-beam lithography) packaging and testing as well as system feasibility experiments. The availability of this research forum centred around the Optoelectronics Technology Center with the basic microelectronic, optoelectronic and photonics knowledge, equipment, and facilities in one location has allowed UCSD to establish itself internationally as one of the leading research institutions in the area of photonics/optoelectronics. Present research focus is on three major areas: Optoelectronic Materials and Devices, Optics for Computing, and Photonic Networks.

The Ultrafast and Nanoscale Optics Group (UNO) is part of the Applied Optics and Photonics Division of the Electrical and Computer Engineering Department at the University of California, San Diego, which primarily focuses on basic and applied research into non-linear and diffractive optics.

Carnegie Mellon University

Department of Materials Science and Engineering
Pittsburgh
PA 15213
USA

This university is involved in research into high resistivity AlGaN layers on sapphire grown by MOVPE. Two departments — the Department of Materials Science and Engineering, and Department of Electrical and Computer Engineering and also in collaboration with Hughes Research Laboratory. It works on research in GaN-InGaN heterostructures on silicon substrates in collaboration with APA Optics and Colorado State University.

University of Colorado

Boulder
CO 80309-0009
USA

This university is working on high temperature HBTs using a SiC and GaN emitter in collaboration with Astralux Inc. A record current gain of ten million has been obtained at room temperature, decreasing to 100 at 535°C. It collaborates with Carnegie Mellon University and APA Optics on research in GaN-InGaN heterostructures on silicon substrates. A collaboration with Astralux involved growth of GaN on 4H-SiC to fabricate a high temperature HBT device at 300°C.

The university received a 3-year DARPA URI grant for the Study of GaN for Short Wavelength Emitters LP-MOVPE fabrication of conducting p-type GaN PN junctions are to be made to emit blue-to-UV light including diode laser structures.

University of Connecticut

Electrical and Systems Engineering Department
University of Connecticut
Storrs
CT 06269-2157
USA
Fax: +1 860 486 2447
This department is working on CdZnSe by MOVPE for blue-green lasers.

Cornell University

School of Electrical Engineering
Ithaca
NY 14853
USA
Tel: +1 607 255 4206
Fax: +1 607 255 5373

Cornell University is involved in many aspects of GaN research. It is home to the CHESS Initiative which grew its first films of a semiconducting gallium nitride in August 1994 by metallorganic MBE.

Researchers have published papers on GaN growth by MBE using a remote RF plasma nitrogen source. The team is also working on the growth of GaN on sapphire substrates by gas source MBE, in order to produce a better substrate for epitaxial growth in device development. MBE GaN research at Cornell has been supported by the National Science Foundation through the Materials Science Center at Cornell, Rome Laboratories, the Office of Naval Research, and Eastman Kodak.

The "Real Time Growth" project is underway with the participation of Cornell's Materials Science Center. A new chamber has been used to study the growth of gallium nitride thin films as a continuation of earlier work. In addition to film growth using conventional gas precursors, the group is can accommodate experiments using ion beam or molecular beam sources. These experiments are carried out in collaboration with faculty members affiliated with the Materials Science Center.

The Cornell group led a multidisciplinary university research initiative (MURI) for high power, linear, broadband solid state amplifiers for ONR applications. Also involved are Renesselaer Polytechnic Institute, and the Northrop Grumman company division (formally part of Westinghouse). The proposed program will focus on development of GaN materials, devices and circuits. Cornell has demonstrated record GaN HFET performances with ft of 40 GHz and f_{max} of 99 GHz in transistors with more than 40 V breakdown. It has also been active in high temperaure electronic devices such as mixed signal ICs based on SiC.

Cornell University researchers have reported significant progress in making a new generation of GaN transistors. Lester F. Eastman, the John L. Given Foundation Professor of Engineering, and James R. Shealy, professor of electrical engineering, say they have tested GaN transistors with output power of up to 2.2 W/mm at a frequency of 4 GHz and expect to see power figures five times higher as soon as improved test equipment is installed. The GaN is grown on a heatsink made of either SiC or sapphire. SiC conducts heat about 10 times as well as sapphire and makes the high power possible, but it is still expensive. Northrop Grumman and Cree Research, two companies also working to develop GaN transistors, supply SiC wafers to the Cornell researchers as a contribution to science education.

The Cornell research is principally supported by a three-year, $1 million per year grant from the Multi-Disciplinary University Research Initiatives (MURI) program of the Office of Naval Research (ONR), with an additional $1 million per year from the Defense Advanced Research Projects Agency (DARPA) and other federal agencies.

Cree Inc

400 Silicon Drive
Durham
NC 27703-8475
USA
Cree (formerly Cree Research Inc) is a world leader in the wide bandgap semiconductor field with an extensive range of products based on SiC and GaN (see Company Profile).

Cree is one of only few companies to have succeeded in developing a continuous wave blue diode laser which operates at room temperature as part of its ongoing development project in this important area. The company has also demonstrated continuous wave laser operation at room temperature. This was achieved by improved high reflectivity facet coatings, which lowered the threshold current for lasing as developed by Arto Nurmikko's group at Brown University. The CW emission lasted for more than fifteen seconds. Cree had been working for two years to develop its laser device with partial funding from the Defense Advanced Research Projects Agency (DARPA).

This was announced on June 9th, 1997, via a demonstration of an electrically pulsed gallium nitride based blue laser. The breakthrough was designed and fabricated using Cree's SiC wafers combined with a proprietary GaN thin film process.

Dr Jan Schetzina, professor of physics at North Carolina State University and a partner in the DARPA effort discovered the initial laser demonstration. Cree retains close ties with the University of N. Carolina, the research department from which it sprang in the early 1990s and with which it retains close ties to this day. Dr Arto Nurmikko, professor of physics at Brown University, also observed blue laser operation from sample devices received from Cree.

The Electro-Optics Center

West Hills Industrial Park
77 Glade Drive
Kittanning
PA 16201
USA
Tel: +1 724 545-9700
Fax: +1 724 545-9797
Website: www.electro-optics.org
Technology areas include fibre optic technology, lasers, materials, design and process technology, night vision and IR.

Projects include 'Fiber Optic Array Electrical Splice' (FOAES) and SPAWAR, Advanced Deployable System (ADS) to develop a more automated, more reliable method for splicing sensors on to multi-conductor, water proof electrical cables to reduce the touch labour, increase yield, and improve array reliability. Also a torpedo tether for the NAVSEA Undersea Weapons Program Office for fibre optic spool winding, spool lubrication, and payout acceptance testing will be developed leading to an efficient manufacturing process to reduce the production cost of the fibre optic tether.

Navy MANTECH Projects include 'Automated Diode Array Manufacturing' (ADAM) to develop an automated manufacturing capability for the laser diode array mounting process with the goal of demonstrating high rate production at a cost of $20 or less per diode bar. HILDA — High Intensity Laser Diode Array for the ARMY Strategic Missile Defense Command (SMDC), to demonstrate the fabrication of high intensity laser diode array with high peak power and pulse rate 10–15% duty cycle at 805 nm.

EMCORE Inc

394 Elizabeth Avenue
Somerset
NJ
USA
Tel: +1 908 271 9090
From its founding as a specialist epitaxy equipment manufacturer in the 1980s, EMCORE has become a vertically-integrated company specialising in the design and production of MOVPE reactors and associated process technology and devices (see Company Profile).

The EMCORE Electronic Materials (E2M) division provides foundry services, including offering GaN-based epitaxial wafers. GE Lighting and EMCORE Corporation announced in January 1999 the formation of a new joint venture GELcore LLC, which will target the market for "white light" LEDs. EMCORE also has a joint venture with Uniroyal Technologies called Uniroyal Optoelectronics LLC, which included a transfer of technology. The company has published a number of papers describing its success in the growth of III-nitride materials for optoelectronics. In particular, growth of GaN, InGaN and AlGaN layers and InGaN/(Al)GaN quantum well structures in a multi-wafer high speed rotating disk reactor. Both n- and p-doping and high quality optical properties have been achieved with excellent uniformity of the structures.

University of Florida

Dept of Materials Science and Engineering
Rhines Hall
Gainesville
FL 32611
USA
Tel: +1 352 846 1087
This group, led by Prof Cammy R Abernathy, regularly reports its findings on wide bandgap semiconductors. It was at the University of Florida that Dr S.

Nakamura began his interest in these materials which ultimately led to his ground-breaking development of blue emitter optoelectronic devices.

The world's first ZnSe blue laser was produced at the University of Florida by Prof Bob Park in the early 1990s.

Today, the University of Florida is investigating a range of topics associated with III-nitrides including epitaxial Er-doping of GaN. It has an on-going collaboration with workers at university departments in Taiwan doing research into GaN Schottky rectifiers.

University of George Mason

George Mason University
Department of Electrical & Computer Engineering
Fairfax
VA 22203
USA
Tel: +1 703 527 5410
Research into electrical contacts for GaN and SiC devices.

Hewlett-Packard Laboratories

Solid State Materials Dept
Solid State Laboratories
Hewlett-Packard Laboratories
3500 Deer Creek Road
Palo Alto
CA 94304
USA
These laboratories were formerly known as H-P but are now under the new name Agilent but their basic corporate R&D role remains largely unchanged (see Company Profile).

The labs are well-known worldwide for many pioneering innovations in optoelectronics.

HP was a member of a consortium that in 1995 was awarded US$4 million for a 2-year contract from the Advanced Research Projects Agency to develop blue semiconductor diode lasers and LEDs. Other members of the consortium were:

- SDL Inc (now JDS Uniphase).
- Xerox American Crystal Technologies.
- Advanced Technology Materials Inc.
- Boston University.
- University of Texas at Austin.

The consortium members provided additional funds of more than US$4 million. Under the programme, the consortium aimed to develop advanced GaN epitaxial wafers for use in the fabrication of blue lasers and LEDs. George Craford, who was largely credited with inventing super-bright transparent-substrate AlGaInP-based LEDs, led the HP team.

In the field of wide bandgap semiconductors the Palo Alto labs have studied structural and optical properties of GaN which was laterally overgrown by MOVPE on Si in collaboration with the University of California, Santa Barbara.

The John Hopkins University

Applied Physics Laboratory
Laurel
MD 20723
USA
This institute has published electrical transport studies on the III-nitrides such as InN, GaInN and AlInN. Other studies include measurement of the optical response so as to to determine compositional information.

Hughes Research Laboratory

Malibu
CA 90265
USA
Hughes Research Laboratory is part of the world famous Hughes company and feeds the various R&D streams for this company which is world leader in aerospace and related application markets.

Workers at the Malibu facility have published work into high resistivity AlxGaN layers grown by MOVPE. Hughes STX Corp has participated with US institutes (University of Maryland & John Hopkins University) on GaN-based UV detectors.

IBM Almaden Research Center

650 Harry Road
San Jose
CA 95120-6099
USA
Tel: + 408 927 1000
The IBM Almaden Research Center is involved in GaN research and has published papers on improved optical activation of ion-implanted Zn acceptors.

University of Illinois at Urbana

Materials Research Laboratory
IL 61801
USA
Work has been published by the Materials Research Lab (in collaboration with CNRS of France) on the growth of prismatic domains in ECR-assisted MBE of GaN/SiC. Using N-rich growth conditions, prismatic domains were formed in the initial stage.

The Microelectronics Lab has also published work on high quality GaN films grown on a- and c-sapphire by atmospheric pressure MOVPE. Double buffer layer, with GaN as the first layer and AlN as the second have been shown to lead to uniform mirror-like films.

The university's Epi-Center Facility enables researchers from Materials Science & Engineering, Electrical Engineering, and Physics to investigate epitaxial growth and synthesize new material configurations for scientific and technical applications. The facility contains six UHV chambers (for crystal growth by MBE) dedicated to: Si and Ge, ceramics, devices, GaAs, and metals. A UHV-CVD reactor has also been brought into operation. The characterization equipment includes surface x-ray scattering, x-ray photoelectron spectroscopy, ultraviolet electron spectroscopy, ion scattering spectroscopy and STM, as well as RHEED and mass spectrometry available during growth.

Recently, As-doped GaN films and GaNAs films have been made using MOVPE grown on sapphire, GaN-coated sapphire, and GaAs substrates.

Kansas State University

Dept. of Chemical Engineering
Durland Hall
Manhattan
KS 66506-5102
USA

This university department has collaborated with other US institutes such as the Oak Ridge National Laboratory in the low temperature CVD of 3C-SiC films grown on Si (100) using special silane mixtures to reduce dislocation density and improve film quality.

Lawrence Livermore National Laboratory

Center for Microelectronics & Optoelectronics
PO Box 5503
M/S L-271
Livermore
CA 94551
USA
Tel +1 510 423 7801

Lawrence Livermore National Laboratory (LLNL) is among the world's largest research institutions. The Center for Microelectronics and Optelectronics (CMO) is just one of the areas of reseach LLNL funds (the annual budget is US$ 1billion, employing 7000 staff). LLNL's areas of interest in microelectronics and optoelectronics arise from high-speed detection, transmission and transient recording requirements for nuclear test programs, specialized computing and instrument interfacing requirements for satellite systems, as well as laser fusion.

University of Massachusetts

Department of Electrical and Computer Engineering
Marcus Hall
Amherst
MA 01003-5220
USA

The university has two semiconductor groups: the Compound Semiconductor Laboratory (CSL) and the Semiconductor Spectroscopy Laboratory (SSL). The two groups research into heterostructure FETs, optical modulators and detectors and semiconductor lasers, as well as OMCVD, QW electronic structure and optical properties.

The CSL, under the direction of Prof Kei May Lau, is an organized research unit within the School of Engineering. Founded in 1983, the laboratory serves as a focus for the R&D of novel compound semiconductor devices for high frequency and photonic applications. It emphasizes experimental studies of III-V compound semiconductor materials and devices. Research activities include: semiconductor device design and modelling, epitaxial growth by MOVPE, materials characterization, device fabrication and circuit applications.

CSL's equipment includes: an in-house built MOVPE system with a specialised "Nitride Addition" which with a 50 mm inner diameter and a water cooling jacket, can ramp from room temperature up to 1350K in less than two minutes; a computer-controlled AIXTRON AIX 200/4 MOVPE system (for growth of III-V QW and other heterostructures), a Hall effect measurement system, commercial CV profiler, double crystal x-ray diffractometer as well as facilities for device fabrication and material characterization.

SSL is concerned with optical characterization of advanced semiconductor materials and devices, using techniques such as photoluminescence, photoreflectance, electroreflectance and photocurrent spectroscopies. SSL developed a new polarized excitation luminescence technique which enables improved identification of optical transition types.

University of Michigan

Solid-State Electronics Laboratory (SSEL)
Department of Electrical Engineering and Computer Science
1301 Beal Avenue
Ann Arbor
MI 48109-1063
USA
The Optoelectronic Research Group is directed by Professor Pallab Bhattacharya, it conducts experimental and theoretical research in various areas of solid-state electronics. The Solid-State Electronics Laboratory (SSEL) of the University of Michigan is well equipped with many facilities, including those for photolithography, CVD, MOVPE and MBE. Comprising the Solid-State Fabrication Facility (opened in 1986), the group is involved in high-speed devices and device technology and optoelectronics.

University of Minnesota

Department of Electrical Engineering
University of Minnesota
200 Union St SE
Minneapolis
MN 55455
USA
This group has published research into Ga surface accumulation and the consequential reduction in the growth rate of GaN.

MIT Lincoln Laboratory

Rm E-118E
244 Wood Street
Lexington
MA 02173-9108
USA
Tel: +1 781 981 7842
Fax: +1 781 981 3867

The centre is a participant in a number of collaborative research programs including in the wide bandgap semiconductors, the GaN Microwave Power Amplifier program.

NASA Lewis Research Center

21000 Brookpark Road
Cleveland
OH 44135
USA
Tel: +1 216 433 8902
Fax: +1 216 433 8643

For some years now NASA has been studying SiC high temperature, high power semiconductor electronics for aerospace applications. SiC epitaxial growth and device fabrication technology needed to realize advanced high temperature, high power SiC semiconductor electronics.

NASA's Lewis Research Center has an EMCORE MOVPE reactor for its SiC R&D efforts and is also studying II-VI semiconductors grown on sapphire.

The MOVPE was tailored by EMCORE specifically for the exceptionally high temperature deposition required in NASA's SiC development efforts. It is equipped with the latest RF heating option that allows operation in excess of $1600°C$. This RF heated MOVPE platform is utilized under the leadership of NASA/Lewis' Dr David J. Larkin.

University of New Mexico

Center for High Technology Materials
University of New Mexico
Albuquerque
NM 87131
USA

Studies at the University of New Mexico include low temperature buffer layer and layer thickness in the optimization of MOVPE growth of GaN on sapphire.

State University of New York at Buffalo

Dept. of Electrical Engineering
Buffalo
NY 14260
USA

This group has investigated the piezoelectric electric field distribution in strained PIN III-nitride MQW structures with the Xerox Palo Alto Research Center.

North Carolina State University

Department of Materials Science and Engineering
Raleigh
NC 276975
USA

Research teams at this university (led by Prof Robert F Davis), are involved in characterization of thin films of wide bandgap semiconductors, such as diamond, SiC, AlN, GaN and InGaN. They have fabricated and completed characterization on working electronic and optoelectronic devices, such as high-power and high-temperature transistors, blue LEDs and high storage capacitors.

ELO MOVPE growth work at NC State has been shown via TEM to bring about a reduction of the dislocation density in overgrown regions.

The University als works closely with Cree Inc (see Company Profile) and Dr Jan Schetzina, professor of physics at NCSU and a partner in the DARPA effort discovered Cree's initial laser demonstration. Cree retains close ties with the University of N. Carolina, the research department from which it sprang in the early 1990s and with which it retains close ties to this day.

Nitres Inc

5655 Lindero Canyon Road
Suite 404
Westlake Village
CA 91362
USA

Nitres Inc (formerly Widegap Technology, LLC [WiTech]), is a privately-owned company which was founded in 1997 as an offshoot of the University of California, Santa Barbara (UCSB). It is mainly research oriented, and has undertaken several US government-funded projects (see Company Profile).

Northwestern University

Center for Quantum Devices
Department of Electrical and Computer Engineering
Evanston
IL 60208
USA

The work underway by this institute has led to publication of many leading edge results in wide bandgap semiconductor devices.

In 1998, The group published a paper claiming to have successfully made a working blue-green diode laser. Published in the *MRS Internet Journal of Nitride Semiconductor Research* diode lasers with very low threshold current densities were said to have been achieved by eliminating Al from the cladding layer of the laser diode structure. The group achieved excellent p-type ohmic contacts and low series resistance. GaInN/GaN multi-quantum well lasers grown on sapphire substrates by low pressure MOVPE.

However, according to industry experts, NWU's achievement of actual lasing in these structures may have appeared to have been premature. The results were perhaps superluminescent LEDs rather than diode lasers.

NWU's centre is supported by a number of US military funding agencies including the ONR, the BMDO, DARPA, and the US Air Force.

Other work includes a collaboration with Nagoya Institute of Technology on the simultaneous growth of two differently oriented GaN epilayers on sapphire by MOVPE.

At the Center for Quantum Devices, Dr M Razeghi and her world famous group of over 40 post graduate students and scientists are engaged in research into a number of material systems and have a capability from in-depth characterization of material up to and including fabrication and life testing of packaged devices.

Oak Ridge National Laboratory

High Temperature Materials Laboratory
Oak Ridge
TN 37831-6064
USA
This national institution has collaborated with other US institutes such as Kansas State University in the low temperature CVD of 3C-SiC films grown on Si (100) using special silane mixtures to reduce dislocation density and improve film quality.

Office of Naval Research (ONR)

800 North Quincy Street
Ballston Tower 1
Arlington
VA 22217-5660
USA
Tel: +1 703 696 4218
Fax: +1 703 696 2611
The ONR research laboratories are responsible for coordinating research programs in advanced semiconductors. For example, the strong piezo-electric coefficient in the nitrides program launched in 1998.

University of Ohio

Department of Physics and Astronomy
Smith Laboratory
Columbus
OH 43210
USA

A group of 9 faculties from Physics, Chemistry, Chemical Engineering and Electrical Engineering at Ohio University were awarded a University Research Initiative Support Program (URISP) grant by the US Department of Defense for research on growth, doping and electrical contact formation on wide bandgap semiconductors. The grant provides US$2 million over 5 years for basic research and development of infrastucture related to materials such as GaN and AlN. The project director will be Prof Martin E Kordesch, who's laboratory and research on diamond, another wide bandgap semiconductor, has been funded since 1990 by the US Department of Defense through the Office of Naval Research and the Office of Innovative Science and Technology of the Ballistic Missile Defense Organization (BMDO).

OSU Center for Laser and Photonics Research

Center for Laser and Photonics Research
413 Noble Research Center
Oklahoma State University
Stillwater
OK 74078
USA
Tel: +1 405 744 6575

Scientists (in collaboration with Oklahoma industry), have a project to develop the use of lasers, enhancing manufacturing of products or production services marketed by the companies involved. An example of this collaboration is the development of blue LEDs and laser diodes with Eagle-Picher Industries Inc.

Pennsylvania State University

Dept of Materials Science & Engineering
221 Steidle Building
University Park
PA 16802
USA
Fax: +1 814 863 8561

This institute has conducted research into ohmic contacts to GaN with Electronic Materials Processing Research Laboratory, EMCORE Corp and Hewlett-Packard Optoelectronics Div.

Purdue University

School of Electrical Engineering
West Lafayette
IN 47907
USA
This group collaborates with Brown University on work into II-VI lasers. It also has an ongoing research activity in III-nitrides using an Aixtron AIX 200 MOVPE system

Rensselaer Polytechnic Institute

Electrical, Computer, and Systems Engineering and Center for Integrated Electronics and Electronics Manufacturing
Room 9017, CII
110 8th Street
Troy
NY 12180-3590
USA
Tel: +1 518 276 2201
Fax: +1 518 276 2990
Rensselaer Polytechnic Institute is a participant in a number of collaborative research programs including semiconductors.

Rice University

Rice Quantum Institute
Rice University
Houston
TX 77251-1892
USA
Tel: +1 713 348 4833
Fax: +1 713 348 5686
Web: www.ruf.rice.edu/~lasersci/
E-mail: fkt@rice.edu
Under Prof Frank K Tittel, this group conducts research and development in quantum electronics, in particular laser spectroscopy applied to sensitive, selective and real-time trace gas detection, laser applications in medicine and biology and deep-UV excimer based microlithography. Also being researched are laser-based mid-IR sensors. Tunable near-infrared diode laser sources are mixed and frequency converted to longer wavelengths using state of the art nonlinear optical crystals such as periodically poled lithium niobate.

Rockwell Science Center

1049 Camino Dos Rios
Thousand Oaks
CA 91358
USA
Tel: +1 805 373 4191
Fax: +1 805 373 4775
Rockwell is a leading developed and supplier of a wide range of electronic devices for the telecoms marketplace. The centre is a participant in a number of collaborative research programs in semiconductors.

Sandia National Laboratories

PO Box 5800
Albuquerque
NM 87185-0603
USA
This group has reported theoretical results on the gain in diode laser structures. It is suggested that the re-absorption in non-homogeneous InGaN quantum well layers may cause an increase in the laser threshold currents.

On Friday the 13th, 1998, SDL (now JDS Uniphase), of San Jose, California announced their recent achievement of nitride-based diodes lasing in the blue-violet.

SDL Inc (now part of JDS Uniphase)

80 Rose Orchard Way
San Jose
CA 95134-1365
USA
SDL Inc which was founded in 1983, is a world leader in the design, manufacture and sale of semiconductor lasers, laser-based systems and fibre optic-related solutions (see Company Profile).

In 1998, SDL was one of the first companies to produce a working blue laser diode. SDL announced in February that it had achieved nitride-based blue-violet laser diodes with wavelengths between 395-408 nm. Typical threshold current densities were 8.5–14 kA/cm^2. The structures are SCHs with 5 QWs, grown on sapphire. A pulsed mode, 150 mW/facet was achieved. However, this is still at the R&D stage.

The power achieved in pulsed mode, although not CW, was said to be in the same ballpark as that from Nichia, the world leader. SDL is perhaps best known for its high power GaAs based pump lasers, and is known to be targeting markets such as DVD by offering short-wavelength lasers.

Earlier, the company was a member of a consortium that in 1995 was awarded an US$4 million, two-year contract from DARPA to develop blue semiconductor lasers and LEDs. Other members of the consortium were Xerox American Crystal Technologies, ATMI, Boston University and the University of Texas at Austin.

The consortium members provided additional funds of more than US$4 million. Under the programme, the consortium aimed to develop advanced GaN epitaxial wafers for use in the fabrication of blue lasers and LEDs.

University of South Carolina

Dept of ECE
Swearingen Engineering Center
Columbia
SC 29208
USA
Tel: +1 803 777 7941
Fax: +1 803 777 8045
University of South Carolina is a participant in a number of collaborative research programs including in the wide bandgap semiconductors, the GaN Microwave Power Amplifier program.

Cree Research (now Cree Inc) was spun out of the University of South Carolina in the early 1990s and has since become the premier supplier of SiC products worldwide. Collaboration between Cree and USC continues with ongoing research into materials and crystal growth.

Stanford University

Solid State Electronics Laboratory
Stanford
CA 94305
USA
Fax: +1 415 723 4659
This group has successfully grown single crystal GaN on sapphire using the trichloride VPE technique. The work was conducted with assistance from ARPA.

Technologies and Devices International Inc

8660 Dakota Dr
Gaithersburg
MD 20877
USA
Tel: +1 301 897 3229
Fax: +1 301 208 8342
TDI is a small start up company which is pursuing the development of bulk crystals, epitaxial structures and devices over a range of materials with applications in short wavelength optoelectronics and high power semiconductor electronics. TDI develops, manufactures and markets electronic components using silicon carbide and III-V nitride semiconductor materials.

University of Texas at Austin

Center for Materials Science & Engineering
Austin
TX 78712
USA

This centre has a long record of publications in semiconductors and related materials. In the wide bandgap semiconductors the group has published simulation studies of crystal growth of SiC for a cold-wall reactor.

The university was a member of a consortium that in 1995 was awarded an US$4 m, 2-year contract from DARPA to develop blue semiconductor lasers and LEDs. Other members of the consortium were SDL, Xerox American Crystal Technologies, Advanced Technology Materials Inc, and Boston University. The consortium members provided additional funds of more than US$4 million. Under the programme, the consortium aimed to develop advanced GaN epitaxial wafers for use in the fabrication of blue lasers and LEDs.

University of Utah

Electrical Engineering Dept
50 S Central Campus
Dr Rm 3280
Salt Lake City
UT 84112-9206
USA
Tel: +1 801 581 6941
Fax: +1 801 581 5281

The Department of Materials Science and Engineering is involved in work on the Pyrolysis of tertiarybutylamine alone and with trimethylgallium for GaN growth. New nitrogen precursors that pyrolyze at low temperatures are desired to replace NH_3 for the MOVPE growth of GaN and, especially, GaInN. Tertiarybutylamine (TBAm) is a candidate having a high vapour pressure that is relatively safe compared with other potential N sources. In order to test the suitability of TBAm for the growth of GaN, the pyrolysis was studied in He and H_3 ambients.

University of Washington

Seattle
WA 98195
USA
Tel: +1 206 543 3063

The Materials Science and Engineering Department is developing a new growth method: van der Waals epitaxy. This method "decouples" the epitaxial film of interest, such as GaN, from the substrate. The Electrical Engineering Department

Wayne State University

Physics and Astronomy Department
666 West Hancock
Detroit
MI 48201
USA
Tel: +1 313 577 2720
Fax: +1 313 577 3932
This department has an interest in wide bandgap semiconductors and has published studies of band structures and related optical properties both at the eV scale in relation to UV and X-ray spectroscopies and at the MeV scale near the band edges.

Wichita State University

Mechanical Engineering Dept
Wichita
KS 67260-0133
USA
This university has collaborated with other US institutes such as Oak Ridge National Laboratory in the low temperature CVD of 3C-SiC films grown on Si (100) using special silane mixtures to reduce dislocation density and improve film quality.

University of Wisconsin

Materials Science Program
1415 Engineering Drive
Madison
WI 53706
USA
This group has been researching MOVPE of ZnSe and ZnTe II-VI materials for visible light emitters. It is involved in the HVPE growth of GaN on sapphire and is also a participant in the ARPA-URA Visible Light Emitters program.

Xerox PARC

Xerox Palo Alto Research Center
3333 Coyote Hill Road
Palo Alto
CA 94304
USA
Tel: +1 415 812 4199
This world-renowned laboratory is within the Xerox company and therefore conducts R&D related to office equipment related products sold by that company worldwide (see Company Profile).

The optoelectronics group, led by Dr Fernando Ponce, is involved in work on the materials properties of III-V nitrides aimed at the resolution of a practical blue-green diode laser for such applications as high-resolution printing.

Xerox has a number of research programs in the arsenide, phosphide and nitride materials and devices. Recent developments include DualSpot red laser diodes for Xerox's announced laser printing systems for both mid-volume (35 pages per minute) and high-speed (180 pages per minute) applications. The two closely spaced beams from the DualSpot laser pass through a single optical system allowing two lines of data to be written simultaneously on the photoreceptor.

Recent work includes a collaboration with the State University of New York at Buffalo to investigate piezoelectric electric field distribution in strained PIN III-nitride MQW structures.

The company was a member of a consortium that in 1995 was awarded an US\$4 million, two-year contract from DARPA to develop blue semiconductor lasers and LEDs. Other members of the consortium were SDL, Xerox American Crystal Technologies, Advanced Technology Materials Inc, Boston University and the University of Texas at Austin. The consortium members provided additional funds of more than US\$4 million. Under the programme, the consortium aimed to develop advanced GaN epitaxial wafers for use in the fabrication of blue lasers and LEDs.

Yale University

PO Box 208267
10 Hillhouse Avenue
New Haven
CT 06520
USA

Researchers at Yale University in collaboration with the University of Illinois and others, have been working on As-doped GaN films and GaNAs films grown by MOVPE on sapphire, GaN-coated sapphire, and GaAs substrates.

8 Directory of Leading Suppliers

AA SA

Opto-Electronics, 6 rue de Versailles, F-78470 St Remy Les Chevreuse, France
Tel: +33 1 30 52 87 71
Fax: +33 1 30 52 78 03
E-mail: aa-opto@teaser.fr
Web: www.a-a.fr
Optoelectronic components manufacturer.

Acrotec Semiconductor Materials

Japan Energy Corp, 10-1 Toranomon 2-chome, Minato-ku, Tokyo 105, Japan
Tel: +81 3 5573 6592
Fax: +81 3 5573 6779
Contact: N Okada
Epiwafer manufacturer.

Acuity Research Inc

3475 Edison Way, Unit P, Menlo Park, CA 94025, USA
Tel: +1 650 369 6782
Fax: +1 650 369 6785
E-mail: rrc@acuityresearch.com
Web: www.acuityresearch.com
Opto research.

Aculight Corp

11805 N Creek Pkwy S, Ste 113, Bothwell, WA 98011, USA
Tel: +1 425 482 1100
Fax: +1 425 482 1101
E-mail: laser@aculight.com
Web: www.aculight.com
Optoelectronics components manufacturer.

ADC Telecommunications Inc

PO Box 1101, Minneapolis, MN 55440-1101, USA
Tel: +1 612 938 8080
Fax: +1 612 946 3292
Opto component manufacturer.

Adlas Lasertechnik GmbH

Seelandstr 9, D-23569 Lübeck, Germany
Tel: +49 451 3909300
Fax: +49 451 395725
Optoelectronic component manufacturer.

Advanced Ceramics Corp

PO Box 94924, 11907 Madison Ave, Cleveland, OH 44107, USA
Tel: +1 216 529 3959
Fax: +1 216 529 3975
E-mail: mannl@advceramics.com
Web: www.advceramics.com
Optoelectronic component manufacturer.

Advanced Optronics Corp

2F1 No 23 Chi-Lin Rd, Chung-Li Ind Area, Tao-Yuan Hsien, Taiwan ROC
Tel: +886 3 452 9689
Fax: +886 3 452 1256
Optoelectronic component supplier.

Advanced Photonics International Inc

54 Plymouth Rd, White Plains, NY 10603, USA
Tel: +1 914 347 7732
Fax: +1 914 347 7732
Optoelectronic component manufacturer.

Advanced Photonic Systems GmbH

Rudower Chaussee 6, D-12489 Berlin, Germany
Tel: +49 30 6392 6520
Fax: +49 30 6392 6521
E-mail: APhS@Compuserve.com
Manufacturers of diode lasers.

Advanced Precision Optics

3760 Yale Way, #2, PO Box 1430, Fremont, CA 94538, USA
Tel: +1 510 490 8850
Fax: +1 510 490 2255
Optoelectronic component manufacturer.

Aero-Laser Gesellschaft für Gasanalytik mbH

Haupstr 44, D-82467 Garmisch-Partenkirchen, Germany
Tel: +49 8821 58670
Fax: +49 8821 58720
Web: www.aero-laser.de
Semiconductor laser manufacturer.

Aeronex

6975 Flanders Drive, San Diego, CA 92121, USA
Tel: +1 858 452 0124
Fax: +1 858 452 0229
Web: www.gaspurifier.com
Control equipment for device manufacturing.

AG Electro-Optics Ltd

Tarporley Business Centre, Tarporley, Cheshire CW6 9UY, UK
Tel: +44 1829 733305
Fax: +44 1829 733679
E-mail: sales@ageo.co.uk
Web: www.ageo.co.uk
Contact: Russell Evans
Optoelectronic component distributor.

Agere Systems Inc

Central Campus, 555 Union Blvd, Allentown, PA 18109, USA
Tel: +1 610 712 4323
Fax: +1 610 391 2849
E-mail: donnac@agere.com
Web: www.agere.com
Contact: Donna Cunningham
Manufacturer of optoelectronic devices for telecommunications applications.

Agilent GmbH

Herrenberger Str 124, D-71034 Böblingen, Germany
Tel: +49 7031 144816
Fax: +49 7031 142216
Web: www.agilent.de
Manufacturer of optoelectronic components including laser diodes and MQW pump lasers.

Agilent Technologies

370 West Trimble Rd, San Jose, CA 95131, USA
Tel: +1 408 435 7400
Fax: +1 408 435 5865
E-mail: northe_osbrink@agilent.com
Web: www.agilent.com
Contact: N Osbrink
Manufacturer of lasers, LEDs, RF and microwave.

AIXTRON AG

Kackerstr 15–17, D-52072 Aachen, Germany
Tel: +49 241 89 09 24
Fax: +49 241 89 09 40
E-mail: info@aixtron.com
Web: www.aixtron.com
Contact: Marc Deschler
Producer of MOCVD equipment for compound semiconductors.

Akzo Nobel Chemicals Inc

Deposition Chemicals, 730 Battleground Rd, PO Box 600, Deer Park, TX 77536,
USA
Tel: +1 281 479 8100
Fax: +1 281 479 4517
E-mail: dick.pearce@akzo-nobel.com
Contact: Dick Pearce
Materials supplier.

Alcatel Cable

Avd. Princesa Juana de Austria, KM 8700, 28021 Madrid, Spain
Tel: +34 9 1330 6623
Fax: +34 9 1330 6946
Contact: Miguel Hernandez
Distributor of electronic components.

Alcatel ITS Japan Ltd

PO Box 5024, Yebisu Garden Place, Tower 20-3, Ebisu 4 chome, Shibuya-ku
Tokyo, Japan
Tel: +81 3 5424 8561
Fax: +81 3 5424 85 81
E-mail: maejima@alcatel.co.jp
Contact: Masakasu Maejima
Manufacturer of optoelectronic components.

Alcatel ITS Inc

12030 Sunrise Valley Drive, Reston, VA 22091-3495, USA
Tel: +1 703 715 39 21
Fax: +1 703 860 11 83
E-mail: jsproul@its.alcatel.com
Contact: Jennifer Sproul
Manufacturer of optoelectronic components.

Alcatel Optronics

Route de Villejust, F-91625 Nozay, France
Tel: +33 1 64 49 49 10
Fax: +33 1 64 49 46 61
E-mail : bruno.vincent@alcatel.fr
Web: www.alcatel.fr
Contact: Bruno Vincent
Manufacturer of optoelectronic components including lasers, photodiodes and
WDM DFB laser modules.

Alcatel Optronics

Loffelholzstr 20, D-90411 Nuremberg, Germany
Tel: +49 911 4230 500
Fax: +49 911 4230 485
Manufacturer of optoelectronic components including WDM DFB laser
modules.

Alfalight Inc

1832 Wright St, Madison, WI 53704, USA
Tel: +1 608 240 4800
Fax: +1 608 240 4801
E-mail: info@alfalight.com
Web: www.alfalight.com
Manufacture of diode laser products.

Alpha Laser Technology GmbH

Hannah-Vogt-Str 1, D-37085 Göttingen, Germany
Tel: +49 551 7706147
Fax: +49 551 7706146
E-mail: info@alphalas.com
Web: www.alphalas.com
Laser and optical component manufacturer.

Alpha Metals Ltd

1/F Block A 21 Tung Yuen St, Yau Tong Bay, Kowloon, Hong Kong
Tel: +852 347 7112
Fax: +852 234 75 301
E-mail: kchow@alphametals.cookson.com
Web: www.alphametals.cookson.com
Contact: Kelvin Chow
Materials and substrate supplier.

American Xtal Technology (AXT)

4311 Solar Way, Fremont, CA 94538, USA
Tel: +1 510 683 5900
Fax: +1 510 683 5901
E-mail: sales@axt-vgf.com
Web: www.axt.com
Contact: Dr Theodore Young
Manufacturer of lasers as well as a supplier of GaAs, GaN, InP wafers and Ge substrate.

Åmic AB

Dag Hammarskjölds väg 52, Uppsala Science Park. SE-75183 Uppsala, Sweden
Tel: +46 18 52 16 40
Fax: +46 18 14 32 50
Email:info@amic.se
Web: www.amic.se
Microsystems manufacturer.

Amistar Inc

353 Cyril Owen Place, Victoria BC V8X 3X1, Canada
Tel: +1 250 544 1451
Fax: +1 250 479 1608
E-mail: amistar@islandnet.com
Contact: Robert Redden
Bulk crystal growth of III-V and II-VI compounds.

Ammo Engineering Ltd

55 Hanevim St, POB 1726, 47116 Ramat-Hasharon, Israel
Tel: +972 3 547 2747
Fax: +972 3 547 2744
E-mail: ammo1@netvision.net.il
Optoelectronic component supplier.

AMS Electronic Ltd

136 Union St, Suite 79, Torquay TQ2 5QG, UK
Tel: +44 1803 200655
Fax: +44 1803 200656
E-mail: UKinfo@ams.de
Contact: Martin Sharratt
Semiconductor laser manufacturer.

AMS OptoTech GmbH

Albrechstr 14, D-80636 Munich, Germany
Tel: +49 89 1268060
Fax: +49 89 12680660
Semiconductor laser manufacturer.

APA Optics Inc

2950 NE 84th Lane, Blaine, MN 55449, USA
Tel: +1 612 784 4995
Fax: +1 612 784-2038
E-mail: info@apaoptics.com
Web: www.apaoptics.com
Contact: K Olsen
Manufacturer of advanced optoelectronic components.

Apex Co Ltd

128 Chuckbuk-ri, Namyi-myun, Chongwon-kun, Chungbuk 363-810, South Korea
Tel: +82 431 260 2000
Fax: +82 431 260 2500
E-mail: htchoi@apexsemi.co.kr
Web: members.namo.co.kr/~htchoi
Contact: Hyuntae Choi
GaAs substrate supplier.

Applied Optoelectronics Inc (AOI)

13111 Jess Pirtle Blvd, Sugar Land, TX 77478, USA
Tel: +1 281 295 1800
Fax: +1 281 295 1888
E-mail: smart@ao-inc.com
Web: www.ao-inc.com
Contact: Sheldon Smart
Manufacturer of single-mode DFB quantum cascade lasers.

Arconium

50 Sims Ave, Providence, RI 02909, USA
Tel: +1 401 456 0800
Fax: +1 401 421 2419
Web: www.arconium.com
Opto materials manufacturer.

ARTAS Advanced Research Technology and Systems Trading Company Ltd

Hopfenweg 22, D-93197 Zeitlarn, Germany
Tel: +49 941 699130
Fax: +49 941 699131
E-mail: artas@csi.com
Manufacturer of crystal optics elements, silicon (optical grade or high resistivity), germanium (optical grade, detector grade), GaAs, windows, lenses prisms, beam splitters, polarises, optical coatings, non-linear optical crystals.

ATC-Semiconductor Devices

PO Box 29, St Petersburg 194156 Russia
Tel: +7 812 244 2532
Fax: +7 812 244 2544
E-mail: ter@atc.rfntr.neva.ru
Web: www.atcsd.neva.ru
Manufacturer of high-power laser diodes.

ATMI Epitaxial Services — Silicon Division

550 West Juanita Avenue, Mesa, AZ 85210, USA
Tel +1 480 668 4000
Fax: +1 480 464 7421
Email: siliconepi@atmi.com
Web: www.epitronics.com

ATMI Epitaxial Services III-V Division

21002 North 19th Avenue, Suite 5, Phoenix, AZ 85027, USA
Tel: +1 623 581 3663
Fax: +1 623 581 3415
Email: episales@atmi.com
Web: www.epitronics.com

ATOS GmbH

Robert-Bosch-Str 14, D- 64319 Pfungstadt, Germany
Tel: +49 6157 95 03 0
Fax: +49 6157 8 59 90
E-mail: info@atos-online.de
Web: www.atos-online.de
Manufacturer of laser components and systems.

Atramet Inc

222 Sherwood Ave, Farmingdale, NY 11735-1718, USA
Tel: +1 516 694 9000
Fax: +1 516 694 9177
Contact: Gary Newman
Supplier of Russian manufactured III-V and II-VI compounds.

August Technology

4900 West 78th St, Bloomington, MN 55435, USA
Tel: +1 952 820 0080
Fax: +1 952 820 0060
E-mail: sales@AugustTech.com
Web: www.AugustTech.com
Optoelectronic component manufacturer.

AUK Corp

2/F, Oh-Sung Bldg, 686-53, Shindaebang-Dong, Tongjak-Gu, Seoul, South Korea
Tel: +82 2 8351014
Fax: +82 2 8351275
E-mail: aukcorp@unitel.co.kr
Optoelectronic component manufacturer.

Avalon Photonics

Badenerstrasse 569, CH-8048 Zürich, Switzerland
Tel: +41 1 497 1414
Fax: +41 1 497 1400
E-mail: info@avap.ch
Web: www.avalon-photonics.com
Research into, and manufacture of VCSELs.

Axcel Photonics Inc

45 Bartlett St, Marlborough, MA 01752, USA
Tel: +1 508 481 9200
Fax: +1 508 481 9261
E-mail: admin@axcelphotonics.com
Web: www.axcelphotonics.com
Start-up developer and manufacturer of next generation opto components for fibre optic telecoms applications.

Ball Aerospace & Technologies Corp

PO Box 1062, Boulder, CO 80306, USA
Tel: +1 303 939 6100
Fax: +1 303 460 6104
E-mail: info@ball.com
Web: www.ballaerospace.com
Manufacturer of optoelectronic components, primarily for the space industry.

BCO Technologies plc

5 Hannahstown Hill, Belfast BT17 0LT, UK
Tel: +44 28 9061 5599
Fax: +44 28 9060 2088
E-mail: marketing@bco-technologies.com
Web: www.bco-technologies.com
Wafer manufacturer.

Bede Scientific Instruments Ltd

Bowburn South Industrial Estate, Bowburn, Co Durham DH6 5AD, UK
Tel: +44 191 377 2476
Fax: +44 191 377 9952
Measurement and characterisation equipment for semiconductor materials and devices.

BFI IBEXSA Elektronik GmbH

Korbinianstr 2, D-85386 Eching, Germany
Tel: +49 89 3197670
Fax: +49 89 3193510
Semiconductor laser manufacturer.

BFi OPTiLAS Ltd

Mill Square, Featherstone Rd, Wolverton Mill South, Milton Keynes MK12 5ZY, UK
Tel: +44 1908 326326
Fax: +44 1908 221110
Contact: Ian Stansfield
Manufacturer of detectors and diode pump solid state lasers.

Big Sky Laser Technologies Inc

PO Box 8100, Bozeman, MT 59715-2001, USA
Fax: +1 406 586 2924
Manufacturer of diode pump lasers.

BiosQuant GmbH

Rudower Chaussee 6, D-12484 Berlin, Germany
Tel: +49 30 63924550
Fax: +49 30 63924833
Manufacturer of semiconductor lasers.

Blue Lotus Micro Devices

7620 Executive Drive, Eden Prairie, MN 55344, USA
Tel: +1 612 934 2100
Fax: +1 612 934 2737
E-mail: jvanhove@blmd.com
Contact: James Van Hove
Manufacturer of epiwafers.

Bookham Technology plc

90 Milton Park, Abingdon, Oxon OX14 4RY, UK
Tel: +44 1235 827200
Fax: +44 1235 827201
E-mail: marketing@bookham.com
Web: www.bookham.com
Manufacturer of optoelectronic components including DWDM transmitters.

Boston Laser Inc (formerly Polaroid Laser Diode)

1 Upland Rd, Norwood, MA 02062, USA
Tel: +1 888 464 7297
Fax: +1 781 386 6426
E-mail: sales@bostonlaserinc.com
Web: www.bostonlaserinc.com
Manufacturer of laser diodes.

Bowei Electroy Co Ltd

91 Dongshan Rd, Nanping, Fujian 353000, China
Tel: +86 591 8832850
Fax: +86 591 8832850
Optoelectronic component manufacturer.

BRAGG Photonics Inc

2270 Saint-Francois Rd, Dorval, Montreal PQ H9P 1K2, Canada
Tel: +1 514 421 6766
Fax: +1 514 421 0560
Manufacturer of fibre Bragg gratings.

Calient Networks Inc

5853 rue Ferrari, San Jose, CA 95138, USA
Tel: +1 408 972 3600
Fax: +1 408 972 3800
E-mail: contactus@calient.net
Web: www.calient.net
Developer of intelligent, all-photonic switching systems and software for networks.

Calient Optical Components

22 Thornwood Drive, Ithaca, NY 14850, USA
Tel: +1 607 257 1525
Fax: +1 607 257 1612
Web: www.calient.net
MEMS subsidiary of Calient (formally Kionix).

CASIX Inc

21822 Lassen St, #G, Chatsworth, CA 91311, USA
Tel: +1 818 709 7636
Fax: +1 818 885 6926
E-mail: casixus@aol.com
Web: www.casix.com
Manufacturer and supplier of crystals, optics and lasers.

CelsiusTech Electronics AB

SE-181 84 Lidingoe, Sweden
Tel: +46 8 731 6000
Fax: +46 8 731 6485
E-mail: sgen@celsiustech.se
Supplier of advanced optical and optronic sights, mainly for military applications.

Changzhou Galaxy Electronics Co Ltd

139 Qing Liang Rd, Changzhou, Jiangsu 213001, China
Tel: +86 519 6646778
Fax: +86 519 6662020
E-mail: yhdggs@public.cz.js.cn
Optoelectronic component manufacturer.

Chiyoda Electronic (S) Pte Ltd

138 Cecil St, #04-02 Cecil Court, Singapore 069538
Tel: +65 3242990
Fax: +65 3241120
Optoelectronic component manufacturer.

Chuo & Co Ltd

3-6-1 Kanda Ogawamachi, Chiyoda-ku, Tokyo 101-0052, Japan
Tel: +81 3 32918423
Fax: +81 3 32935850
E-mail: lily@candcl.co.jp
Optoelectronic component manufacturer.

Cleveland Crystals Inc

19306 Redwood Ave, Cleveland, OH 44110-2738, USA
Tel: +1 216 486 6100
Fax: +1 216 486 6103
E-mail: sales@clevelandcrystals.com
Web: www.clevelandcrystals.com
Grower and fabricator of linear, nonlinear, and electro-optic crystals.

CMK Ltd

Sandricka 30, Zarnovica 966 81, Slovakia
Tel: +421 858 681 21 41
Fax: +421 858 681 21 25
E-mail: sales@cmk.sk
Contact: Alex Murin
Substrate manufacturer.

Coherent Inc

Semiconductor Group, 5100 Patrick Henry Drive, Santa Clara, CA 95054, USA
Tel: +1 408 764 4983
Fax: +1 408 988 3638
E-mail: tech_sales@cohr.com
Web: www.diodes.coherentinc.com
Manufacturer of high power laser diodes.

Coherent Tutcore Ltd

Kauhakorvenkatu 52, PO Box 48, Tampere 33721, Finland
Tel: +358 50 67 580
Fax: +358 3 571 600
E-mail: Harry.Asonen@tutcore.fi
Contact: Harry Asonen
Manufacturer of Al-free epiwafers for semiconductor lasers.

Coherent (UK) Ltd

Cambridge Science Park, Milton Rd, Cambridge CB4 4FR, UK
Tel: +44 1223 424065
Fax: +44 1223 420073
E-mail: Cohr_UK_CLG@cohr.com
Web: www.coherentinc.com
Contact: Roger Beaman
Manufacturer of diode pumped solid state lasers.

Cohu Inc

Electronics Division, 5755 Kearny Villa Rd, San Diego, CA 92123-1111, USA
Tel: +1 619 277 6700
Fax: +1 619 277 0221
E-mail: sales@cohu.com
Web: www.cohu.com/cctv
Optoelectronic component manufacturer.

Columbia Tech Inc

Qing Wang, 33E, West Building, Yi Hai Square, Chuang Ye Rd, Nan Shan District, Shen Zen 518054, China
Tel: +86 755 6643 190, 6643 191
Fax: +86 755 6643 187
E-mail: coltek@ihw.com.cn
Optoelectronic component supplier.

Commonwealth Scientific Corp

500 Pendleton St, Alexandria, VA 22314-1974, USA
Tel: +1 703 548 0800
Fax: +1 703 548 7405
E-mail: csc@ionbeam.com
Manufacturer of ion beam sources and systems.

Compound Semiconductor Technologies (CST)

Block 7, Kelvin Campus, West of Scotland Science Park, Maryhill Rd, Glasgow G20 0TH, UK
Tel: +44 141 579 3000
Fax: +44 141 579 3040
Web: www.compoundsemi.co.uk
Optoelectronic start-up company for high-bandwidth fibre-optic datacoms.

ConOptics Inc

19 Eagle Rd, Danbury, CT 06810-4127, USA
Tel: +1 800 748 3349
Fax: +1 203 790 6145
E-mail: conoptic@aol.com
Optoelectronic components manufacturer.

Cookson India Ltd

PO Box 2296, Calcutta 1, India
Tel: +91 33 467 0434
Fax: +91 33 467 0433
E-mail: rkoshy@alphametals.cookson.com
Materials supplier.

Coors Ceramics Co

16000 Table Mountain Pkwy, Golden, CO 80403, USA
Tel: +1 303 278 4000
Fax: +1 303 277 4779
E-mail: elecsales@coorsceramics.com
Web: www.corrsceramics.com
Opto materials supplier.

Cotco Luminant Device Ltd

Unit 1-5 13/F Vanta Ind Ctr, 21 Tai Lin Pai Rd Kwai Chung, New Territories,
Hong Kong
Tel: +852 24248228
Fax: +852 24222737
Optoelectronic component manufacturer.

Craft Data GmbH

PO Box 1163, D-24559 Kaltchkirchen, Germany
Tel: +49 4191 2755
Fax: +49 4191 4495
Optoelectronic component distributor.

Cree Inc

4425 Silicon Drive, Durham, NC 27703, USA
Tel: +1 919 313 5300
Fax: +1 919 313 5452
E-mail: sales@cree.com
Web: www.cree.com
Optoelectronic component manufacturer, including LEDs.

Crystacomm Inc

1599 North Shoreline Blvd, Mountain View, CA 94043, USA
Tel: +1 650 961 4311
Fax: +1 650 961 4364
E-mail: gantypas@aol.com
Contact: George Antypas
Manufacturer of InP substrates for optoelectronic devices.

Crystal GmbH

Ostendstraße 2-14, D-12459 Berlin, Germany
Tel: +49 30 695 387 0
Fax: +49 30 635 043 6
Contact: Steffan Sander
Substrate supplier.

Crystallod Inc

25 Fourth St, Somerville, NJ 08876, USA
Tel: +1 908 575 0803
Fax: +1 908 575 0794
E-mail: wabonner-crystallodinc@worldnet.att.net
Contact: William Bonner
Single crystal growth research.

Crystal Specialties Inc

2853 Janitell Rd, Colorado Springs, CO 80906-4104, USA
Tel: +1 719 540 0990
Fax: +1 719 540 0994
E-mail: crystal.spec@pcisys.net
Contact: Thomas M Stavish
Manufacturer of wafers for the optoelectronic industry.

CSEM

Microsystems Division, rue Jaquet-Droz 1, CH-2007 Neuchâtel, Switzerland
Tel: +41 32 720 5111
E-mail: hvandenvlekkert@csem.ch
Web: www.csem.ch
Contact: Hans van den Vlekkert
Manufacturer of VCSELs.

CTA Centrotherm

Halbleitertechnologie GmbH, Joseph von Fraunhofer Str 7, D-52477 Alsdorf, Germany
Tel: +49 2404 90680
Fax: +49 2404 906868
E-mail: fritz@centrotherm.de
Contact: U Fritz
Specialists in LPE, develops epi reactors and technologies.

Cutting Edge Optronics

20 Point West Blvd, St Charles, MO 63301, USA
Tel: +1 314 916 4900
Fax: +1 314 916 4994
Laser diode manufacturer.

CVD Equipment Corp

1881 Lakeland Ave, Ronkonkona, NY 11779, USA
Tel: +1 516 981 7081
Fax: +1 516 981 7095
E-mail: info@cvdequipment.com
Contact: Karen Hamberg
Manufacturer of custom deposition systems.

CVI Laser Corp

200 Dorado Place SE, PO Box 11308, Albuquerque, NM 87123, USA
Tel: +1 505 296 541
Fax: +1 505 298 9908
Manufacturer of optoelectronic components, including photodiodes.

Da Jeon Electric Co

#175-1 Poongsan-Dong, Hanam-Si, Kyunggi-Do, South Korea
Tel: +82 347 7947510
Fax: +82 347 7927510
Optoelectronic component manufacturer.

Daktronics Inc

331 32nd Ave, Brookings, SD 57006, USA
Tel: +1 605 697 4000
E-mail: Webmaster@daktronics.com
Web: daktronics.com
Optoelectronic component manufacturer.

Delta Developments

PO Box 115, Liss, Hampshire GU33 6DF, UK
Tel: +44 1730 827404
Fax: +44 1730 827407
E-mail: admin@Delta-dev.co.uk
Web: www.delta-dev.co.uk
Contact: Dr J G Edwards
Optoelectronic component distributor.

Denton Vacuum

1259 North Church St, Moorestown, NJ 08057, USA
Tel: +1 856 439 9100
Fax: +1 856 439 9111
E-mail: info@dentonvacuum.com
Web: www.dentonvacuum.com
Makers of evaporators and related process equipment for compound semi-conductor industry, eg laser facet AR coatings.

Deposition Sciences Inc

386 Tesconi Court, Santa Rosa, CA 95401-4653, USA
Tel: +1 800 231 7390
Fax: +1 707 579 0731
E-mail: lora.sansom@depsci.com
Supplier of materials for opto devices.

Digital Optics Corp

5900 Northwoods Parkway, Suite J, Charlotte, NC 28269, USA
Tel: +1 704 599 9191
Fax: +1 704 599 4997
E-mail: bacon@doc.com
Web: www.doc.com
Contact: Allan Bacon
Optoelectronic components manufacturer.

DILAS Diodenlaser GmbH

Galileo-Galilei-Str 10, D-55129 Mainz, Germany
Tel: +49 6131 506840
Fax: +49 6131 506855
Manufacturer of high power laser diodes.

Dongsung Moolsan Co Ind

342-1 Gakpyung-ri, Majang-myun, Icheon-gun, Kyonggi, South Korea
Tel: +82 336 322020
Contact: D H Kim
Manufacturer of optoelectronic components.

Dora Texas Corp

2642, Shelby Drive, Pearland, TX 77584, USA
Tel: +1 713 334 7277
Fax: +1 713 334 5543
E-mail: Digilov@neosoft.com
Contact: Michael Digilov
Substrate supplier.

Dowa Mining Co Ltd

New Materials Division, Semiconductor Department, 8-2, Marunouchi 1-Chome, Chiyoda-ku, Tokyo 100, Japan
Tel: +81 3 3201 1067
Fax: +81 3 3201 1098
E-mail: yuji@dicny.com
Contact: T Suzuki
Supplier of semiconducting wafers and AlGaAs LEDs.

Dynatex International

5577 Skylane Boulevard, Santa Rosa, CA 95403, USA
Tel: +1 707 579 4227
Fax: +1 707 542 8599
E-mail: sales @dynatex.com
Contact: Leanne Schmidt
Manufacturer of dry process dicing equipment.

E2O Communications Corp

26679 W Agoura Rd, Calabasas, CA 91302, USA
Tel: +1 818 466 2800
Fax: +1 818 878 9163
Web: www.e2oinc.com
Manufacturer of VCSELs.

Eagle-Picher Technologies

ESAT Department, 200 BJ Tunnell Blvd, Miami, OK 74354, USA
Tel: +1 918 542 1801
Fax: +1 918 542 3223
E-mail: esat@galstar.com
Contact: Bob Brown
Epitaxy of II-VI compounds by MBE or MOVPE.

Eagle-Picher Technologies

PO Box 47, Joplin, MO 64802, USA
Tel: +1 417 623 8000
Fax: +1 417 781 1910
E-mail: bharsch@epi-tech.com
Contact: Bill Harsch
Substrate manufacturer.

Ealing Electro-Optics plc

Greycaine Rd, Watford WD2 4PW, UK
Tel: +44 1923 242261
Fax: +44 1923 234220
Optoelectronic components supplier.

Edinburgh Instruments Ltd

Research Park, Riccarton, Currie, Edinburgh EH14 4AS, UK
Tel: +44 131 449 5844
Fax: +44 131 449 5848
E-mail: sales@edinst.com
Web: www.edinst.com
Contact: Dr Richard Dennis
Optoelectronic component supplier.

Edmund Scientific Company

Industrial Operations Division, 101 E Gloucester Pike, Barrington, NJ 08007-1380, USA
Tel: +1 609 547 3488, Ext 6885
Fax: +1 609 573 6840
E-mail: indopt@edsci.com
Web: www.edsci.com
Optoelectronic component manufacturer, including lasers.

EHD Imaging GmbH

Bahnhofstr 22, D-49401 Damme, Germany
Tel: +49 5491 2090
Fax: +49 5491 2098
E-mail: info@ehd.de
Web: www.ehd.de
Manufacturer of optical products.

Eksma Co

Mokslininku Str 11, 2600 Vilnius, Lithuania
Tel: +370 2 729900
Fax: +370 2 7279715
E-mail: raimis@eksma.elnet.it
Optoelectronic component supplier.

Electron Transfer Technologies

155 Campus Plaza, Edison, NJ 08818-5812, USA
Tel: +1 732 225 3995
Fax: +1 732 225 3580
E-mail: wmaett@aol.com
Gas control equipment suppliers.

Elekon Industries

3882 Del Amo Blvd, Suite 601, Torrance, CA 90503-2162, USA
Tel: +1 310 370 8022
Fax: +1 310 370 8079
E-mail: sales@elekon.com
Web: www.elekon.com
Optoelectronic component manufacturer.

Elektronska Industrija

Bul Viljka Vlahovica 80–82, NIS (Beograd), Russia
Tel: +7 18 55 583
Fax: +7 18 32 5726
Manufacturer of optoelectronic devices.

Electro-Optics Technology Inc

1030 Hastings St, Suite 140, Traverse City, MI 49686-3470, USA
Tel: +1 616 935 4044
Fax: +1 616 935 4046
E-mail: eot@gtii.com
Web: www.eotech.com
Manufacturer of Faraday rotators, optical isolators and InGaAs diodes.

Electrox Ltd

Avenue One, The Business Park, Letchworth, Herts SG6 2HB, UK
Tel: +44 1462 472400
Fax: +44 1462 472444
Contact: Baz Hartnell
Optoelectronic component distributor.

Elliot Scientific Ltd

Gladstone Place, 36–38 Upper Marlborough Rd, St Albans, Herts AL1 3US, UK
Tel: +44 1727 847900
Fax: +44 1727 847922
Web: www.elliotscientific.com
Contact: Mike Elliot
Component distributor.

EMCORE Corp

394 Elizabeth Ave, Somerset, NJ 08873, USA
Tel: +1 732 271 9090
Fax: +1 732 271 9686
E-mail: info@emcore.com
Web: www.emcore.com
Manufacturer of optoelectronic components including VCSELs.

EMF Ltd

71 Panton St, Cambridge CB2 1HL, UK
Tel: +44 1223 364 080
Fax: +44 1223 354 701
E-mail: jd@emf.co.uk
Contact: Jim Dixon
Epiwafer manufacturer.

Epichem

Power Rd, Bromborough, Wirral, Merseyside L62 3QF, UK
Tel: +44 151 334 2774
Fax: +44 151 334 6422
Web: www.epichem.com
Contact: Sarah Leese
Opto materials manufacturer.

EPIGAP Optoelektronik GmbH

Köpenicker Str 325 b, Haus 201, D-12555 Berlin, Germany
Tel: +49 30 6576 2543
Fax: +49 30 6576 2545
E-mail: sales@epigap.de
Web: www.epigap.de
Manufacturer of devices and epiwafers.

Epigress AB

Ideon Science & Technology Park, SE-22370 Lund, Sweden
Tel: +46 46 286 89 90
Fax: +46 46 286 89 89
E-mail: info@epigress.se
Manufacturer of equipment for epitaxial growth.

Epitaxial Technologies LLC

1450 South Rolling Rd, Baltimore, MD 21227, USA
Tel: +1 410 455 5594
Fax: +1 410 455 5595
E-mail: epiwafers@rols.com
Web: www.erols.com/epiwafers/
Contact: Dr Leye Aina
Manufacturer of epitaxial wafers.

EpiWorks Inc

1606 Rion Drive, Champaign, IL 61822, USA
Tel: +1 217 373 1590
Fax: +1 217 373 1591
E-mail: epiworks@epiworks.com
Web: www.epiworks.com
Epiwafer manufacturer.

Equin SA

Ulises 96, 28043 Madrid, Spain
Tel: +34 1 388 0898
Fax: +34 1 388 6078
Optoelectronic components distributor.

Ericsson Microelectronics AB

Isafjordgatan 16, SE-164 81 Kista, Sweden
Tel: +46 8 757 4700
Fax: +46 8 757 4776
Web: www.microe.ericsson.se
Contact: B Callmer
Manufacturer of optoelectronics components, include PIN photodiodes and digital lasers.

Ericsson Components AB

Fiber Optics Research Center, KI/EK/SF, S-164 81 Kista, Sweden
Tel: +46 8 757 5000
Fax: +46 8 757 4764
E-mail: ekahed@eka.ericsson.se
Contact: Hans Eklund
Manufacturer of optoelectronic devices.

Eurodis Enatechnik Electronics GmbH

Pascalkehre 1, D-25451 Quickborn, Germany
Tel: +49 41 06 701201
Fax: +49 41 06 701391
Laser manufacturer.

ExceLight Communications

4021 Stirrup Creek Drive, Suite 200, Durham, NC 27703, USA
Tel: +1 919 3611600
Fax: +1 919 3611619
E-mail: info@excelight.com
Web: www.excelight.com
Manufacturer of optoelectronic components.

Excellence Optoelectronic Inc

Li-Hsing Rd Science-Based Ind Park, 8F, No 10, Hsin-Chu 300, Taiwan ROC
Tel: +886 3 5679000
Fax: +886 3 5679999
Optoelectronic component manufacturer.

Exitech Ltd

Hanborough Park, Long Hanborough, Oxford OX8 8LH, UK
Tel: +44 1865 3883324
Fax: +44 1865 3883334
Contact: M C Gower
Optoelectronic component distributor.

Fermionics Lasertech Inc

1153 Lawrence Drive, Newbury Park, CA 91320, USA
Tel: +1 805 375 0999
Fax: +1 805 375 0339
E-mail: YZLiu@pacbell.net
Contact: Dr Yet-Zen Liu
Manufacturers of standard laser products, edge emitting light diodes, super-luminescent diodes, semiconductor optical amplifiers and custom wavelength lasers.

Fifth Dimension Technologies (5DT)

PO Box 5, Persequor Park 0020, South Africa
Tel: +27 123 491 400
Fax: +27 123 491 404
E-mail: 5dt@pixie.co.za
Contact: Paul Olkers
Optoelectronic component manufacturer.

Firebird Semiconductors Ltd

2950 Highway Drive, Tail, BC V1R 2T3 Canada
Tel: +1 604 364 5605
Fax: +1 604 364 5643
Contact: Grant Fines
Optoelectronic substrate supplier.

Fisba Optik AG

Rorschacher Str 268, 9016 St Gallen, Switzerland
Tel: +41 712 823131
Fax: +41 712 823130
E-mail: info@fisba.ch
Manufacturer of diode pumped solid state lasers.

Fo Sheng Photoelectron Co Ltd

3F-1 No 9 Prosperity Rd, 1 Science Based Ind Park, Taipei 300, Taiwan ROC
Tel: +886 3 5787746
Fax: +886 3 5787748
Optoelectronic component manufacturer.

Fotec Inc

151 Mystic Ave, Medford, MA 02155-4615, USA
Tel: +1 781 396 6155
Fax: +1 781 396 6395
E-mail: info@fotec.com
Optoelectronic component supplier.

Fraunhofer-Institut für Lasertechnik ILT

Steinbach Str 15, D-52074 Aachen, Germany
Tel: +49 241 89060
Fax: +49 241 8906121
Research and prototyping of semiconductor lasers and laser diodes.

Freiberger Compound Materials GmbH

Am Junger Löwe Schacht 5, D-09599 Freiberg, Germany
Tel: +49 3731 2 80 0
Fax: +49 3731 2 80 106
E-mail: info@fcm-germany.com
Web: www.fcm-germany.com

Freiberger Elektronikwerkstoffe GmbH

PO Box 211, D-09584 Freiberg, Germany
Tel: +49 3731 278 572
Fax: +49 3731 278 233
Contact: V Geidel
Optoelectronic substrate supplier.

FTS I KINETICS

3538 Main St, Stone Ridge, NY 12484, USA
Tel: +1 914 687 0071
Fax: +1 917 687 7481
Manufacturer of optoelectronic devices.

Fujitsu Compound Semiconductor Inc

2355 Zanker Rd, San Jose, CA 95131-1138, USA
Tel: +1 408 232 9500
Fax: +1 408 428 9111
Manufacturer of fibre optic products and lightwaves.

Fujitsu Ltd

Marunouchi Center Building, 6-1 Marunouchi, Chiyoda-ku, Tokyo 100, Japan
Tel: +81 3 3216 3211
Fax: +81 3 3216 9365
Manufacturer of optoelectronic components.

Fujitsu Quantum Devices Ltd

1000 Kamisukiahara, Showa-cho, Nakakomagun, Yamanashi 409-3883, Japan
Tel: +81 552 75 4411
Fax: +81 552 75 9461
Contact: K Hayashi
Manufacturer of optoelectronic components, including laser diodes.

Furukawa Electric Ltd

Compound Semiconductor Department, New Business Development Division, 6-1, Marunouchi 2-Chome, Chiyoda-Ku, Tokyo 100, Japan
Tel: +81 3 3286 3219
Fax: +81 3 3286 3965
Web: www.furukawa.co.jp
Contact: H Matsushita
Supplier of GaAs MOCVD wafers.

Furukawa Electric Technologies Inc

900 Lafayette St, Suite 401, Santa Clara, CA 95050, USA
Tel: +1 408 248 4884
Fax: +1 408 248 8815
Contact: Ranjit Mand
Optoelectronic substrate supplier.

Galaxy Compound Semiconductors

9922 East Montgomery Ave, #7 Spokane, WA 99206, USA
Tel: +1 509 892 1114
Fax: +1 509 892 1116
Opto component manufacturer.

GCA Fibreoptics Ltd

GCA House, Building 19, Thorney Leys Business Park, Witney, Oxon OX8 7GE, UK
Tel: +44 1993 700800
Fax: +44 1993 700444
Laser and LED manufacturer.

GEO Gallium

903 Sheehy Drive, Suite E Babylon Business Center, Horsham, PA 19044-1231,
USA
Tel: +1 519 3883, Ext 122
Fax: +1 215 773 9310
Materials supplier.

Geola Technologies Ltd

Shaw House, Pegler Way, Crawley RH11 1AF, UK
Tel: +44 1293 763 078
Fax: +44 1293 527 20
Manufacturer of optical products, RGB lasers and holographic materials.

Geola Uab

41 Naugarduko, PO Box 343, 2006 Vilnius, Lithuania
Tel: +370 2 232 737
Fax: +370 2 232 838
Web: www.geola.com
Manufacturer of optical products, RGB lasers and holographic materials.

Gerhard Franck Optronik GmbH

Kuehnstr 75, D-22045 Hamburg, Germany
Tel: +49 40 6696220
Fax: +49 40 66962230
Optoelectronic component distributor.

Gore Photonics

425 Commerce Ave, Lompoc, CA 93436, USA
Tel: +1 805 737 7391
Fax: +1 805 737 7393
E-mail: vjayaram@wlgore.com
Web: www.wlgore.com
Contact: V Jayaram
Manufacturer of optoelectronic components, including VCSELs.

Groupe Arnaud Electronics

68 ave General Michel Bizo, F-75012 Paris, France
Tel: +33 1 4473 1070
Fax: +33 1 4473 1053
E-mail: Arnaud.Electronics@A-Arnaud.fr
Web: www.arnaudelectronics.com
Distributor of materials for the opto industry.

GTI Technologies Inc

6 Armstrong Rd, PO Box 433, Shelton, CT 06484, USA
Tel: +1 203 929 2200
Fax: +1 203 926 0074
E-mail: bob@gti-USA.com
Web: www.gti-USA.com
Component manufacturer.

Hamamatsu Corp

360 Foothill Rd, Bridgewater, NJ 08807, USA
Tel: +1 800 524 0504
Fax: +1 908 231 1218
Manufacturer of laser diodes and other components.

Hamamatsu Photonics K K

Solid State Division, 11261 Ichino-cho, Hamamatsu-shi, Shizouka 435, Japan
Tel: +81 53 434 3311
Fax: +81 53 434 5184
Contact: T Hiruma
Manufacturer of optical devices.

Hamamatsu Photonics UK Ltd

Lough Point, 2 Gladbeck Way, Windmill Hill, Enfield, Middlesex EN2 7JA, UK
Tel: +44 181 367 3560
Fax: +44 181 367 6384
E-mail: info@hamamatsu.co.uk
Contact: Tim Stokes
Manufacturer and supplier of optoelectronic components.

Hanvac Corp

1675-4 Shinil-Dong, Taedok-Gu, Taejon 306-230, South Korea
Tel: +82 42 935 4900
Fax: +82 42 935 4905
E-mail: sales @hanvac.co.kr
Web: www.hanvac.co.kr
MOCVD systems for epitaxial growth of III-nitrides for blue LEDs etc.

Hanyoung Electronic Co Ltd

Hanyoung Bldg, #40-11, 2ga, Munlae-dong, Youngdeungpo-ku, Seoul, South Korea
Tel: +82 2 6788781
Fax: +82 2 6333332
Optoelectronic component manufacturer.

Harris Diamond Corp

100 Stierli Court, Suite 106, Mt Arlington, NJ 07856, USA
Tel: +1 201 770 1420
Fax: +1 201 770 1549
Manufacturer of optoelectronic devices.

Heraeus Med GmbH

Heraeusstr 12-14, D-63450, Germany
Tel: +49 6181 355514
Fax: +49 6181 355960
Manufacturer of lasers.

Hitachi Cable Ltd

Chiyoda Building, 2-1-2 Marunouchi, Chiyoda-ku, Tokyo 100, Japan
Tel: +81 3 5252 3686
Fax: +81 3 3213 0402
E-mail: shouhei_uwai@cc.hitachi-cable.co.jp
Web: www.hitachi-cable.co.jp
Contact: Shouhei Uwai
Manufacturer of substrate materials as well as laser diodes etc.

Hitachi Europe GmbH

Herrenbergerstr 140/G7, D-71034 Böblingen, Germany
Tel: +49 7031 14 7206
Fax: +49 7031 14 3931
Optoelectronics manufacturer.

Hitachi Europe Ltd

Whitebrook Park, Lower Cookham Rd, Maidenhead, Berks SL6 8YA, UK
Tel: +44 1628 585000
Fax: +44 1628 778322
Web: www.hitachi-eu.com
Contact: M Jones
Optoelectronics manufacturer.

Hitachi Ltd

6 Kanda Surugadai, 4-chome, Chiyoda-ku, Tokyo 101, Japan
Tel: +81 3 2851111
Contact: K Mita
Manufacturer of optoelectronic devices, mainly laser diodes.

Hitachi Ltd

Komoro Plant, 190 Kashiwagi, Komoro-shi, Nagano 384, Japan
Tel: +81 267 22 4111
Manufacturer of laser diodes and infrared emitting diodes.

Hitachi Semiconductor (KEDAH) SDN BHD

Plot 54 Kulim Industrial Estate, 09000 Kulim, Kedah Darulaman, Malaysia
Tel: +60 4 4892241
Fax: +60 4 4891122
Contact: Chng Peik Hock
Manufacturer of optoelectronic devices.

Honeywell International

102-721 Vanalman Ave, Victoria, BC V8Z 3B6, Canada
Tel: +1 250 479 9922
Fax: +1 250 479 2734
E-mail: fsteeds@crystarinc.com
Web: www.sapphireproducts.com
Component manufacturer.

Honeywell Micro Switch Division

830 East Arapaho Rd, Richardson, TX 75081, USA
Tel: +1 972 234 4271
Fax: +1 972 470 4417
Contact: J Staley
Manufacturer of optoelectronic components, including IREDs and VCSELs.

IICO Customized Wafer Services

3050 Oakmead Village Drive, Santa Clara, CA 95051, USA
Tel: +1 408 727 2547
Fax: +1 408 727 1322
E-mail: arnold@iico.com
Contact: Arnold Framption
Ion inplantation services for production and research.

III/V Reclaim

Eisenfelden 92, D-84543 Winhoring, Germany
Tel: +49 8671 73442
Fax: +49 8671 73443
E-mail: Reclaim@T-Online.de
Contact: Jorg Schwar
Reclaimed/recycled GaAs and InP wafers.

II-VI Inc

375 Saxonburg Boulevard, Saxonburg, PA 16056, USA
Tel: +1 412 352 1504
Fax: +1 412 352 4980
Web: www.optics.org/II-VI
Laser manufacturer.

ILEE AG

Schützenstr 29, CH-8902 Urdorf, Switzerland
Tel: +41 1 7342777
Fax: +41 1 7342722
LEDs, semiconductor lasers and laser systems.

Indium Corp of America

1676 Lincoln Ave, Utica, NY 13502, USA
Tel: +1 315 853 4900
Fax: +1 315 853 1000
E-mail: askus@indium.com
Web: www.indium.com
Contact: David Minckler
Materials supplier.

Infineon Technologies AG

iGr, PO Box 800949, D-81609 Munich, Germany
Tel: +49 89 23424497
Fax: +49 89 234 28482
E-mail: sales@infineon.com
Web: www.infineon.com
Optoelectronic component manufacturer, including lasers.

Institute of Photonics, University of Strathclyde

Wolfson Centre, 106 Rottenrow, Glasgow G4 0NW, UK
Tel: +44 141 553 4120
Fax: +44 141 552 1575
E-mail: photonics@strath.ac.uk
Web: www.photonics.ac.uk
Collaborative research partnerships undertaking strategic long-term research, industrial contracts and consultancy.

Instruments SA (ISA)

3880 Park Ave, Edison, NJ 08820-3012, USA
Tel: +1 732 494 8660
Fax: +1 732 549 5125
E-mail: ReCarWalk@aol.com
Web: www.instrumentssa.com
Manufacturer and distributor of instrumentation and components for optical spectroscopy.

Integrated Optical Components Ltd

3/4 Waterside Business Park, Eastways, Witham CM8 3YQ, UK
Tel: +44 1376 502110
Fax: +44 1376 502125
E-mail: sales@ioc.co.uk
Web: www.ioc.co.uk
Contact: M Powell
Optoelectronic component manufacturer and supplier.

Intense Photonics Ltd

Kelvin Campus, West of Scotland Science Park, Glasgow G20 0SP, UK
Tel: +44 141 589 7000
Fax: +44 141 589 7039
E-mail: enquiries@intensephotonics.com
Web: www.intensephotonics.com
Optoelectronics component manufacturer.

Intercrystal

Assenovgradsko shosse, Plovdiv 4009, Bulgaria
Tel: +359 32 62 30 60
Fax: +359 2 62 30 62
E-mail: interc@plovdiv.inetg.bg
Web: www.inetg.bg
Contact: Uriy Grigorev
Opto substrate supplier.

INTRACO Ltd

230 Victoria St #12-00, Bugis Junction Towers, Singapore 188024
Tel: +65 337 0011
Fax: +65 337 7200/337 7300
E-mail: lowm@intraco.com.sg
Contact: Maureen Low
Optoelectronic component distributor

IPG Laser GmbH

im IGZ Adlershof, Rudower Chaussee 5, D-12489 Berlin, Germany
Tel: +49 30 639250
Fax: +49 30 63925041
Manufacturers of high-power fibre lasers and solid-state lasers.

IQE Inc (formerly Quantum Epitaxial Designs [QED])

119 Technology Drive, Bethlehem, PA 18015, USA
Tel: +1 610 861 6930
Fax: +1 610 861 5273
E-mail: jmckeown@iqep.com
Web: www.iqep.com
Contact: James McKeown
Manufacturer of quantum well infrared photodetectors.

IQE plc (formerly Epitaxial Products International [EPI])

Cypress Drive, St Mellons, Cardiff CF3 0EG, UK
Tel: +44 2920 839400
Fax: +44 2920 779929
E-mail: mscott@iqep.com
Web: www.iqep.com
Contact: Dr Mike Scott
Manufacturer of custom GaAs and InP MOCVD wafers for optoelectronic applications.

Japan Energy Corp

Compound Semiconductor Marketing Dept, 10-1 Toranomon 2-Chome, Minato-ku, Tokyo 105, Japan
Tel: +81 3 5573 6592
Fax: +81 3 5573 6779
E-mail: interc@plovdiv.inetg.bg
Web: www.inetg.bg
Contact: M Suzuki
Manufacturer of detectors, GaAs and InP wafers.

JDS Uniphase GmbH

Arbeostrasse 5, D-85386 Eching, Munich, Germany
Tel: +49 89 3196026
Fax: +49 89 3193002
Contact: I Duckminor
Manufacturer of semiconductor lasers for telecoms applications.

JDS Uniphase (formerly SDL Inc)

80 Rose Orchard Way, San Jose, CA 95134-1365, USA
Tel: +1 408 943 9411
Fax: +1 408 943 1070
E-mail: sales@sdli.com
Manufacturer of semiconductor lasers, laser-based systems and fibre optic-related solutions.

JDS Uniphase Laser Division

163 Baypointe Parkway, San Jose, CA 95134, USA
Tel: +1 408 434 1800
Fax: +1 408 954 0405
Web: www.jdsuniphase.com
Manufacturer of semiconductor lasers for telecoms applications.

JDS Uniphase Ltd

2 Viewpoint, Babbage Rd, Stevenage SG1 2EQ, UK
Tel: +44 1438 745055
Fax: +44 1438 742490
Web: www.jdsuniphase.com
Contact: S Ebdon
Manufacturer of semiconductor lasers for telecoms applications.

JDS Uniphase Netherlands BV

Prof Holstlaan 4, 5656 AA Eindhoven, the Netherlands
Tel: +31 40 27 43038
Fax: +31 40 27 43859
Contact: Wim Nijman
Manufacturer of semiconductor lasers for telecoms applications.

Jenoptec

12 rue Jean Baptise Huet, F-78350 Les Metz, Jouy en Josas, France
Tel: +33 1 34 659 102
Fax: +33 1 34 651 863
E-mail: marketing@jenoptec.com
Contact: Pascal Slobadzian
Optoelectronic component manufacturer.

Jenoptik Laserdiode GmbH

Prüssingstraße 41, D-07745 Jena, Germany
Tel: +49 3641 654300
Fax: +49 3641 654391
E-mail: joldoo@ibm.net
Manufacturer of laser diodes.

Kaiser Systems Inc

126 Sohier Rd, Beverly, MA 01915, USA
Tel: +1 508 922 9300
Fax: +1 508 922 8374
E-mail: sales@kaisersys.com
Web: www.kaisersys.com
Manufacturer of high-performance power supplies for lasers.

Kamelian Ltd

Block 7, Kelvin Campus, West of Scotland Science Park, Maryhill Rd, Glasgow
G20 0TH, UK
Tel: +44 141 579300
Fax: +44 141 573099
E-mail: enquiries@kamelian.com
Web: www.kamelian.co.uk
Manufacturer of opto components.

Kaye Dee Ltd

Stamford St, New Hall Rd Trading Estate, Sheffield, S Yorkshire S9 2TX, UK
Tel: +44 114 256 0222
Fax: +44 114 256 0019
Contact: F Vine
Optoelectronic component supplier.

Kelvin Nanotechnology

Rankine Building, University of Glasgow, Glasgow G12 8LT, UK
Tel: +44 141 330 4869
Fax: +44 141 330 3726
E-mail: S.Hicks@elec.gla.ac.uk
Web: www.elec.gla.ac.uk/groups/knt/KNTMBE.html
Contact: Dr Simon Hicks
Fabrication of optoelectronic materials as well as modulators and laser diodes.

Kimmon Electric Co Ltd

Laser Division, 1-53-2 Itabashi-ku, Tokyo 173, Japan
Tel: +81 3 5248 4811
Fax: +81 3 5248 0018
E-mail: lasers@kimmon.com
Web: www.kimmon.com
Laser manufacturer.

Kingmax Optoelectronics

No 1 Kuang Fu North Rd, Hsin Chu Industrial Park, Hukou, Hsin Chu 303, Taiwan ROC
Tel: +886 3 5970888
Fax: +886 3 590999
E-mail: sales@kingmax-opto.com.tw
Web: www.kingmax-opto.com.tw
Opto component manufacturer.

Korea Optoelectronics Corp

53-10 Kee Dong, Sungdong ku, Seoul, South Korea
Tel: +82 2 446 0002
Fax: +82 2 452 4614
Manufacturer of LEDs, IREDs, photodiodes, phototransistors and laser detectors.

Kymata Ltd

Starlaw Park, Livingston EH54 8SF, UK
Tel: +44 1506 426000
Fax: +44 1506 460066
E-mail: sales@kymata.com
Web: www.kymata.com
Contact: David Plekenpol
Manufacturer of integrated optical components.

Labsphere Inc

PO Box 70, Shaker St, North Sutton, NH 03260, USA
Tel: +1 603 927 4266
Fax: +1 603 927 4694
Web: www.labsphere.com
Manufacturer of reflectance material for optical components and laser targets.

Labsphere Inc (UK)

1st Floor, Lower Washford Mill, Mill St, Buglawton, Cheshire CW12 2AD, UK
Tel: +44 1788 562828
Fax: +44 1788 562929
Web: www.labsphere.com
Contact: Robert Yeo
Manufacturer of reflectance material for optical components and laser targets.

Lambda Photometrics Ltd

Lambda House, Batford Mill, Harpenden, Hertfordshire AL5 5DZ, UK
Tel: +44 1582 764334
Fax: +44 1582 712084
E-mail: info@lambdaphoto.co.uk
Web: www.lambdaphoto.co.uk
Contact: Bob Carless
Optoelectronic component manufacturer.

Lambda Physik GmbH

Hans-Böckler-Straße 12, D-37079 Göttingen, Germany
Tel: +49 551 69380
Fax: +49 551 68691
E-mail: salesgermany@lambdaphysik.com
Manufacturer of solid-state lasers.

Lambda Physik Inc

3201 West Commercial Boulevard, Fort Lauderdale, FL 33309, USA
Tel: +1 954 486 1500
Fax: +1 954 486 1501
E-mail: marcom@lambdaphysik.com
Contact: Lynn Kobel
Manufacturer of lasers.

Lambda Physik Japan Co Ltd

German Industry Center, 1-18-2, Hakusan, Midori-ku, Yokohama 226, Japan
Tel: +81 45 939 7848
Fax: +81 45 939 7849
Manufacturer of solid state lasers.

Laser 2000 GmbH

Argelsrieder Feld 14, D-82234 Wessling, Germany
Tel: +49 8153 4050
Fax: +49 8153 40533
Manufacturer of optoelectronic components including semiconductor lasers.

Laser Analytical Systems (LAS) GmbH

Ruhlsdorfer Str 95, D-14532 Stahnsdorf, Germany
Tel: +49 33 29 63870
Fax: +49 33 29 638733
E-mail: info@las.de
Web: www.las.de
Manufacturer of laser systems.

Laser Analytical Systems (LAS) Inc

3333 Bowers Ave, Suite 130, Santa Clara, CA 95054, USA
Tel: +1 408 253 8350
Fax: +1 408 253 7288
E-mail: info@lasinc.com
Web: www.las.com
Manufacturer of laser systems.

Laser Components GmbH

Werner von Siemensstr 15, D-82140 Olching, Germany
Tel: +49 8142 28640
Fax: +49 8142 286411
E-mail: info@lasercomponents.de
Optoelectronic component manufacturer.

Laser Diode Inc

4 Olsen Ave, Edison, NJ 08820, USA
Tel: +1 732 549 9001
Fax: +1 732 906 1559
E-mail: sales@laserdiode.com
Web: www.laserdiode.com
Contact: M Robertson
Manufacturer of high-powered pulsed and CW lasers, solar cells.

Laser Diode Inc

Crystal Products Division, 205 Liberty St, Metuchen, NJ 08840, USA
Tel: +1 908 549 9222
Fax: +1 908 549 9897
E-mail: sales@laserdiode.com
Web: www.laserdiode.com
Contact: Steve Lerner
Manufacturer of optoelectronic substrates.

Laser Graphics GBR

LG Laser Technologies GmbH
Hanauer Straße 59, D-63801 Kleinostheim, Germany
Tel: +49 6027 4662-0
Fax: +49 6027 4662 33
E-mail: info@lg-lasertechnologies.com
Web: www.lg-lasertechnologies.com
Manufacturer of diode lasers.

Laser Lines Ltd

Beaumont Close, Banbury, Oxon OX16 7TQ, UK
Tel: +44 1295 267755
Fax: +44 1295 269651
E-mail: corp@laserlines.co.uk
Web: www.laserlines.co.uk
Contact: Steve Knight
Laser supplier.

Lasermate Corp

1977 W Holt Ave, Pomona, CA 91768, USA
Tel: +1 909 868 6818
Fax: +1 909 868 0948
E-mail: info@lasermate.com
Web: www.lasermate.com
Manufacturer of optoelectronic components including laser diode modules, diode pumped solid-state lasers, VCSELs.

Laseroptics SA

Lavalle 1634, 1048 Buenos Aires, Argentina
Tel: +54 1 372 7547
Fax: +54 1 372 7531
E-mail: laser@laseroptics.com.ar
Optoelectronic component manufacturer.

Laser Precision Corp

109 N Genessee St, Utica, NY 13502-2509, USA
Tel: +1 315 797 4449
Fax: +1 315 798 4038
Manufacturer of fibre-optic test and measurement instruments and systems utilizing advanced optical design and microprocessor technology.

Laser Power Optics

12777 High Bluff Drive, San Diego, CA 92130, USA
Tel: +1 619 755 0700
Fax: +1 619 259 9093
Laser manufacturer.

Laser SOS Ltd

Unit 3, Burrel Rd, St Ives Industrial Estate, St Ives, Cambridgeshire PE17 4LE, UK
Tel: +44 1480 460990
Fax: +44 1480 469978
Contact: Antoni Koszykowski
Laser distributor.

Laser Support Services Ltd

38 James St, Pittenweem, Fife KY10 2QN, UK
Tel: +44 1333 311938
Fax: +44 1333 312082
Contact: Grahame Rogers
Provider of 300 mW solid state diode pumped lasers.

Lasertel

7775 N Casa Grande Highway, Tucson, AZ 85743, USA
Tel: +1 520 744 5700
Fax: +1 520 744 5766
E-mail: info@lasertel.com
Web: www.lasertel.com
Manufacturer of diode lasers.

Lasertron

37 North Ave, Burlington, MA 01803, USA
Tel: +1 617 2726462
Fax: +1 617 2732694
Optoelectronic component manufacturer.

LasIRvis Technology (Europe) Ltd

26 Gosforth Cose, Middlefield Industrial Estate, Sandy SG19 1RB, UK
Tel: +44 1767 692727
Fax: +44 1767 692626
E-mail: lasirvis@kbnet.co.uk
Contact: Ian Bulavs
Manufacturer of optoelectronic components, including laser diodes.

Lasotronic AG

Blegistr 13, CH-6340 Baar-Zug, Switzerland
Tel: +41 42 330033
Fax: +41 42 330030
Manufacturer of lasers for medical applications.

Leadwell CNC Machines Mfg Corp

No 23, Gong 33 Rd, Taichung Industrial Park, Taichung 407, Taiwan ROC
Tel: +886 4 23591880
Fax: +886 4 23592555, 23593875
Web: www.leadwell.com.tw
Laser cutting specialists.

LiCONiX

3281 Scott Blvd, Santa Clara, CA 95054-3014, USA
Tel: +1 408 496 0300
Fax: +1 408 492 1303
E-mail: mdowley@LiCONIX.com
Web: www.LiCONIX.com
Manufacturer of diode laser systems.

Lightchip

27 Northwestern Drive, Salem, NH 03079, USA
Tel: +1 603 894 7165
Fax: +1 603 870 9890
E-mail: Info@lightchip.com
Web: www.lightchip.com
DWDM technology supplier.

LightLogic Inc

8674 Thornton Ave, Newark, CA 94560, USA
Tel: +1 510 578 5600
Fax: +1 510 578 5900
E-mail: info@lightlogic.com
Web: www.lightlogic.com
Contact: Gary Wiseman
Lightlogic designs, manufactures and sells advanced optoelectronic modules.

Lite-On Computer Tech (Dong Guan) Ltd

Hastelweg 232, 5652 CM Eindhoven, the Netherlands
Tel: +31 40 295 7500
Fax: +31 40 295 7565
E-mail: Chris_Wang@liteontc.com
Contact: Chris Wang
Optoelectronic component manufacturer.

Lite-On (Europe) Ltd

Blegistrasse 11B, CH-6340 Baar-Walterswill, Switzerland
Tel: +41 41 761 38 44
Fax: +41 41 763 14 34
Web: www.liteon-europe.com
Contact: Hans Heusser
Optoelectronic component manufacturer.

Lite-On Technology

5F 16 Sec 4, Nanking E Rd, Taipei, Taiwan ROC
Tel: +886 2 2570 6999
Fax: +886 2 2570 6888
E-mail: Edward_Chen@liteontc.com
Web: www.liteontc.com.tw
Contact: Edward Chen
Optoelectronic component manufacturer.

Lite-On Technology (M) SDN BHD

Jalan Rozhan, Off Jalan Alma, 14000 Bukit, Mertajam, Penang, Malaysia
Tel: +60 4 5516711
Fax: +60 4 5511051
E-mail: KJ_Liu@liteontc.com
Contact: K J Liu
Optoelectronics component manufacturer.

LITESPEC Optical Fiber LLC

Alexander Drive, Research Triangle Park, NC 27709, USA
Tel: +1 919 541 8420
Manufacturer of optical fibres.

LOT-Oriel Ltd

1 Mole Business Park, Leatherhead, Surrey KT22 7AU, UK
Tel: +44 1372 378822
Fax: +44 1372 375353
Contact: S Parr
Optoelectronic component distributor.

Loughborough University of Technology

Dept of Mechanical Engineering, Loughborough, Leics LE11 3TU, UK
Tel: +44 1509 223222
Fax: +44 1509 268013
E-mail: J.R.Tyrer@Lboro.ac.uk
Web: www.lboro.ac.uk
Contact: Dr John Tyrer
Contract R&D supplier.

LS Laser Systems GmbH

Gollierstr 70, D-80339 Munich, Germany
Tel: +49 89 5026867
Fax: +49 89 5004509
Manufacturer of semiconductor-, solid-state- and diode lasers.

Lynton Lasers Ltd

Lindow House, Beech Lane, Wilmslow, Cheshire SK9 5ER, UK
Tel: +44 1625 536646
Fax: +44 1625 530633
E-mail: sales@lyntlaz.demon.co.uk
Web: www.lyntlaz.demon.co.uk
Contact: Paul Vernon
Laser manufacturer.

LYNX Comerico E Importacao

Rua Alves Guimaraes, 1426-Jardim America, 05410-010-Sao Paolo-SP, Brazil
Tel: +55 11 3871 1996
Fax: +55 11 3871 1803
E-mail: lynx@dialdata.com.br
Optoelectronic component supplier.

M/A-COM III-V Materials Group

100 Chelmsford St, Lowell, MA 01853, USA
Tel: +1 978 656 2630
Fax: +1 978 656 2800
E-mail: kobar@tycoelectronics.com
Contact: Richard Koba
Web: www.macom-gaaswafers.com
Materials manufacturer.

Maintech Inc

1957 D Pioneer Rd, Huntingdon Valley, PA 19006, USA
Tel: +1 215 328 0820
Fax: +1 215 328 0810
E-mail: cschweriner@maintechinc.com
Web: www.maintechinc.com
Contact: Craig Schweriner
Opto components manufacturer.

Marconi Optical Components (formerly Caswell Technology)

Caswell, Towcester NN12 8EQ, UK
Tel: +44 1327 356751
Fax: +44 1327 356775
E-mail: oc.enquiries@marconi.com
Web: www.oc.marconi.com
Contact: Ray Taylor
Manufacturer of optoelectronic components.

Marconi Optical Components (formerly Marconi Applied Technologies)

Waterhouse Lane, Chelmsford, Essex CM1 2QU, UK
Tel: +44 1245 493493
Fax: +44 1245 492492
E-mail: mtech.uk@marconi.com
Web: www.oc.marconi.com
Contact: Neil O'Brien
Manufacturer of optical components.

Marketech International

5869 Beacon St, Pittsburgh, PA 15217, USA
Tel: +1 412 421 3103
Fax: +1 412 421 1826
Contact: J Spieckerman
Optoelectronic substrate supplier.

Marubun Corp

Marubun Daiya Bldg, 8-1 Nihonbashi Odenmacho, Chuo-ku, Tokyo 103, Japan
Tel: +81 3 3639 9811
Fax: +81 3 3662 1349
Manufacturer of solid state lasers.

Matheson Tri-Gas

625 Wool-Creek Drive, San Jose, CA 95112, USA
E-mail: info@matheson-trigas.com
Web: www.mathesontrigas.com
Materials supplier.

Matsushita Electronics Corp

Semiconductor Group, 1-1 Saiwai-cho, Takatsuki-shi, Osaka 569-1193, Japan
Tel: +81 726 82 5521
Fax: +81 726 82 3093
Manufacturer of laser diodes, LEDs and other optoelectronic devices.

Matsushita Electronics Corp of North America

1 Panasonic Way, Secaucus, NJ 07094, USA
Tel: +1 201 348 7000
Fax: +1 201 348 5310
Manufacturer of optoelectronics, including LEDs.

MBE Technology Pte

14 Science Park Drive, #04-03 The Maxwell, Singapore Science Park, Singapore 0511
Tel: +65 7735211
Fax: +65 7735068
E-mail: mbetech@pacific.net.sg
Web: www.mbetech.com
Contact: Zhong Wang
Optoelectronic substrate manufacturer.

MBE Technology USA Inc

1170 Ninth St, #3, Alameda, CA 94501, USA
Tel: +1 510 523 0288
Fax: +1 510 864 1308
E-mail: srschoenly@email.msn.com
Contact: R Schoenly
Optoelectronic substrate manufacturer.

Meller Optics Inc

120 Corliss St/PO Box 6001, Providence, RI 02940, USA
Tel: +1 800 821 0180
Fax: +1 401 331 0519
E-mail: steve@melleroptics.com
Web: www.melleroptics.com
Contact: Steve Lydon, Marketing
Manufacturer of epi polished sapphire substrates for GaN, III-V, and II-VI thin-film deposition.

Melles Griot

Laser Group, 2251 Rutherford Rd, Carlsbad, CA 92008, USA
Tel: +1 760 438 2131
Fax: +1 760 438 5208
E-mail: sales@carlsbad.mellesgriot.com
Web: www.mellesgriot.com
Manufacturer of diode lasers.

Melles Griot

2 Pembroke Ave, Waterbeach, Cambridge CB5 9QR, UK
Tel: +44 1223 203300
Fax: +44 1223 203311
Web: www.mellesgriot.com
Contact: N Pratt
Manufacturer of diode lasers.

Metalorganics

60 Willow St, North Andover, MA 01845, USA
Tel: +1 978 557 1700
Fax: +1 978 557 1701
E-mail: bstennic@rohmhaas.com
Materials supplier.

Microlase Optical Systems

141 St James Rd, Glasgow GL4 0LT, UK
Tel: +44 141 552 8205
Fax: +44 141 552 3906
E-mail: lasers@microlase.co.uk
Web: www.microlase.co.uk
Contact: Dr Graeme Malcolm
Laser manufacturer.

Mitsubishi Cable America Inc

520 Madison Ave, New York, NY 10022, USA
Tel: +1 212 888 2270
Fax: +1 212 888 2276
Contact: M Scott
Optoelectronic component manufacturer.

Mitsubishi Chemical Corp

Compound Semiconductor Dept, 5-2 Marunouchi 2-Chome, Chiyoda-ku, Tokyo
100, Japan
Tel: +81 3 283 4673
Fax: +81 3 3283 4485
Contact: Katsuyuki Simakawa
Manufacturer of optoelectronic substrates.

Mitsubishi Electric Corp

Optoelectronic & Microwave Devices Laboratory, Kita-Itami Works, 4-1 Mizuhara, Itami City, Hyogo 664, Japan
Tel: +81 727 84 7384
Fax: +81 727 80 2694
Contact: Dr Yasuo Mitsui
Manufacturer of optoelectronic devices, including GaAs FETs and optoelectronic modules at its Kamakura Works.

Mitsubishi Electric Europe

Semiconductor Business Unit, Travellers Lane, Hatfield, Herts AL10 8XB, UK
Tel: +44 1707 276 100
Fax: +44 1707 278 997
E-mail: christine.warren@meUKmee.com
Web: www.mitsubishichips.com
Contact: C Warren
Manufacturer of optoelectronic components.

MMR Technologies

1400 N Shoreline Blvd, Ste A 5, Mountain View, CA 94043-1346, USA
Tel: +1 415 962 9620
Fax: +1 415 962 9647
Manufacturer of optoelectronic devices.

Mochem

Hannah-Arendt-Strasse 3-7, D-35007 Marburg, Germany
Tel: +49 6421 350 250
Fax: +49 6421 350 251
E-mail: greiling@mochem.de
Web: www.mochem.de
Contact: Arnd Greiling
Materials supplier.

MODE Division of EMCORE

5741 Midway Park Place NE, Albuquerque, NM 87109, USA
Tel: +1 505 343 1111
Fax: +1 505 343 8300
E-mail: Bill_Leasure@Emcore.com
Contact: Bill Leasure
Manufacturer of VCSELs.

Modulight Inc

PO Box 770, FIN-33101 Tampere, Finland
Tel: +358 3 3399 1200
Fax: +358 3 3399 1210
E-mail: sales@modulight.com
Web: www.modulight.com
Optical component manufacturer.

MOELLER-WEDEL GmbH

Rosengarten 10, D-22880 Wedel, Germany
Tel: +49 4103 709 01
Fax: +49 4103 709 355
E-mail: info@moeller-wedel.com
Web: www.moeller-wedel.com
Leading manufacturer of operating microscopes and eye examination equipment.

Molecular OptoElectronics Corp

877 25th St, Wastervliet, NY 12189, USA
Tel: +1 518 270 8208
Fax: +1 518 273 5701
E-mail: sales@moec.com
Web: www.moec.com
Component manufacturer.

MTI Corp

5327 Jacuzzi St, Unit 3G, Richmond, CA 94804, USA
Tel: +1 510 525 3070
Fax: +1 510 234 5235
E-mail: mtixpj@aol.com
Contact: XP Jiang
Optoelectronic substrate manufacturer.

Multiplex Inc

115 Corporate Blvd, South Plainfield, NJ 07080, USA
Tel: +1 908 757 8817
Fax: +1 908 757 8910
Web: www.multiplexinc.com
Contact: Steven Kukoda
Manufacturer of high-end opto-electronic components.

Murata Electronics (UK) Ltd

Oak House, Ancells Rd, Ancells Business Park, Fleet, Aldershot GU13 8UN, UK
Tel: +44 1252 811666
Fax: +44 1252 811777
E-mail: enquiry@murata.co.uk
Web: www.murata.co.jp
Contact: R Kingdom
Optoelectronics supplier.

Mütek Infrared Laser Systems

Arzbergerstr 10, D-82211 Herrsching, Germany
Tel: +49 8152 5245
Manufacturer of diode lasers.

Nanovation Technologies Inc

601 Brickell Key Drive, Suite 802, Miami, FL 33131, USA
Tel: +1 305 373 1664
Fax: +1 305 373 2203
E-mail: Roberttatum@worldnet.att.net
Web: www.nanovation.com
Contact: Robert Tatum
Research into and manufacture of microcavity semiconductor lasers and LEDs.

NEC Electronics

Electronic Devices Group, 7-1 Shiba 5-chome, Minato-ku, Tokyo 1098-01, Japan
Tel: +81 3 3454 1111
Fax: +81 3 3798 1510
Manufacturer of laser diodes for telecoms and data storage applications.

NEC Electronics (Europe) GmbH

Oberrather Str 4, D-40472 Düsseldorf, Germany
Tel: +49 211 650301
Fax: +49 211 6503327
Web: www.euronec.com
Contact: Mr Tachimoto
Manufacturer of optoelectronic components.

NEL

Compound Semiconductor Device Division, 3-1 Wakamiya, Morinosato, Atsugi-shi, Kanagawa 243-01, Japan
Tel: +81 462 40 4051
Fax: +81 462 50 2488
Contact: Tohru Takada
Manufacturer of high speed fibre-optic transmission systems.

Neraeus Noblelight Ltd

Cambridge Science Park, Milton Rd, Cambridge CB4 4GQ, UK
Tel: +44 1223 423324
Fax: +44 1223 423999
Optoelectronic component supplier.

New Dimension Research & Instrument Inc

400 West Cummings Park, Suite 6950, Woburn, MA 01801, USA
Tel: +1 781 933 1165
Fax: +1 781 933 1214
Web: www.newdri.com
Developer of VCSEL technology.

New Focus Inc

1275 Reamwood Ave, Sunnyvale, CA 94089, USA
Tel: +1 408 7348988
Fax: +1 408 7348882
Web: www.newfocus.com
Manufacturers of diode lasers and photodiodes.

New Focus Inc

5215 Hellyer Ave, San Jose, CA 95138-1001, USA
Tel: +1 408 284 5039
Fax: +1 408 284 4829
E-mail: cdavidson@newfocus.com
Web: www.newfocus.com
Contact: Cathy Davidson
Manufacturer of fibre optic products for next-generation optical networks.

New Focus Inc

2630 Walsh Ave, Santa Clara, CA 95051-0905, USA
Tel: +1 408 9808088
Fax: +1 408 9808883
E-mail: Contact@NewFocus.com
Web: www.newfocus.com
Manufacturer of high-performance photonics tools for laser applications.

Newport Corp

1791 Deere Ave, Irvine, CA 92606, USA
Tel: +1 949 863 3144
Fax: +1 949 253 1800
E-mail: gspiegel@newport.com
Web: www.newport.com
Manufacturer of optoelectronic components.

Newport Ltd

4320 First Ave, London Rd, Newbury Business Park, Newbury RG14 2PZ, UK
Tel: +44 1635 521757
Fax: +44 1635 521348
E-mail: uk@newport.com
Web: www.newport.com
Contact: Dr David Welsh
Optoelectronic component distributor.

Nichia Corp

491 Oka, Kaminaka-Cho, Anan-Shi, Tokushima-Ken, 774 Japan
Tel: +81 884 22 2311
Fax: +81 884 210 148
E-mail: laser@nichia.co.jp
Web: www.nichia.co.jp
Manufacturer of optoelectronic components, including blue LEDs and lasers.

Nichia America Corp

3775 Hempland Rd, Mountville, PA 17554, USA
Tel: +1 717 285 2323
Fax: +1 717 285 9378
E-mail: info@nichia.com
Web: www.nichia.com
Manufacturer of optoelectronic components.

nLight Photonics Corp

800 Maynard Ave South, Suite 100, Seattle, WA 98134, USA
Tel: +1 206 336 5567
fax: +1 206 336 5558
E-mail: info@nlightphotonics.com
Web: www.nlightphotonics.com
Start-up company developing pump modules for DWDM.

NOK EG&G Optoelectronics Corp

Parale Mitsui Bldg, 18th Floor, 8 Higashida-cho, Kawasaki-ku, Kawasaki-shi, Kanagawa 210, Japan
Tel: +81 44 200 9150
Fax: +81 44 200 9160
Web: www.egginc.com
Optoelectronics component manufacturer.

Nortel Networks

Optical Components, 2745 Iris St, Ottawa, ON K2C 3V5, Canada
E-mail: sales-opticalcomponents@nortelnetworks.com
Web: www.nortelnetworks.com/opticalcomponents
Manufacturer of optoelectronics components.

Nortel Networks plc

Brixham Rd, Paignton, Devon TQ4 7BE, UK
Tel: +44 1803 662000
Fax: +44 1803 662801
E-mail: smburns@nortelnetworks.com
Web: www.nortelnetworks.com/optoelectronics
Contact: J Duffey
Manufacturer of semiconductor laser emitters, IREDs and photodiode detectors and receivers.

Nortel Networks GmbH

Bereich OptoElectronics, Munsterstr 100A, D-40476 Düsseldorf, Germany
Tel: +49 211 9488850
Fax: +49 211 9488853
Manufacturer of optoelectronic components.

Nortel Networks

Kalvebod Brygge 33, 1560 Copenhagen V, Denmark
Tel: +45 33 74 22 22
Fax: +45 33 74 22 44
Web: www.nortelnetworks.com
Manufacturer of optoelectronic components for telecoms applications.

Nortel Networks

33 Quai Paul Doumer, Paris La Défense, F-92415 Courbevoie Cedex, France
Tel: +33 1 41 99 15 15
Fax: +33 1 41 99 15 00
Web: www.nortelnetworks.com
Manufacturer of optoelectronic components for telecoms applications.

Novotech Inc

916 Main St, Acton, MA 02135, USA
Tel: +1 978 929 9458
Fax: +1 978 929 9459
E-mail: novohm@fiam.net
Web: www.novotech.net
Component manufacturer.

NSG/Fiber Optics Div

8-1 5-Chome, Nishi Hashimoto, Sagamihara-shi, Kanagawa 229-11, Japan
Tel: +81 354 460 159
Fax: +81 354 430 160
E-mail: nsg10797@tax2.nsg.co.jp
Contact: Dr Oikawa
Optoelectronic component manufacturer, especially fibre optics.

Oak Technology

139 Kifer Court, Sunnyvale, CA 94086, USA
Tel: +1 408 737 0888
Fax: +1 408 737 3838
Contact: R Simone
Optical storage products.

Ocean Optics Europe

Soerense Zand 4, NL-6961 LL Eerbeek, the Netherlands
Tel: +31 313 651 978
Fax: +31 313 655 783
E-mail: ooe@oceanoptics.com
Fibre optic spectrometer manufacturer.

Ocean Optics Inc

380 Main St, Dunedin, FL 34698, USA
Tel: +1 727 733 2447
Fax: +1 727 733 3962
E-mail: Info@OceanOptics.com
Web: www.oceanoptics.com
Manufacturer of miniature fibre optic spectrometers.

Okab-Roederstein AB

Ellipsvagen 12, S-14175 Huddinge, Sweden
Tel: +46 8 740 8100
Fax: +46 8 740 1177
Component distributor.

Oki Electric Industry Co Ltd

7–12 Toranomon 1-chome, Minato-ku, Tokyo 105, Japan
Tel: +81 3 3501 3111
Fax: +81 3 3581 5522
Web: www.oki.co.jp
Manufacturer of optoelectronic components.

Optec GmbH

Siemensstr 7, D-57299 Burbach, Germany
Tel: +49 2736 44090
Fax: +49 2736 440915
Manufacturer of lasers for medical applications.

Optigain Inc

1174 Kingstown Rd, PO Box 3732, Peace Dale, RI 02883, USA
Tel: +1 401 783 9222
Fax: +1 401 783 9224
E-mail: wulmschneider@optigain.com
Web: www.optigain.com
Manufacturer of erbium doped fibre amplifiers for telecommunication, CATV and photonics research.

Optikzentrum NRW

Universitätsstr 142, D-44799 Bochum, Germany
Tel: +49 234 970700
Fax: +49 234 9707070
Optoelectronics component manufacturer.

Optilas GmbH

Boschstr 12, D-82178 Puchheim, Germany
Tel: +49 89 890135
Fax: +49 89 8002561
Optoelectronic component manufacturer and supplier.

OptoCom Innovation

1 rue de Langonaval, F-22304 Lannion, France
Tel: +33 2 96 372699
Fax: +33 2 96 464689
Manufacturer of high-power optical amplifiers.

Optometrech Systems/Applied Ocean Physics

9899 Hilbert, Suite A, San Diego, CA 92131, USA
Tel: +1 619 566 572
Fax: +1 619 695 1197
Manufacturer of optoelectronic devices.

Opto Power Corp

3321 East Global Loop, Tucson, AZ 85706, USA
Tel: +1 520 746 1234
Fax: +1 520 294 3300
E-mail: opc-info@optopower.com
Web: www.optopower.com
Manufacturer of diode laser arrays.

OptoSci Ltd

141 St James Rd, Glasgow G4 0LT, UK
Tel: +44 141 552 7020
Fax: +44 141 552 3886
Web: www.optosci.co.uk
Optoelectronic component manufacturer.

Opto Speed Holding SA

Via Cantonale, CH-6805 Mezzovico, Switzerland
Tel: +41 91 935 52 52
Fax: +41 91 935 52 62
E-mail: ptenti@optospeed.com
Web: www.optospeed.com
Contact: Paolo Tenti
Producer of high-end optical components.

Optotek Ltd

62 Steacie Drive, Kanata, ON K2K 2A9, Canada
Tel: +1 613 591 0336
Fax: +1 613 591 0584
Web: www.optotek.com
Contact: P Wareberg
Manufacturer of advanced optoelectronics, including LED displays.

Optotel AS

Gert Rasmussen, Skovlytoften 4, Box 124, Denmark
Tel: +45 45 41 05 06
Fax: +45 45 41 07 30
Web: www.optotel.dk
Optoelectronic component distributor.

Oriel Instruments

150 Long Beach Blvd, PO Box 872, Stratford, CT 06497-0872, USA
Tel: +1 203 377 8282
Fax: +1 203 378 2457
E-mail: res_sales@oriel.com
Web: www.oriel.com
Optical component supplier.

Oxford Applied Research

Crawley Mill, Witney, Oxon OX29 9SP, UK
Tel: +44 1993 773 575
Fax: +44 1993 702 326
E-mail: sales@oaresearch.co.uk
Web: www.oaresearch.co.uk
Supplier of thin film deposition equipment, eg RF atom sources for oxides and nitrides.

Oxford Lasers Ltd

Abingdon Science Park, Barton Lane, Abingdon, Oxon OX14 3YR, UK
Tel: +44 1235 554211
Fax: +44 1235 554311
E-mail: oxlasers@rl.ac.uk
Web: www.oxfordlasers.com
Contact: Keith Errey
Laser manufacturer.

Oxley Inc

25 Business Park Drive, PO Box 814, Branford, CT 06405, USA
Tel: +1 203 488 1033
Fax: +1 203 481 6971
Web: www.oxleyinc.com
Manufacturer of optoelectronic components.

Panasonic Industrial Europe GmbH

Neukeferloh, Bretonischer Ring, D-85630 Grasbrunn, Germany
Tel: +49 89 46007 119
Fax: +49 89 46007 223
Web: www.panasonic.de
Contact: Norbert Spring
Manufacturer of optoelectronic components.

Panasonic Industrial Europe Ltd

Willoughby Rd, Bracknell RG12 8FP, UK
Tel: +44 1344 862444
Fax: +44 1344 861656
Web: www.panasonic.co.jp
Contact: Henry Ohashi
Manufacturer of optoelectronic components, including LED displays.

PD-LD Inc

3451 E Harbour Drive, Phoenix, AZ 85034-7229, USA
Tel: +1 602 437 85000
Fax: +1 602 437 8555
Manufacturer of standard visible laser diodes.

Performance Materials

4 Park Ave, Hudson, NH 03051-3927, USA
Tel: +1 603 598 9122
Fax: +1 603 598 9126
E-mail: jdecosta@performancematerial.com
Web: www.performancematerial.com
Materials supplier.

PerkinElmer Optoelectronics (formerly EG&G Inc)

2175 Mission College Blvd, Santa Clara, CA 95054, USA
Tel: +1 408 565 0707
Fax: +1 408 565 0793
Web: www.opto.perkinelmer.com
Contact: Shannyn Roberts
Manufacturer of optoelectronics components.

Phoenicon GmbH

Bismarckstr 58, D-10627 Berlin, Germany
Tel: +49 30 3421088
Fax: +49 30 3429055
Component manufacturer.

Photodigm Inc

3026 Mockingbird Lane #155, Dallas, TX 75205, USA
Tel: +1 972 747 7721
E-mail: jmattis@photodigm.com
Web: www.photodigm.com
Contact: Jack Mattis
Start-up company to commercialise optical grating and semiconductor laser technology.

Photonic Materials Ltd

6 Mallard Way, Strathclyde Business Park, Bellshill ML4 3BF, UK
Tel: +44 1698 73810
Fax: +44 1698 73811
E-mail: info@photonicmaterials.com
Web: www.photonicmaterials.com
Manufacturer of GaN crystals for emitters.

Photon Inc

1115 Space Park Drive, Santa Clara, CA 95054, USA
Tel: +1 408 492 9449
Fax: +1 408 492 9659
E-mail: info@photon-inc.com
Web: www.photon-inc.com
Manufacturer of laser beam profiling equipment.

Picogiga

Place Marcel Rebuffat, Parc de Villejust, 91971 Courtaboeuf 7 Cedex, Villejust
France
Tel: +33 169 31600
Fax: +33 169 316178
Web: www.picogiga.com
Supplier of epiwafers.

Picogiga

5 rue de la Reunion, ZA de Courtaboeuf, F-91940 Les Ulis, France
Tel: +33 1 6907 1950
Fax: +33 1 6907 3208
Web: www.picogiga.com
Contact: Linh Nuyen
Supplier of epiwafers.

Picogiga Inc

122 Calistoga Road # 330, Santa Rosa, CA 95409, USA
Tel: +1 707 539 2508
Fax: +1 707 539 2508
Web: www.picogiga.com

Picopolish SA

CD 56, Route des Michels, F-13790 Peynier-Rousset, France
Tel: +33 442 29 19 79
Fax: +33 442 29 19 70

Picopolish Switzerland GaAs

Chapons des Prés 13, Bevaix, CH-2022, Switzerland
Tel: +32 847 05 55
Fax: +32 847 05 56
E-mail: picopolish@compuserve.com
Web: www.picipolish.com
Contact: Gerard Gilles
Materials supplier.

PicoQuant GmbH

Rudower Chaussee 29 (IGZ), D-12489 Berlin, Germany
Tel: +49 30 6392 6560
Fax: +49 30 6392 6561
E-mail: photonics@pq.fta-berlin.de
Web: www.picoquant.com
Supplier of small picosecond pulsed diode lasers.

Pilkington Optronics

Barr & Stroud Ltd, 1 Linthouse Rd, Glasgow G51 4BZ, UK
Tel: +44 41 954 4000
Fax: +44 41 954 4001
Supplier of electro-optical equipment for both defence and civilian applications.

Planar Plus Ltd

PO Box 155, Nijegorodskaya 23, Office 307, Novosibirsk 630092, Russia
Tel: +7 3832 664689
Fax: +7 3832 664689
E-mail: misha@planar.nsk.ru
Optoelectronic component manufacturer.

Polaron CVT Ltd

4 Carters Lane, Kiln Farm, Milton Keynes MK11 3ER, UK
Tel: +44 1908 563267
Fax: +44 1908 568354
E-mail: rdavies@polaron-group.co.uk
Web: www.polaron-group.co.uk/cvt/index.html
Contact: Dr Richard Davies
Manufacturer of custom and off-the-shelf vacuum equipment.

Polytec GmbH

Polytec-Platz 5-7, D-76337 Waldbronn, Germany
Tel: +49 7243 6040
Fax: +49 7243 69944
Manufacturer of optoelectronic components including semiconductor lasers.

Precision-Optical Engineering

42 Wilbury Way, Hitchin, Herts SG4 0TP, UK
Tel: +44 1462 440328
Fax: +44 1462 440329
Web: www.p-oe.co.uk
Contact: Colin Freeland
Optoelectronic equipment manufacturer.

Princeton Lightwave Inc

2601 Route 130 South, Cranbury, NJ 08512, USA
Tel: +1 609 925 8100
Fax: +1 609 409 7023
E-mail: info@princetonlightwave.com
Web: www.princetonlightwave.com
Manufacturer of next-generation optical networking components.

Princeton Scientific Corp

PO Box 143, Princeton, NJ 08542, USA
Tel: +1 609 924 3011
Fax: +1 609 924 3018
E-mail: psi1 @idt.net
Web: www.princetonscientific.com
Makers of a wide range of III-V materials — Inp, InSb, InAs, GaP, GaSb, GaAs, etc, epi-ready and range of sizes.

PROXITRONIC

Robert-Bosch-Straße 34, D-64625 Bensheim, Germany
Tel: +49 6251 1703 0
Fax: +49 6251 1703 90
E-mail: info@proxitronic.de
Web: www.proxitronic.de
German manufacturer of high quality image intensifiers and image converters.

PRP Optoelectronics Ltd

Wood Burcote Way, Towcester NN12 6TF, UK
Tel: +44 1327 359135
Fax: +44 1327 359602
E-mail: geoffl@prpopto.demon.co.uk
Contact: Dr Geoff Lidgard
Manufacturer of optoelectronics components.

Quality Technologies UK Ltd

10 Prebendal Court, Oxford Rd, Aylesbury HP19 3EY, UK
Tel: +44 1296 394499
Fax: +44 1296 392432
E-mail: nspencer@qtopto.com
Web: www.qtopto.com
Contact: N Spencer
Optoelectronic component manufacturer, including LEDs and optocouplers.

Quarton Inc

9F, No 185, Sec 1, Ta-Tung Rd, Hsi-Chih,Taipei Hsien,Taiwan ROC
Tel: +886 2 2643 1000
Fax: +886 2 2643 2000
E-mail: quarton@quarton.com.tw
Web: www.quarton.com.tw
Manufacturer of laser sights, pointers and modules.

Quarton USA Ltd Co

7042 Alamo Downs Parkway, Suite 250, San Antonio, TX 78238-4518, USA
Tel: +1 210 520 8430
Fax: +1 210 520 8433
Manufacturer of laser modules.

Qusion Technologies

Suite 200, 100 Canal Pointe Boulevard, Princeton, NJ 08540, USA
Tel: +1 609 951 4270
Fax: +1 609 987 0254
Start-up company manufacturing optical components.

QWIP Technologies

2400 Lincoln Ave, Ste 217, Altadena, CA 91001, USA
Tel: +1 626 296 6316
Fax: +1 626 296 6323
E-mail: jllorens@qwip.com
Web: www.qwip.com
Opto components manufacturer.

Raboutet SA

250 ave Louis-Armand, ZI des Grands-Prés, BP 31, F-74301 Cluses Cedex, France
Tel: +33 4 50 98 15 18
Fax: +33 4 50 98 92 57
E-mail: info@raboutet.fr
Web: www.raboutet.fr
Manufacturer of molybdenum components for MBE. Specialized cleaning and degassing available.

Recon Optical Inc

Pacific Optical Div, 2660 Columbia St, Torrance, CA 90503-3802, USA
Tel: +1 310 328 5840
Fax: +1 310 320 0550
E-mail: allie_baker@ROI.BOURNS.com
Designer and custom manufacturer of lenses.

Regisbrook Group Ltd

Units 1 and 2 Suffolk Way, Drayton Rd, Abingdon, Oxon OX14 5JY, UK
Tel: +44 1235 534909
Fax: +44 1235 528971
Contact: Chris Williams
Distributor of optoelectronic components.

Remcor Products Company

500 Regency Drive, Glendale Hights, IL 60139, USA
Tel: +1 708 980 6900
Fax: +1 708 980 8511
Manufacturer of optoelectronic devices.

Renishaw PLC

Transducer Systems Division, Old Town, Wotton-under-Edge GL12 7DH, UK
Tel: +44 1453 844302
Fax: +44 1453 844236
E-mail: pitt_david/gb@renishaw.co.uk
Contact: Dr David Pitt
Manufacturer of transducers and related components.

Rep-Tec Inc

2245 Valwood Parkway, Dallas, TX 75234, USA
Tel: +1 972 488 8383
Fax: +1 972 488 8384
E-mail: gary@rep-tec.com
Web: www.rep-tec.com
Contact: Gary Moss
Optoelectronic component supplier.

RF Nitro Communications

10420-F Harris Oaks Boulevard, Charlotte, NC 28269 USA
Tel: +1 704 596 9060
Fax: +1 704 596 0950
E-mail: sales@rfnitro.com
Web: www.rfnitro.com
Opto components manufacturer.

Riber SA

92503 Rueil-Malmaison, France
Tel: +33 1 47 08 84 55
Fax: +33 1 47 16 02 55
Web: www.riber.com

RIFOCS Corp

833 Flynn Rd, Camarillo, CA 93012, USA
Tel: +1 805 389 9800
Fax: +1 805 389 9808
Manufacturers of diode lasers and LEDs.

RM Photonics Ltd

Hatirosh St, PO Box 1507, Ramat Modiim/Hashmonaim, DN Modiin 73127, Israel
Tel/Fax: +972 8 9762021
E-mail: rmphoton@inter.net.il
Contact: Meir Chazon
Optoelectronic component distributor.

Roditi International Corp Ltd

Carrington House, 130 Regent St, London W1, UK
Tel: +44 171 439 4390
Fax: +44 171 434 0896
E-mail: kenmore@roditi.co.uk
Web: www.roditi.co.uk
Contact: Chris Kenmore
Supplier of optoelectronic materials.

ROHM Co Ltd

21 Saiin Mizosaki-cho, Ukyo-ku, Kyoto 615-8585, Japan
Tel: +81 75 311 2121
Fax: +81 75 315 0172
Web: www.rohm.com
Contact: J Hikita
Manufacturer of LEDs and laser diodes.

ROHM Electronics (UK) Ltd

Whitehall Ave, Kingston, Milton Keynes MK10 0AD, UK
Tel: +44 1908 282666
Fax: +44 1908 282528
Web: www.rohm.com
Manufacturer of optoelectronic components, including LEDs.

ROHM Electronics

10145 Pacific Heights Blvd, Suite 1000, San Diego, CA 92121, USA
Tel: +1 858 625 3630
Fax: +1 858 625 3670
Web: www.rohmelectronics.com.
Manufacturer of optoelectronic components, including LEDs.

ROHM Korea Corp

371–11 Kasan-Dong, Kumchon-ku, Seoul 152-023, South Korea
Tel: +82 2 8182 600
Fax: +82 2 837 0039
Web: www.rohm.com
Manufacturer of opto components.

ROHM Wako Co Ltd

100 Tomioka, Kasaoka, Okayama 714-8585, Japan
Tel: +81 865 67 0111
Fax: +81 865 67 2551
Web: www.rohm.com
Manufacturer of LEDs, LED displays and laser diodes.

Roithner Lasertechnik

Fleischmanngasse 9, A-1040 Vienna, Austria
Tel: +43 1 586 52 43
Fax: +43 1 586 41 43
E-mail: rlt@mcb.at
Manufacturer of optoelectronic components, including VCSELs.

Roxburgh Electronics Ltd

Roxburgh House, Foxhills Industrial Park. Scunthorpe, Lincs DN15 8QJ, UK
Tel: +44 1724 281770
Fax: +44 1724 281650
Optoelectronic component supplier.

Royal Philips Electronics NV

Afdeling Consumentenbelagen, Antwoordnummer 500, 5600 VB Eindhoven, The Netherlands
Tel: +31 40 2732305
Fax: +31 40 788399
Web: www.philips.com
Manufacturer of optoelectronic components.

Samsung Electronics

250, 2-ga, Taepyung-ro, Chung-gu, Seoul 100-742, South Korea
Tel: +82 2 727 7692
Fax: +82 2 727 7826
Research into, and manufacture of, semiconductor lasers.

San-Es Trading

6-15 Sangenjaya, 1-Chome Setagaya-Ku 154, Tokyo, Japan
Tel: +81 337 958 121
Fax: +81 337 958 008
E-mail: pxc05457@niftyserve.or.jp
Contact: Masao Takakura
Optoelectronic component distributor.

Sanyo Electric Co Ltd

5-5 Keihan Hondori 2-chome, Moriguchi City, Osaka 570-8677, Japan
Tel: +81 6 991 1181
Fax: +81 6 991 6566
Web: www.sanyo.co.jp
Optoelectronic component manufacturer, including lasers.

Schäfter & Kirchhoff

Celsiusweg 15, D-22761 Hamburg, Germany
Tel: +49 40 8512562
Fax: +49 40 8503137
Optoelectronic component supplier.

Scientific Measurement Systems Inc

2527 Foresight Circle, Grand Jct, CO 81505, USA
Tel: +1 800 7473308
Fax: +1 303 246 6618
Laser equipment manufacturer.

Semco Laser

18007 Cortney Court, City of Industry, CA 91748, USA
Tel: +1 626 581 0211
Fax: +1 626 581 0511
Web: www.semcolaser.com
Products include epiwafers, diode laser chips and devices.

SEMEFAB plc

Coventry Rd, Lutterworth LE17 4JB, UK
Tel: +44 1455 552505
Fax: +44 1455 5562612
Contact: C Robins
Manufacturer of optoelectronics/arrays.

Sharp Corp

22-22 Nagaike-cho, Abeno-ku, Osaka 545, Japan
Tel: +81 6 621 1221
Fax: +81 6117 725300
Web: www.sharp.co.jp
Manufacturer of optoelectronic components, including lasers.

Sharp Electronics (Europe) GmbH

Microelectronics Division, Sonnistr 3, D-20097 Hamburg, Germany
Tel: +49 40 23760
Fax: +49 40 2376 2510
Web: www.sharp.com
Contact: Peter Thiele
Manufacturer of optoelectronic components.

Sharp Microelectronics

5700 Northwest, Pacific Rim Blvd, #20 Camas, WA 98607, USA
Tel: +1 360 834 2500
Fax: +1 360 834 8903
Web: www.sharp.co.jp
Optoelectronic component manufacturer.

Shenai Semiconductor Co Ltd

Bldg 1, Guangxianxiaoqu, Baguasan Rd, Shenzhen, Guangdong 518028, China
Tel: +86 755 2260060
Fax: +86 755 240042
Optoelectronic component manufacturer.

Shinkosha Co Ltd

2-4-1 Kosugaya, Sakae-ku, Yokohama City 247-0007, Japan
Tel: +81 45 892 2174
Fax: +81 45 892 2986
E-mail: msales@shinkosha.com
Web: www.shinkosha.com
Sapphire substrates supplier.

Shiny Opto Electronics Co Ltd

Tzu Ching St, 4F No 4, Lane 11, Tu Chen City, Taipei 106, Taiwan ROC
Tel: +886 2 22684950
Fax: +886 2 22684374
E-mail seco1 @ms31.hinet.net
Optoelectronic component manufacturer.

Showa Denko America Inc

951 Mariners Island Blvd, Suite 680, San Mateo, CA 94404, USA
Tel: +1 415 345 1338
Fax: +1 415 345 5403
Contact: Bettye Garrett
Supplier of epiwafers.

Showa Denko KK

13-9 Shiba Diamon 1-chome, Minato-ku, Tokyo 105, Japan
Tel: +81 3 5470 3503
Fax: +81 3 3435 1034
Contact: Mr Yamamoto
Manufacturer of optoelectronic substrates.

SiCrystal AG

Heinrich-Hertz-Platz 2, D-92275 Eschenfelden, Germany
Tel: +49 96 65 91 370/+49 96 65 91 3790
E-mail: info@sicrystal.de
Web: www.sicrystal.de
Contacts: Dr Robert Eckstein — r.eckstein@sicrystal.com; Christian Diehl — c.diehl@sicrystal.com
Supplier of SiC wafers.

Siemens AG

Wittelsbacherplatz 2, D-80333 Munich, Germany
Tel: +49 89636 28480
Fax: +49 89 63628482
E-mail: katja.schlendorf@UKsiemens.de
Web: www.siemens.de
Optoelectronic component manufacturer.

Siemens

10950 North Tantau Ave, Cupertino, CA 95015, USA
Tel: +1 408 777 4968
Web: www.siemens.de
Optoelectronic component manufacturer.

Sirus Technology Corp

No 156 Kao-Shy Rd. Yang-Mei Chen, Tao Yuan 326, Taiwan ROC
Tel: +886 3 4852687
Fax: +886 3 4751625
E-mail: sales@sirustech.com
Manufacturer of LD Modules and semiconductor lasers.

SLI Inc

500 Chapman St, Canton, MA 02021, USA
Tel: +1 781 828 2948
Fax: +1 781 828 2012
Web: www.sli-lighting.com
Manufacturer of optoelectronic components.

SMK Corp

5-5 Togoshi 6-chome, Sinagawa-ku, Tokyo 142, Japan
Tel: +81 3 3785 1105
Fax: +81 3 3785 1878
Optoelectronic component distributor.

SOITEC SA

Parc Technologique des Fontaines, F-38190 Bernin, France
Tel: +33 4 76 92 75 00
Fax: +33 4 76 92 75 01
E-mail: sales@soitec.fr
Web: www.soitec.com
Thin film substrate manufacturer.

Solid Circuits Inc

Olivarez Compound, Dr A Santos Ave, Paranaque, Metro Manilla,
the Philippines
Tel: +63 2 827 5715
Fax: +63 2 819 0981
Subcontract assembler of optoelectronic devices.

Solid State Optronics Inc

2 North First St, San Jose, CA 95113, USA
Tel: +1 408 293 4600
Fax: +1 408 293 4848
Manufacturer of optoelectronic devices.

Solkatronic Chemicals

351 Philadelphia Ave, Morrisville, PA 19067, USA
Tel: +1 215 736 0700
Fax: +1 215 736 3666
E-mail: info@solkatronic.com
Web: www.solkatronic.com
Contact: Nancy McGrath
Materials supplier.

Sony Corp

4-14-1 Asahi-cho, Atsugi-shi, Kanagawa 243, Japan
Tel: +81 462 30 5111
Fax: +81 462 30 5160
Web: www.sony.co.jp
Manufacturer of optoelectronic components, including lasers.

Sony Electronics Inc

1 Sony Place, San Antonio, TX 78245, USA
Tel: +1 210 681 9000
Fax: +1 210 647 6492
Web: www.sony.com
Optoelectronic component manufacturer.

Sopra Inc

33 Nagog Park, Acton, MA 01720, USA
Tel: +1 508 263 2520
Fax: +1 508 263 2790
Optoelectronic component supplier.

Spectra-Physics Inc

Siemensstr 20, D-64289 Darmstadt, Germany
Tel: +49 6151 708240
Fax: +49 6151 79120
Manufacturer of optoelectronic components including diode lasers.

Spectra-Physics Inc

1330 Terra Bella Ave, PO Box 7013, Mountain View, CA 94039-7013, USA
Tel: +1 650 966 5435
Fax: +1 650 969 3546
E-mail: aheld@splasers.com
Web: www.spectra-physics.com
Contact: Andy Held
Manufacturer of optoelectronic components including diode lasers.

Spectra-Physics Inc (formerly Optical Research Associates)

3280 E Foothill Blvd, Suite 300, Pasadena, CA 91107-2193, USA
Tel: +1 626 795 9101
Fax: +1 626 795 9102
E-mail: service@opticalres.com
Web: www.opticalres.com
Optical engineering services.

Spectrum Technologies Ltd

Western Ave, Bridgend CF31 3RT, UK
Tel: +44 1656 655437
Fax: +44 1656 655920
E-mail: e.hardy@spectrum-technologies.co.uk
Web: www.spectrum-technologies.co.uk
Contact: Elaine Hardy
Manufacturer of fibre-Bragg gratings.

Spindler & Hoyer GmbH

Königsallee 23, D-37081 Göttingen, Germany
Tel: +49 551 69350
Fax: +49 551 6935166
E-mail: sales@spindlerhoyer.de
Manufacturer of optoelectronic components including diode lasers.

Spindler & Hoyer Inc

459 Fortune Blvd, Milford, MA 01757-1745, USA
Tel: +1 508 334 5678
Fax: +1 508 478 5980
E-mail: info@spindlerhoyer.com
Manufacturer of optoelectronic components including diode lasers.

Spindler & Hoyer UK Ltd

2 Drakes Mews, Crownhill, Milton Keynes MK8 0ER, UK
Tel: +44 1908 262525
Fax: +44 1908 262526
E-mail: sales@SpindlerHoyer.co.uk
Manufacturer of optoelectronic components including diode lasers.

Spire Corp

One Patriots Park, Bedford, MA 01730-2396, USA
Tel: +1 617 275 6000
Fax: +1 617 275 7470
E-mail: spire.corp@channel1.com
Contact: Kurt Linden
Manufacturer of epiwafers and opto components, including: high-power diode bars, solar cells, laser power converters and thermophotovoltaic converters.

Spiricon Inc

2600 North Main Logan, UT 84341, USA
Tel: +1 435 753 3729
Fax: +1 435 753 5231
E-mail: sales@spiricon.com
Web: www.spiricon.com
Laser manufacturer.

Staib Instruments Inc

PO Box 120484, Newport News, VA 23612, USA
Tel: +1 757 872 8827
Fax: +1 757 872 8838
E-mail: staib-us@staib-instruments.com
Web: www.staib-instruments.com.
Equipment supplier.

Stanley Electric (Asia Pacific) Ltd

Unit 1605 16/F, Silvercord Tower 1, 30 Canton Rd, Taimshatsui, Kowloon,
Hong Kong
Tel: +852 2730 1738
Fax: +852 2730 1933
Web: www.stanley.co.jp
Manufacturer of optoelectronic components.

Stanley Electric US Co Inc

420 E High St, London, OH 43140, USA
Tel: +1 614 852 5200
Fax: +1 614 852 5201
Web: www.stanleyelec.com
Manufacturer of optoelectronic components, especially LEDs.

Sterling Semiconductor Inc

22660 Executive Drive, Suite 101, Sterling, VA 20166-9535, USA
Tel: +1 703 834 7535
Fax: +1 703 834 7537
E-mail: sales@sterlingsemiconductor.com
Web: www.sterlingsemiconductor.com
SiC substrate and epiwafer supplier.

Strem Chemicals Inc

7 Mulliken Way, Newburyport, MA 01950-4098, USA
Tel: +1 978 462 3191
Fax: +1 978 465 3104
E-mail: info@strem.com
Web: www.strem.com
Contact: Ephraim S Honig
Materials supplier.

Sumitomo Chemical Co Ltd

Electronic Materials Division, 27-1 Shinkawa 2-chome, Chuo-ku, Tokyo 104-8260, Japan
Tel: +81 3 5543 5816
Fax: +81 35543 5934
Manufacturer and supplier of epiwafers and source materials for opto- and electronic devices.

Sun Opto Inc

20793 E Valley Blvd #C, Walnut, CA 91789, USA
Tel: +1 909 598 8266
Fax: +1 909 598 4966
Web: www.sun-led.com
Manufacturer of optoelectronic devices.

Switchtec Electronics Ltd

Brooms Rd, Stone Business Park, Stone ST15 0SH, UK
Tel: +44 1785 818600
Fax: +44 1785 811900
Optoelectronic component distributor.

Syscom Instrument AG

Imfeldstrasse 29, 8037 Zurich, Switzerland
Tel: +41 136 20500
Fax: +41 136 20650
Contact: Markus Blattler
Supplier of optoelectronic components.

Taiwan Dacoms

No 168, Sec 1, Chung Shan Rd, Yung-Ho, Taipei, Taiwan ROC
Tel: +886 22 920 7500
Fax: +886 22 920 7077
Optoelectronic component distributor.

Teem Photonics

46 ave Felix Viallet, F-38031 Grenoble Cedex, France
Tel: +33 476 57 47 89
Fax: +33 476 57 45 84
Manufacturer of fibre optic devices.

Tellium Inc

2 Crescent Place, PO Box 901, Oceanport, NJ 07757-0901, USA
Tel: +1 732 923 4100
E-mail: sales@tellium.com
Web: www.tellium.com
Manufacturer of high capacity optical switches.

Temescal

2700 Maxwell Way, PO Box 2529, Fairfield, CA 94533-0252, USA
Tel: +1 707 423 2100
Fax: +1 707 425 1706
E-mail: temescal.sales@ct.boc.com
Web: www.boc.com/vacuum/coating
Contact: Anita Allison
Provider of optoelectronic etching and deposition equipment.

Terahertz Photonics Ltd

Rosebank Park, Livingston EH54 7EJ, UK
Tel: +44 1506 818 500
Fax: +44 1506 818 501
E-mail: info@thzonline.com
Web: www.thzonline.com
Opto start-up company — specializing in thin-film coatings.

Thales Optics (formerly THOMSON-CSF Laser)

7 rue du Bois Chaland, CE 2906 Lisses, F-91029 Evry Cedex, France
Tel: +33 1 69 11 39 39
Fax: +33 1 64 97 52 03
Web: www.laser-diodes.thomson-csf.com
Manufacturer of lasers (including diode-pumped) and optical sensors.

Thales Optics (formerly THOMSON-CSF Laser Diodes)

Route Départementale 128, BP 46, F-91401 Orsay Cedex, France
Tel: +33 1 69 33 00 00
Fax: +33 1 69 33 03 21
Web: www.thomson-csf.com
Manufacturer of laser diodes.

Thales Optronique (formerly THOMSON-CSF Optronique)

rue Guynemer, BP 55, F-78283 Guyancourt Cedex, France
Tel: +33 1 30 96 70 00
Fax: +33 1 30 96 75 50
Optoelectronics conveyor.

The Ninth Semiconductor Device Factory of Beijing

33, JuEr Hu Tong Jiao Dao Kou, Beijing 100009, China
Tel: +86 10 64031664
Fax: +86 10 64013887
Optoelectronic component manufacturer.

Thermo VG Semicon

The Birches Industrial Estate, Imberhorne Lane, East Grinstead RH19 1XZ, UK
Tel: +44 1342 325011
Fax: +44 1342 315800
Web: www.vgsemicon.com

TLC Precision Wafer Technology Inc

661 5th Avenue North, Suite 160, Minneapolis, MN 55404, USA
Tel: +1 612 341 2795
Fax: +1 612 341 2799
Contact: Timothy Childs
Supplier of advanced custom GaAs and InP epiwafers.

TOPTICA Photonics AG (formerly TuiOptics GmbH)

Fraunhoferstr14, D-82152 Martinsried/Munich, Germany
Tel: +49 89 899969 0
Fax: +49 89 899969 35
Web: www.toptica.com
Manufacturer of diode lasers.

Toshiba Corp

1-1 Shibaura 1-chome, Minato-ku, Tokyo 105, Japan
Tel: +81 3 3457 4511
Fax: +81 3 3502 3979
Web: www.toshiba.com
Optoelectronic component manufacturer.

Toshiba Electronics Europe GmbH

Hansaallee 181, D-40549 Düsselsorf, Germany
Tel: +49 211 52960
Fax: +49 211 5296400
Web: www.toshiba.com
Manufacturers of optoelectronics components including semiconductor lasers, photodiodes, and optocouplers.

Toyoda Gosei Co Ltd

1 Nagahata, Ochiai, Haruhi-Cho, Nishikasugai-Gun, Aichi 452, Japan
Tel: +81 52 400 5120
Fax: +81 52 400 5881
Web: www.toyoda-gosei.co.jp
Manufacturer of optoelectronics devices.

TrueLight Corp

2F-2 No 21 Prosperity Rd, Science-Based Industrial Park, Hsin Chu 300, Taiwan ROC
Tel: +886 3 5780080
Fax: +886 3 5780555
E-mail: sales@truelight.com.tw
Web: www.truelight.com.tw
Manufacturer of optoelectronic components, including LEDs, diodes, infrared, photodiodes and semiconductor lasers.

TuiLaser Aktiengesellschaft

Industriestraße 15, D-82110 Germering, Munich, Germany
Tel: +49 89 89407 0
Fax: +49 89 8545610
Email: info@tuilaser.com
Web: www.tui-laser.de
Excimer laser technology and solutions for medical, industrial and scientific applications, with a special focus on laser vision correction.

TwinStar

325 Scarlet Boulevard, Oldsmar, FL 34677, USA
Tel: +1 813 854 2930
Fax: +1 813 854 2765
E-mail: sales@starstar.com
Web: www.starstar.com
Component manufacturer.

u2t Innovative Optoelectronic Components GmbH

Tangermünder Weg 18, D-13583 Berlin, Germany
Tel: +49 171 6431849
Fax: +49 30 31002 558
E-mail: contact@u2t.de
Web: www.u2t.de
Manufacturer of optoelectronic components.

U-Jin Co Ltd

478-17 Mangwon Dong, Mapo-ku, Seoul, South Korea
Tel: +82 2 336 8351
Fax: +82 2 333 0336
Contact: W J Yoo
Manufacturer of optoelectronic devices.

Ulm Photonics

Lise-Meitner Str 13, D-89081 Ulm, Germany
Tel: +49 731 550194 014
Fax: +49 731 550194 026
Web: www.ulmphotonics.de
ULM photonics has been set up by a group of entrepreneurs from the University of Ulm, Department of Optoelectronics to provide leading edge VCSEL light sources.

UM Electro-Optic Materials

12 Channel St #702, Boston, MA 02210, USA
Tel: +1 617 960 5901
Fax: +1 617 345 9271
E-mail: Nick.Sink@um.be
Web: www.um.be
Contact: Nick Sink
Materials supplier.

United Epitaxy Co Ltd

9F No 10 Li-Hsin Rd, Science-Based Industrial Park, Hsinchu, Taiwan ROC
Tel: +886 3 567 8000
Fax: +886 3 567 8753
E-mail: UECCOM@mail.ttn.com.tw
Manufacturer of epiwafers and custom epilayers.

United Mineral & Chemical Corp

1100 Valley Brook Ave, Lyndhurst, NJ 07071, USA
Tel: +1 201 507 3300
Fax: +1 201 507 1506
E-mail: enquiry@umccorp.com
Web: www.umcorp.com
Contact: Irwin Drangel
Supplier of high purity metals (arsenic, red phosphorous, indium, gallium, aluminium, antimony trioxide, antimony, boron oxide, intermettalic compounds) for MBE, bulk crystal growth and dopants.

USHA India Ltd

12/1 Delhi Mathura Rd, Faridabad (Haryana), India
Tel: +91 129 277641
Fax: +91 129 277679
Contact: Y P Singh
Manufacturer of optoelectronic components.

Vacuum Process Technology Inc

Corporate Park, Ste 840, 300 Oak St, Pembroke, MA 02359, USA
Tel: +1 781 826 0087
Fax: +1 781 826 7047
E-mail: vpt@tiac.net
Equipment supplier.

Valpey Fisher Corp

Ultrasound & Optics Division
75 South St, Hopkinton, MA 01748, USA
Tel: +1 508 435 6831
Fax: +1 508 435 5289
E-mail: vfisher@kersur.net
Equipment supplier.

VBR Inframetrics

Mechelse Steenweg 277, B-1800 Vilvoorde, Belgium
Tel: +32 2 257 9030
Fax: +32 2 257 9039
Manufacturer of infrared thermal imaging and measurement systems.

VBS Technology Ltd

Unit 2, Croft Ct, Sandall Carr Rd, Kirk Sandall, Doncaster, South Yorkshire DN3 1QL, UK
Tel: +44 1302 884702
Fax: +44 1302 883105
Equipment manufacturer.

Veco Custom Optics Inc

3350 Scott Blvd, Ste 38, Santa Clara, CA 95054, USA
Tel: +1 408 727 1303
Fax: +1 408 727 1380
E-mail: veco@vecooptics.com
Web: www.vecooptics.com
Equipment supplier.

Vector-Optics Ltd

27/2 Yartsevskaya St, Moscow 121552, Russia
Tel: +7 095 140 5903
Fax: +7 095 140 5903
Materials and device supplier.

Vector Technology Ltd

9 & 10 Roseheyworth Business Park, Abertillery, Gwent NP3 1SP, UK
Tel: +44 1495 320222
Fax: +44 1495 320484
E-mail: sales@vector-technology.co.uk
Web: www.vector-technology.co.uk
Gas processing equipment.

Vicon Infrared

4 Seneca Court, Acton, MA 01720, USA
Tel: +1 978 263 4248
Opto component manufacturer.

Viratec Thin Films Inc

2150 Airport Drive, Faribault, MN 55021, USA
Tel: +1 507 334 0051
Fax: +1 507 334 0059
E-mail: sales@viratec.com
Component manufacturer.

Virginia Semiconductor Inc

1501 Powhatan St, Fredericksburg, VA 22401, USA
Tel: +1 540 373 2900
Fax: +1 540 371 0371
E-mail: vasemi@juno.com
Opto component manufacturer.

Vishay Intertechnology Inc

63 Lincoln Highway, Malvern, PA 19355-2120, USA
Tel: +1 610 644 1300
Fax: +1 610 296 0657
Web: www.vishay.com
Manufacturer of optoelectronic components.

Vitana Corp

2500 Don Reid Dr, Ottawa ON K1H 1E1, Canada
Tel: +1 613 247 1211
Fax: +1 613 247 2001
E-mail: guest@vitana.com
Web: www.vitana.com
Opto component manufacturer.

Vixel Corp

11911 North Creek Parkway South, Bothell, WA 98011, USA
Tel: +1 425 806 4317
Fax: +1 425 806 4050
E-mail: marketing@vixel.com
Manufacturer of laser diodes.

VTI Inc

4200 Colonel Glenn Hwy, Ste 800, Dayton, OH 45431-1663, USA
Tel: +1 937 427 1104
Fax: +1 937 427 1107
E-mail: vtiinc@compuserve.com
Supplier of materials and devices.

Wafer Technology Ltd

34 Maryland Rd, Tongwell, Milton Keynes MK15 8HJ, UK
Tel: +44 1908 210 444
Fax: +44 1908 210 443
E-mail: Klamb@wafertech.co.uk
Web: www.wafertech.co.uk
Contact: Karen Lamb
Optoelectronic substrate manufacturer.

Welt Electronic Srl

Via R, Bardazzi 62/68, 50127 Firenze, Italy
Tel: +39 055 437 9933
Fax: +39 055 435 728
Optoelectronic component distributor.

Westron Inc

4-5-6 Shibuya, Shibuya-ku, Tokyo 150-0002, Japan
Tel: +81 3 34862751
Fax: +81 3 34861549
E-mail: westron@sco.bekkoame.or.jp
Optoelectronic component manufacturer.

XEROX Palo Alto Research Center (PARC)

3333 Coyote Hill Rd, Palo Alto, CA 94304, USA
Tel: +650 812 4000
Fax: +1 650 812 4973
Contact: D Hofstetter
Laser research including violet diode lasers.

Xinetics Inc

410 Great Rd #A6, Littleton, MA 01460, USA
Tel: +1 508 486 0181
Fax: +1 508 486 0281
Manufacturer of optoelectronic devices.

Xmr Inc

5303 Betsy Ross Drive, Santa Clara, CA 95054, USA
Tel: +1 408 988 2426
Fax: +1 408 970 0742
Product/service company.

Yellow Stone Corp

9 Lane 113, Chih-Yuan 2nd Rd, Taipei, Taiwan ROC
Tel: +886 2 28221522
Fax: +886 2 28222309
Manufacturer of LED displays.

Yokogawa Electric Corp

2-9-32 Naka-machi, Musashino-shi, Tokyo 180-0006, Japan
Tel: +81 422 525543
Fax: +81 422 557311
Optoelectronic component manufacturer.

Zarlink Semiconductor (formerly Mitel Semiconductor)

400 March Rd, Ottawa K2K 3H4, Canada
Tel: +1 613 592 0200
Fax: +1 613 591 2317
E-mail: corporate@zarlink.com
Web: www.zarlink.com
Manufactuer of optoelectronic devices.

Zarlink Semiconductor AB

Bruttovagen 1, Box 520, S17526 Jarfalla, Sweden
Tel: +46 8 58 02 45 00
Fax: +46 8 58 01 67 15
Manufacturer of optoelectronic devices, including VCSELs.

ZETEX

Fields New Rd, Chadderton, Oldham OL9 8NP, UK
Tel: +44 161 622 4444
Fax: +44 161 622 4446
Contact: C Greene
Manufacturer of optoelectronic devices.

9 Appendices

Country	Currency	2000
AUSTRALIA	Australian Dollar	A$ 1.73
AUSTRIA	Schilling	Sch 14.93
BELGIUM	Franc	BF 43.77
BRAZIL	Cruzerio	Crz 1.83
CANADA	Canadian dollar	C$ 1.49
CHINA	Yuan	Yn 8.28
DENMARK	Krone	DKr 8.09
EURO	Euro	€ 1.09
FINLAND	Markka	Fmk 6.45
FRANCE	Franc	FFr 7.12
GERMANY	Deutsch mark	DM 2.12
HONG KONG	HK Dollar	HK$ 7.79
INDIA	Rupee	Rup 45.00
INDONESIA	Rupiah	Rp 8415.00
IRELAND	Irish Punt	£ 0.85
ISRAEL	Shekel	Shk 4.09
ITALY	Lira	Lira 2100.00
JAPAN	Yen	¥ 107.90
MALAYSIA	Ringgit	Rt 3.80
MEXICO	Peso	Ps 9.47
NETHERLANDS	Guilder	Gd 2.39
NORWAY	Krone	NKr 8.80
PHILIPPINES	Philippine Peso	Peso 44.30
PORTUGAL	Escudo	Esc 217.50
SINGAPORE	Singapore Dollar	S$ 1.72
SOUTH AFRICA	Rand	Rd 6.94
SOUTH KOREA	Won	Wn 1131.00
SPAIN	Peseta	Pts 180.50
SWEDEN	Krona	Swkr 9.17
SWITZERLAND	Franc	SFr 1.69
TAIWAN	New Taiwan Dollar	NT$ 31.26
THAILAND	Baht	Bt 40.20
UK	Pound Sterling	£ 0.66

9.2 Mergers and Acquisitions

Recent major mergers and acquisitions include:

- JMAR Technologies of San Diego, CA, USA, completed the purchase of Semiconductor Advanced Lithography (SAL), a supplier of X-ray lithography stepper systems based in South Burlington, Vermont, USA, in August 2001. JMAR manufactures a collimated laser plasma source that directs 25 W X-rays onto a mask/wafer target. Known as the picosecond X-ray light source (PXS), this short wavelength (1 nm) beam can be used in X-ray lithography (XRL) steppers to directly write patterns with line-widths smaller than 0.13 μm, making it suitable for high-frequency GaAs circuits.

- Vishay Intertechnology announced in June 2001 that, subject to government approvals, it had acquired the infrared components business of Infineon Technologies for approximately US$120 million. The deal includes facilities in San Jose, CA, USA and Malaysia.

- PerkinElmer announced in May 2001 that it had acquired Sonoran Scanners' proprietary Laser Direct Imaging (LDI) and Computer-to-Plate (CTP) technologies. Sonoran was founded in 1997 and is based in Tucson, Arizona. This technology complements PerkinElmer's Digital Imaging and Lithography businesses which manufacture Large Area Electronic Substrate exposure machines. Sonoran Scanners' LDI technology uses direct CTP printing to produce a precise image, while increasing yields at reduced cost.

- Bookham Technology announced in January 2001 that it had acquired privately held Measurement Microsystems A-Z Inc of Quebec, Canada, for approximately US$47.5 million. Measurement Microsystems develops innovative algorithms and DSP solutions, which enable optimized signal extraction using DWDM optical components.

- In March 2001 Vitesse announced that it had agreed to acquire fabless semiconductor company, Exbit Technology A/S of Herlev, Denmark in exchange for 4.4 million shares of Vitesse common stock.

- During April 2001 Infineon Technologies entered into a definitive agreement to acquire Catamaran Communications Inc of San Jose, CA, USA, for US$250 million in common stock. Catamaran specializes in ICs for next-generation 40 Gb/s and 10 Gb/s segments of the optical networking market.

- NEC Corp and Toshiba Corp announced in April 2001 the incorporation of NEC TOSHIBA Space Systems Ltd, a joint venture company for space-related business. The new company will be headed by Hiromi Hayashi, formerly a senior vice president of NEC Corp. The two partners have agreed to transfer their space operations to the new company, ready for it to start operation in October 2001.

- NEC Corp announced in April 2001 that it would transfer its laser printer business to Fuji Xerox (a leader in the colour laser engine market), with the intention that the details would be finalized by end August 2001. NEC will transfer its laser printer business including R&D, manufacturing, marketing and maintenance service operations currently performed by NEC and its subsidiaries to Fuji Xerox. NEC will also transfer all shares of NEC

Niigata Ltd, (NEC's major development and manufacturing centre for laser printers), to Fuji Xerox.

- The Business Systems Communications division of Mitel Corp was sold for C$350 million in early 2001. It was bought by a company controlled by Dr Terrance H Matthews (one of the original founders of Mitel), and renamed Mitel Networks.

- JDS Uniphase announced in February 2001 that it had acquired Optical Process Automation Inc (OPA) of Melbourne, FL, USA, for an undisclosed sum. OPA designs and manufactures automated and semi-automated systems for the manufacture of fiberoptic components and modules.

- Nortel Networks announced in February 2001 that it would acquire JDS Uniphase's Zurich, Switzerland subsidiary for approximately 65.7 million of Nortel's common shares. The agreement also included related assets in Poughkeepsie, New York, USA.

- Also in February 2001, JDS Uniphase and SDL Inc completed their merger. SDL is a manufacturer of laser diodes and optoelectronics ICs. Each outstanding share of SDL Inc common stock was exchanged for 3.8 shares of JDS Uniphase common stock. SDL became a wholly-owned subsidiary of JDS Uniphase.

- Optenia was officially launched on 15 January 2001, A new start-up company, Optenia (investors include Mitel, the National Research Council of Canada [NRC] and others) has been formed, based on joint research on DWDM between Mitel and NRC. Optenia has a manufacturing agreement with Mitel encompassing foundry arrangements and joint development of microfabrication techniques.

- NEC and Samsung SDI announced the formation of their Won 9.4 billion, joint venture company in January 2001, Samsung NEC Mobile Display Co Ltd, based in Pusan, South Korea. The company will develop, manufacture and market displays based on organic electroluminescent technology.

- In January 2001 JDS Uniphase acquired a minority equity investment in Avantas Networks Corp, a privately-held telecommunications network testing equipment company based in Montreal, Quebec, Canada. The terms of the investment were not disclosed.

- San Jose-based photonics systems developer, Calient Networks acquired privately-owned Kionix Inc in January 2001 for an undisclosed sum. Kionix has facilities in New York and California. The Kionix subsidiary has been renamed Calient Optical Components. Kionix's non-optical MEMS product lines were spun off into a separate company, Cayuga Sensors Inc.

- Cree completed its acquisition of UltraRF, formerly a division of Spectrian Corp in December 2000 for 1.8 million shares of Cree common stock plus additional Cree shares worth US$30 million. UltraRF designs, manufactures and markets a complete line of LDMOS and bipolar RF power semiconductors; it will operate as a wholly-owned Cree subsidiary.

- In October 2000, French semiconductor company, Picogiga, made a capital investment in Finish opto start-up company Modulight Inc, in exchange for a 40% share of Modulight.

- Kymata Ltd followed its acquisition of BBV Design and Software by acquiring MEMS manufacturer, Total Micro Products of Enschede, The Netherlands, for an undisclosed amount, in September 2000.

- In a July 2000 US$175 million transaction, Alcatel Optronics acquired privately-held Innovative Fibers Inc of Gatineau, Quebec, Canada, the world leader in DWDM optical Fibre Bragg Grating technology.

- During FY 2000, JDS Uniphase acquired AFC Technologies, Ramar Corp, EPITAXX Inc, SIFAM Ltd, Oprel Technologies Inc, IOT Ltd, Optical Coating Laboratories Inc, Cronos Integrated Microsystems Inc, Fujian Casix Lasers Inc and E-TEK Dynamics Inc.

- In 2000, Vishay sold its 65% share of the joint venture formed in 1997 with Taiwanese company, Lite-On (Vishay/Lite-On Power Semiconductor of Singapore); it is now 100% owned by Lite-On.

- In summer 2000, Mitel acquired privately-held Vertex Networks Inc (a fabless semiconductor company providing high-performance network packet processing, switching and routing silicon solutions for the enterprise and WAN access markets) for a stock transaction for 11 million Mitel shares, valued at approximately CDN$300 million.

- In June 2000 Nortel Networks announced it had acquired CoreTek, a manufacturer of tunable lasers, for US$1.43 billion in Nortel Networks common shares.

- Nortel also acquired Xros Inc in June 2000, a leader in second-generation, large-scale, fully photonic switching, for US$3.25 billion in Nortel Networks common shares.

- In June 2000 Lucent Technologies (now Agere) acquired privately-held Herrmann Technology Inc of Dallas, TX, USA, for Lucent shares valued at approximately US$438 million. Herrmann is a leading supplier of devices for next-generation DWDM optical networks.

- Uniroyal Technology in May 2000 announced that it acquired Sterling Semiconductor Inc (a leading developer of SiC technology and materials), through a merger for approximately 1.5 million shares of Uniroyal common stock.

- In May 2000, Kymata Ltd acquired BBV Design and Software of Enschede, The Netherlands, for an undisclosed amount. BBV is a leading provider of passive optical device design and simulation tools for integrated optic device manufacturers.

- Nortel Networks in May 2000 announced the acquisition of Photonic Technologies, a pioneer in strategic optical components based in Sydney, Australia, for up to US$35.5 million in cash. Nortel already owned approximately one third of the privately held company, through its 1998 investment in the company.

- In April 2000 Lucent (Agere) acquired Ortel Corp of Alhambra, CA, USA, a leading developer of optoelectronic components for cable TV. Ortel was founded in 1980 and employs 700 people at two facilities. Each share of Ortel was converted into 3.135 shares of Lucent, which gave an approximate value of US$2.95 billion.

9.3 Joint Ventures and Agreements

Recent major joint ventures and agreements include:

- In October 2001 DuPont and Cambridge Display Technology (CDT) signed an agreement to advance the commercialization of light emitting polymer displays. As part of the agreement, DuPont will receive a non-exclusive license to technology and intellectual property assets held by CDT. DuPont and CDT entered into a joint development agreement to bring light-emitting polymer (LEP) technology invented by CDT into commercial use in flexible displays in May 1998. CDT also has a technology licensing agreement with Hoechst AG (of Frankfurt, Germany), whereby, Hoechst's key customers are granted licences from CDT for LEP technology.

- Osram in May 2001 acquired a licence from Cambridge Display Technology (CDT) to enable them to use CDT's proprietary Light Emitting Polymer display technology to develop products for mobile, cellular and automotive applications.

- Sharp and Sony Corp began licensing Version 1.0 of "OP i.LINK", (a digital interface using optical fibres for connecting audio, video, and IT products aimed at home networks in the era of broadband communications) from April 2001.

- Agilent Technologies and Mitel in April 2001 signed a MSA for a common standard for next-generation 4-channel parallel fibre-optic modules operating at data rates of up to 10 Gbaud.

- Vitesse and Nortel Networks in April 2001 announced a packaging agreement, utilising Nortel Networks OPTera Packet Edge resilient packet ring technology to develop complete 2.5 Gb/s and 10 Gb/s chipset platforms.

- In April 2001 Vixel and Vitesse announced a partnership to jointly develop and market Fibre Channel Loop switch semiconductors for use in stand-alone switches or in embedded applications such as disk arrays and RAID subsystems. In July 2001 the partnership announced that they will bring the technology to the market as the VSC7192. The VSC7192 is a 12 Channel Fibre Channel Arbitrated Loop (FC-AL) switch IC which delivers advanced diagnostics and low latency for embedded applications, such as disk arrays, RAID subsystems and standalone switches. General availability for production volumes will be November 2001.

- In April 2001, Sony and Universal Display Corp announced a joint development agreement for high efficiency active matrix OLED display devices for use in large area monitor applications.

- ROHM and Eastman Kodak announced in April 2001 that they had formed a technical agreement for the manufacture of OEL displays, with products scheduled for 2002.

- Agere Systems, Alcatel Optronics and OpNext in March 2001 expanded their MSA for 10 Gb/s transponder modules used in optical networking systems to include Agilent, Ericsson Microelectronics, Mitsubishi Electric, JDS Uniphase, NEC and ExceLight.

- JDS Uniphase and Alcatel Optronics in March 2001 announced an extension to the agreement begun in 1999 to supply 980 nm laser chips and pump stabilizers manufactured by the former SDL (now part of JDS Uniphase) to Alcatel Optronics. The new agreement lasts until the end of 2003.

- Agere Systems and Agilent Technologies in March 2001 announced a multi-source agreement for 10 Gigabit Ethernet fibre-optic transceivers. The agreement is the first developed to support the proposed IEEE 10 Gigabit Ethernet interoperability standard, and specifies a uniform form factor, size, connector type and electrical pin-outs. Agere and Agilent will independently develop 10 Gigabit Ethernet transceivers.

- Oki and Matsushita in March 2001 announced an alliance to develop and market personal identification products for gate management and financial systems based on the tint patterns of the human iris.

- In February 2001 Tohoku Pioneer Corp, Semiconductor Energy Laboratory Co Ltd and Sharp Corp announced an agreement to establish a joint venture (ELDis Inc, in Tochigi Prefecture) for the manufacture and marketing of CG Silicon TFT substrates for organic EL displays. Commencement of operations is scheduled for Autumn 2002, with initial production capacity expected to be 500k units/month of 2-inch panel equivalents.

- Toyoda Gosei, Toshiba America Electronic Components Inc (TAEC) and Toshiba Corp announced in February 2001 that they had jointly developed a white LED that points the way to a stable white light source with the luminosity required for future replacement for incandescent lamps. Volume production is scheduled for end 2001.

- Mitel Corp, WL Gore & Associates and Agilent Technologies in February 2001 announced they had signed a MSA that will set the industry standard for next-generation parallel fibre-optic modules. The modules are designed to meet demand for network capacity in proprietary terabit switch/router links as well as very short reach OC-192 and Infiniband connections.

- Agilent Technologies, Mitel Corp and WL Gore & Associates in February 2001 announced they had signed a Multi-Source Agreement that will set the industry standard for next-generation parallel fibre-optic modules. Designed for a variety of fibre-optic communications applications, the parallel fibre optic modules include transmitters and receivers featuring 12 channels at 1–2.7 Gbaud/channel for an aggregate bandwidth of 12–32.4 Gbaud.

- As part of its January 2001 acquisition by Cree, UltraRF entered into a two-year supply agreement with its former owner, Spectrian, under which Spectrian will purchase UltraRF products representing a significant portion of its anticipated needs. Cree, UltraRF and Spectrian also entered into a one-year development agreement under which Cree and UltraRF will work to develop high gain driver modules and high efficiency power modules based on LDMOS materials and high power MESFET components based on SiC materials.

- Bookham Technology announced in January 2001 that it had signed a multi-million dollar agreement to supply additional ASOC optical components to Fujitsu Telecommunications Europe Ltd. Under the terms of the arrangement, Bookham will supply Fujitsu with up to 10k WDM bi-directional transceiver modules per month in 2001, for use in Fujitsu's FBX-Access platform.

- Cree Inc announced an alliance with ROHM Co Ltd in December 2000. The agreement included a licence agreement under which ROHM granted Cree a five-year exclusive licence to several US patents and a non-exclusive licence under a Japanese patent for optoelectronic devices. The companies have also signed a non-binding memorandum of understanding for the co-operative development of a packaged blue laser diode for consumer applications. In addition, Cree and ROHM have entered into an annual supply agreement for the purchase by ROHM of LED chips manufactured by Cree. ROHM and Kyoto University have a joint development agreement researching into ZnO.

- In November 2000 IBM licenced its pump laser technology and silicon oxynitride (SiON) technology to Kymata Ltd. SiON will enable Kymata to manufacture new chip designs for both themselves and IBM.

- Infineon, JDS Uniphase, Nortel and LightLogic in September 2000 announced a multi-source agreement (MSA) for common mechanical, electrical, and optical specifications for small footprint OC-192 (10 Gb/s) SDH/SONET transponders in order to support the optical networking market. These devices provide a space and cost saving optical-to-electrical conversion solution for LAN to metropolitan networks that is four times faster than today's standard OC-48 (2.5 Gb/s) products optical transponder. A full drop-in solution is specified: a 200-pin connector and pinout that provides customers with an electrical interface compliant with Optical Internetworking Forum specifications.

- Cree Inc in September 2000 licenced and completed a technology transfer of Osram's ATON technology, which enhances brightness of blue InGaN-LEDs.

- Applied Optoelectronics in July 2000 received a licence from Lucent/(Agere Systems) giving Applied Optoelectronics the right to manufacture and distribute the quantum cascade (QC) laser for applications in industries other than telecommunications. This was the first time this technology has been licensed.

- EMCORE and JDS Uniphase in June 2000 formed a joint development manufacturing and marketing agreement to develop, manufacture and market a family of fibre optic array transceivers (based on EMCORE's laser technology), for fibre optic communications products.

- Spectrolab Inc announced in May 2000 that as part of a technical agreement that lasts until the end of 2002 with Dornier Satellitensysteme GmbH (DSS) of Germany, it would supply solar cells for DSS' space solar array systems.

- In January 2000, Agilent and EMCORE entered into a three-year supply agreement whereby EMCORE manufactures Gigarray® VCSEL arrays for use in parallel optical transceivers.

- Sanyo Electric and Eastman Kodak have a comprehensive alliance agreement to jointly develop next generation organic EL panel displays on organic electroluminescence display technology. The agreement was signed in February 1999.

- JDS Uniphase (SDL), San Jose, CA, has an agreement until the end of 2002 to provide high volumes of multi-mode pump lasers to IPG Photonics of Sturbridge, MA, for cladding-pumped fibre products.

9.4 Alphabetical List of Acronyms

A	Ampere
Å	Angstrom
A/D	Analog to Digital
ACIA	Asynchronous Communications Interface Adapter (=UART*)
ADC	Analog to Digital Converter
ADM	Add Drop Multiplexers
ADPCM	Adaptive Differential Pulse Code Modulation
ADSL	Asymmetric Digital Subscriber Line
ADTV	Advanced Definition TV
AFD	All Format Decoder
AGC	Automatic Gain Control

ALDA	Addressable Laser/Detector Array
ALE	Atomic Layer Epitaxy
AlGaAs	Aluminium Gallium Arsenide
AIN	Aluminium Nitride
ALVT	Advanced Low Voltage Technology
AM	Amplitude Modulation
AMBA	Advanced Microcontroller Bus Architecture
AMLCD	Active Matrix LCD (*q.v.*)
Amp	Amplifier (eg op-amp — Operational Amplifier)
AMPS	Advanced Mobile Phone System
AND	logic term — output is high if and only if all inputs are high
AO/DI	Always On/Dynamic ISDN
AOI	logic term — And-Or-Invert
APA	Audio Power Amplifiers
APC	Automatic Power Control
APD	Avalanche Photodiode
APS	Active Pixel Sensor
ARPA	Advanced Research Project Agency
ASCP	Application Specific Custom Product
ASD	Application Specific Discrete
ASIC	Application Specific Integrated Circuit
ASK	Amplitude Shift Keying
ASP	Average Selling Price
ASSP	Application Specific Standard Product
ATA	Advanced Technology Attachment
ATM	Asynchronous Transfer Mode
AVP	Audio/Video Processor
AWDM	Analog Wavelength Division Multiplexing
AWG	Arrayed Waveguide Grating
Baud	Level changes per second (often confused with BPS *q.v.*)
BESOI	Bond & Etch Back SOI
BGA	Ball Grid Array
BGO	BackGround Operation
BIPS	Billion Instructions Per Second
BIST	Built-In Self Test
BPS	Bits Per Second
BRS	Buried Ridge Stripe
BSC	Base Station Controllers
BST	Barium Strontium Titanate
BTS	Base Transceiver Systems
CAD	Computer Aided Design
CAM	Computer-Aided Manufacturing
CAN	Controller Area Network
CARS	Collision Avoidance Radar System
CATV	CAble TeleVision
CAV	Constant Angular Velocity
CBE	Chemical Beam Epitaxy
CBGA	Chip Scale Ball Grid Array
CCD	Charge Coupled Device
cd	Candela
CDMA	Code Division Multiple Access
CDR	Clock and Data Recovery
CG	Continuous Grain (silicon)
CGBA	Ceramic Ball Grid Array
CHMSL	Centre High Mounted Stop Lamp
CIF	Common Intermediate Format

CIM	Computer Integrated Manufacturing
CIS	Copper Indium Selenide
CIS	Contact Image Sensors
CMMR	Common Mode Rejection Ratio
CMP	Chemical Mechanical Polishing
CMP	Chemical Mechanical Planarization
CMT	Cadmium Mercury Telluride
CMOS	Complementary Metal Oxide Semiconductor
COA	Chip-on-ALIVH (*q.v.*)
COB	Chip On Board
COD	Catastrophic Optical Damage
codec	Compression/decompression
CPGA	Ceramic Pin Grid Array
CPU	Central Processing Unit
CRT	Cathode Ray Tube
CSIC	Customer Specific Integrated Circuit
CSR	Cell Switch Router
CTR	Current Transfer Ratio
CW	Continuous Wave
CWDM	Coarse Wavelength Division Multiplexing
DAA	Data Access Arrangement
DAB	Digital Audio Broadcasting
DAC	Digital to Analog Converter
DARPA	Defence ARPA (*q.v.*)
DBR	Distributed Bragg Reflector
DBS	Direct Broadcast Satellite
DCFL	Direct Coupled FET Logic
DD	Double Density
DDP	Double Density Package
DDR DRAM	Double Data Rate Dynamic Random Access Memory
DECT	Digital Enhanced Cordless Telephone
DELTT	Double Electron Layer Tunnelling Transistor
Demux	Demultiplexer
DES	Data Encryption Standard
DFB	Distributed Feed Back
DIB	Duel Independent Bus
DIL	Dual-In-Line
DIP	Dual-In-Line Package
DMA	Direct Memory Access
DME	Direct Memory Execution
DMIC	Digital Monolithic Integrated Circuit
DML	Directly Modulated Laser
DMT	Discrete MultiTone (standard)
DOE	Diffractive Optical Elements0
DOIT	Direct Optical Interface Technology
DPSS	Diode-Pumped Solid-State (laser)
DPSSFD	Diode-Pumped Solid-State Frequency-Doubled (laser)
DPST	Double Pole, Single Throw (switch)
DRAM	Dynamic Random Access Memory
DS	Direct Sequence
DS	Double Sided (floppy disk)
DSC	Digital Still Camera
DSL	Digital Subscriber Line
DSL	Dynamic Single Longitudinal
DSP	Digital Signal Processing
DSVD	Digital Simultaneous Voice and Data

DSS	Digital Satellite System
DSS	Digital Spread Spectrum
DSSS	Direct Sequence Spread Spectrum
DTS	Digital Tuning System
DTTV	Digital Terrestrial Television
DTV	Digital Television
Dual	Containing 2 equal devices
DUART	Dual* Universal Asynchronous Receiver Transmitter (=2 × UART*)
DUV	Deep UV
DVB	Digital Video Broadcasting
DVC	Digital Video Compression
DVD	Digital Video Disk
DWDM	Dense Wavelength Division Multiplexing
EA	Electroabsorption
EAM	Electro-Absorption Modulator
ECC	Elliptic-Curve Cryptosystems
EDA	Electronic Design Aid
EDA	Electronic Design Automation
EDC	Extended Data Cut
EDFA	Erbium Doped Fibre Amplifier
EDO	Extended Data Out
EDP	Electronic Data Processing
EMI	ElectroMagnetic Interference
EMILD	Electroabsortion Modulator Integrated (DFB) Laser Diode
EMILM	Electroabsorption-Modulated Isolated Laser Module
EMAS	European Eco-Management and Audit Scheme
EMC	Electromagnetic Compatibility
EPLC	Electrically Programmable Logic Device
EPLD	Erasable PLD (*q.v.*)
ERA	Electrically Reconfigurable Arrays
ESPRIT	European Strategic Program for R&D in Information Technology
ETC	Enhanced Throughput Cellular
EUV	Extreme Ultra Violet
FA	Factory Automation
FAP	Fibre Array Package
FCSS	Fibre Channel Storage Systems
FDC	Floppy Disc Controller
EDFA	Erbium Doped Fibre Amplifier
FEC	Forward Error Correction
FED	Field Emissive Display
FET	Field Effect Transistor
FFT	Fast Fourier Transforms
FGL	Fibre Grating Laser
FHSS	Frequency Hopping Spread Spectrum
FIB	Focussed Ion Beam
FIFO	First In First Out — or — Fan In, Fan Out
FLOPS	Floating Point Instructions Per Second (MFLOPS, GFLOPS)
FM	Frequency Modulation
FP	Fabry Perot
FPA	Focal Plane Array
FPAA	Field Programmable Analog Array
FPM	First Page Mode
fps	Frames Per Second
FPSC	Field Programable System Chips

FRED	Fast REcovery Diode
FSK	Frequency Shift Keying
FS-PAL	Field Sequential Phase Alternation Line
FTTC	Fibre-To-The-Curb
FTTH	Fibre-To-The-Home
GaAlAs	Gallium Aluminium Arsenide
GaAs	Gallium Arsenide
GaAsP	Gallium Arsenide Phosphide
GaInAsP	Gallium Indium Arsenide Phosphide
GaN	Gallium Nitride
GaP	Gallium Phosphide
GbE	Gigabit Ethernet
GBIC	GigaBit Interface Converter
Gbps	Gbit per second
GLUE	General 7400-series chips to glue (connect) ICs
GMR	Giant Magneto Resistive
GND	Ground (voltage level)
GRINSCH	Graded INdex Semiconductor Confinement Heterostructure
GSMBE	Gas Source MBE (*q.v.*)
GSM	Global Systems for Mobile (communications)
HAD	Hole — Accumulation — Diode
HBLED	High Brightness LED
HBT	Heterostructure Bipolar Transistor
HD	High Density (floppy disk)
HDD	Hard Disk Drive
HDSL	High bit rate Digital Subscriber Line
HDTV	High Definition TV
HED	High-Efficiency Diode
HEMT	High Electron Mobility Transistor
Hex	Containing 6 equal devices *or* hexadecimal base notation
HFD	Hybrid Fibre Coax
HIC	Hybrid Integrated Circuits
HJBT	Hetero Junction Bipolar Transistor
HJFET	Hetero Junction FET
HPDL	High Power Diode Laser
HVD	High Voltage Differential
I/O	Input and Output
IC	Integrated Circuit
ICE	In-Circuit Emulator (system to simulate a processor)
ICP	Inductively coupled Plasma
I2C	(IIC) Intra Integrated circuit (-bus) (Philips)
IF	Intermediate Frequency
IGBT	Insulated Gate Bipolar Transistor
IHT	Information Handy Terminal
I2L	Integrated Injection Logic
ILM	Isolated Laser Modules
IMOS	Integrated Micro-Optical Systems
IMS	Intelligent Manufacturing System
InAs	Indium Arsenide
InN	Indium Nitride
InP	Indium Phosphide
InSb	Indium Antimonide
INTC	Interrupt Controller
IO	Input and Output
IP	Intellectual Property
I-PAL	Improved PAL

IRLED	InfraRed Light Emitting Diode
ISDN	Integrated Service Digital Network
ISM	Industrial, Scientific & Medical band
ISP	In-System Programming/Internet Service Provider
ISR	In-System Reprogrammable
ITSEC	Information Technology Security Evaluation Criteria
JBOD	Just a Bunch of Disks
JEDEC	Joint Electron Device Engineering Council
JEIDA	Japan Electronic Industry Development Association
JFET	Junction FET (*q.v.*)
JIT	Just-In-Time
JPEG	Joint Photographic Experts Group
KGD	Known Good Die
Kilo	Prefix 1024 or 1000
λ_p	Peak sensitivity wavelength
LB	Low Back (reflection)
LCAS	Line Card Access Switch
LCD	Liquid Crystal Display
LCOS	Liquid Crystal on Silicon
LD	Laser Diode
LDO	Low DropOut (regulators)
LDSD	Low Dimensional Structures & Devices
LED	Light Emitting Diode
LEO	Lateral Epitaxial Overgrowth
LEP	Light Emitting Polymer
LER	Lissajous Electron Plasma
LIU	Line Interface Unit
LMDS	Local Multipoint Distribution System
LNA	Low Noise Amplifier
LO	Local Operating (frequency)
LOCMOS	Local Oxidization CMOS (*q.v.*)
LOG	Epitaxially Laterally Overgrown GaN
LS	Low-power Schottky (type of TTL)
LSB	Least Significant Bits
LSI	Large Scale Integration
LVD	Low Voltage Differential
LVDS	Low Voltage Differential Signalling
LVT	Low Voltage Technology
MAC	Media Access Controller
MAC	Multiplexed Analogue Components
MAX	Multiple Array Matrix
MB	Megabyte
MBE	Molecular Beam Epitaxy
Mbit	Megabit
Mbps (or Mb/s)	Mbit per second
Msps	Mega samples per second
mCBIC	Micro Cell-Based Integrated Circuit
mcd	Millicandela
MCM	Multi Chip Module
MCM-C	MCM-ceramic
MCM-D	MCM-dielectric
MCP	Multi-Chip Package
MCSN	Multimedia Cable Network System
MCT	Mercury Cadmium Telluride (also CMT)
MCT	MOS Controlled Thyristors
MCU	Microcontroller

MD	Mini Disk
MDS	Multicore Development System
Mega	Prefix — 1024 × 1024 or 1000 × 1000 or 1000 × 1024
MEMS	Microelectromechanical Systems
MESFET	Metal Semiconductor FET (*q.v.*)
MFMIS FET	Metal Ferroelectric Metal Insulator Semiconductor FET
MGM	Mostly Good Memory
MID	Moulded Interconnect Device
MIDI	Musical Instrument Digital Interface
MIM	Media Interface Modules
MIPS	Million Instructions Per Second
MISFET	Metal Insulator Semiconductor FET
MLC	Multi-Level Cell
MLLD	Mode-Locked Laser Diode
MMIC	Microwave Monolithic Integrated Circuit
MMU	Memory Management Unit
mmwave	Millimetre Wave
MOCVD	Metal Organic Chemical Vapour Deposition
MODE	Micro-Optical Devices
MODEM	Modulator (and) Demodulator
MOEMS	Micro-Opto-Electro-Mechanical systems
MOPS	Mega Operations Per Second
MOS	Metal Oxide Semiconductor
MOSFET	Metal Oxide Semiconductor Field Effect Transistor
MOVPE	Metal Organic Vapour Phase Epitaxy (also OMVPE [*q.v.*])
MP@ML	Main Profile at Main Level
MPEG	Motion Picture Experts Group
MPL	Multi-level Pass transistor Logic
MPU	Microprocessor
MPW	Multi-Product Wafer
MQW	Multi Quantum Wells
MR	Magneto Resistive
MRAM	Magnetic RAM (*q.v.*)
MSA	Multi-Source Agreement
MSI	Medium Scale Integration
MSOP	Mini Small Outline Package
MSPS	Mega Symbols per Second
MSS	Modulated Semiconductor Structure
MTP	Multi-Time Programmable
MUART	Multiple Universal Asynchronous Receiver/Transmitter
Mudem	Multiplexer (and) Demultiplexer (Not an official term)
Mux	Multiplexer
MVI	Motion Video Instruction
NAM	Non-Absorbing Mirrors
NAND	Logic term — Not-AND (result is NOT-operation after an AND-operation)
NIC	Network Interface Cards
NMOS	N-channel MOS
NoBL	No Bus Latency
NOR	Logic term — Not-OR (result is NOT*-operation after an OR-operation)
NOT	Logic term (output is high if the input is low and low if the input is high)
NPT	Non-Punch Through
nm	Nanometre
ns	Nansecond

NTSC	National Television System Committee or 'Never The Same Colour'	
NV	Non Volatile (sustains content without power-connection)	
NVG	Negative Voltage Generator	
NVM	Non-Volatile Memory	
NVRAM	Non Volatile RAM	
NXOR	Logic term — Not-XOR (result is a NOT-operation after an XOR-operation)	
OADM	Optical Add-Drop Multiplexer	
OC	Optocoupler	
OCM	Optical Channel Monitor	
Octal	Containing 8 equal devices (Octuple)	
ODP	Ozone Depletion Potential	
OEIC	Optoelectronic IC (*q.v.*)	
OEL	Organic Electroluminscence	
OEM	Original Equipment Manufacturer	
OFA	Optical Fibre Amplifier	
OFC	Open Fibre Control	
OHCI	Open Host Controller Interface	
OI	Opto Isolator	
OLED	Organic LED (*q.v.*)	
OMVPE	Organometallic VPE (*q.v.*)	
OpAmp	Operational Amplifier	
OPIC	Optical Chip	
OPO	Optical Parametric Oscillator	
OR	Logic term (output is high if at least one input is high)	
ORCA	Optimized Reconfigurable Cell Array	
OS	Operating System, eg DOS (Disk Operating System)	
OSA	Optical Sub-Assembly	
OSL	Organic Semiconductor Laser diode	
OTDM	Optical Time Division Multiplexing	
OTP	One Time Programmable	
PAL	Programmable Array Logic	
PBGA	Plastic Ball Grid Array	
PBSRAM	Pipeline Burst SRAM (*q.v.*)	
PC	Personal Computer	
PCB	Port Bypass Circuit	
PCB	Printed Circuit Board	
PCBA	Printed Circuit Board Assembly	
PCI	Peripheral Component Interface	
PCM	Pulse Code Modulation	
PCMCIA	PC Memory Card International Association	
PCN	Personal Communication and Network	
PCS	Personal Communications Services	
PD	Photo Diode	
PDA	Personal Digital Assistant	
PDP	Plasma Display Panel	
PDP	Parallel Data Processing	
PDIP	Plastic Dual In-line Packaging	
PERM	Pre-Embossed Rigid Magnetic	
PGA	Pin Grid Array	
PHY	Physical Layer	
PIC	Photonic IC (*q.v.*)	
PIN	Positive Intrinsic Negative	
PLA	Programmable Logic Array	
PLC	Planar Light Circuits	

PLCC	Plastic Leaded Chip Carrier
PLD	Programmable Logic Device
PLL	Phase-Locked-Loop
PLM	Pump Laser Module
PM	Polarization Maintaining
pm	picometres
PMD	Physical Media Dependent
PMF	Polarization Maintaining Fibre (pigtails)
PMT	Photo-Multiplier Tube
PNP	P-type, N-type, P-type (transistor)
PPI	Precise-Pixel Interpolation
POF	Plastic Optical Fibre
Poly-Si	Polysilicon
POTS	Plain Old Telephone System
PPRAM	Parallel Processing Random Access Memory
PPV	Poly-p-phenylenevinylene
PQFP	Plastic Quad Flat Package
PRML	Partial Response/Maximum Likelihood
PROM	Programmable Read Only Memory
PSD	Position Sensitive Detector
PSRR	Power Supply Rejection Ratio
PV	Photo-Voltaic (ie solar cell)
PZT	Lead-zirconium-titanate
QAM	Quadrature Amplitude Modulation
QCL	Quantum Cascade Laser
QCW	Quasi-Continuous Wave
QML	Qualified Manufacturers Listing
QPA	Quad Port Acceleration
QPLC	Quad Physical Layer Controller
QPSK	Quadrature Phase-Shift-Keying
QW	Quantum Well
QWI	Quantum Well Intermixing
QWIP	Quantum Well Infrared Photodetectors (QWIP)
RAID	Redundant Array of Independent Disks
RAM	Random Access Memory
RAS	Remote Access Server
RBSOA	Reverse Bias Safe Operating Area
RCLED	Resonant-Cavity LED
RDIMM	Registered Dual In-line Memory Modules
RDRAM	Rambus Dynamic Random Access Memory
RF	Radio Frequency
RFIC	Radio Frequency Integrated Circuit
RFID	Radio Frequency IDentification
RIE	Reactive Ion Etching
RIN	Relative Intensity Noise
RISC	Reduced Instruction Set Computing
RM	Resource Management
ROADM	Reconfigurable Optical Add-Drop Multiplexer
ROM	Read Only Memory
ROS	Record-On-Silicon
RTOS	Real Time Operating System
Rx	Receiver
SAM	Self-Aligned by MBE (*q.v.*)
SAN	Storage Area Networks
SAR	Segmentation and Reassembly
SAW	Surface Acoustic Wave

SBP	Serial Bit Protocol
SBR	Saturable Bragg Reflector
SBS	Stimulated Brillouin Scattering
SCALPEL	Scattering with Angular Limitation in Projection Electron-beam Lithography
SCSI	Small Computer System Interface
SDH	Synchronous Digital Hierarchy
SDL	Simplified Data Link
SDTV	Standard-Definition Digital Television
SEC	Single Edge Contact
SERDES	Serialization/Deserialization
SEU	Single Event Upset
SFF	Small Form Factor
SFFP	Small Form Factor Pluggable
SHG	Second Harmonic Generation
SI	Semi-Insulating
SIA	Semiconductor Industry Association
SiC	Silicon Carbide
SiGe	Silicon Germanium
Si3N4	Silicon Nitride
SIU	System Interface Unit
SLAC	Subscriber Line Audio processing Circuits
SLI	System-Level Integration
SLIC	Standard-function Linear Integrated Circuit
SLIC	Subscriber Line Interface Circuit
SLM	Spatial Light Modulator
SLO	Scanning Laser Ophthalmoscope
SLMQW	Strained Layer Multiple Quantum Well
SMBus	System Management Bus
SMD	Surface Mount Device
SMF	Single Mode Fibre
SMID	Single Instruction Multiple Data
SMIF	Standard Mechanical InterFace
SMPS	Switch Mode Power Supply
SMR	Surface Mount Radial
SMT	Surface Mount Technology
SOA	Semiconductor Optical Amplifier
SO	Small-Outline
SOC	System On a Chip
SOD	Silicon On Diamond
SO DIMM	Small Outline Dual-In-line Memory Module
SOI	Silicon-On-Insulator
SOIC	Small Outline Integrated Circuit
SON	Small Outline Non-lead
SONET	Synchronous Optical Network
SOP	Small Outline Package
SOQ	Silicon-On-Quartz
SOS	Silicon on Sapphire
SPD	Single Photon Detection
SPGA	Staggered Pin Grid Array (package)
SPGA	System Programmable Gate Array
SPI	Serial Peripheral Interface
SPICE	Simulation Program with Integrated Circuit Emphasis
SPLD	Standard Programmable Logic Device
SPLICS	Special-Purpose Linear Integrated Circuits
SQW	Single Quantum Well

SSCI	Synchronous Serial Communication Interface
SSFDC	Solid-State Floppy-Disk Card
SSI	Small Scale Integration
SSOP	Shrink Small Outline Package
SSRAM	Synchronous (cache) Static Random Access Memory
SSTL	Solid State Transistor Logic
STA	Static Timing Analysis
STB	Set-Top Box
SWAC	Single Wide Area Connection
TCP	Tape Carrier Package
TDMA	Time Division Multiple Access
TEC	Theremoelectric Cooled
TFT	Thin-Film Transistor
TIA	Transimpedance Amplifier
TMOS	T-Metal Oxide Semiconductor
TMR	Triple Modular Redundancy
TO	Transistor Outline
T_{Opr}	Operating Temperature
TOPS	Texture Optimized Pyramidal Surface
TOS	Thermo-Optic Switch
TPV	ThermoPhotoVoltaic
TRAC	Totally Reconfigurable Analog Circuit
TRON	The Realtime Operation System Nucleus (Japanese R&D program)
TSSOP	Thin Shrink SOP (*q.v.*)
T_{Sol}	Soldering Temperature
T_{Str}	Storage Temperature
TTL	Transistor Transistor Logic
UBC	User Break Controller
μBGA	Micro Ball Grid Array
UHB-LED	Ultra-High Brightness LED (*q.v.*)
ULSI	Ultra Large Scale Integrated Circuit
UQFP	Ultra Quad Flat Package
USB	Universal Serial Bus
USIC	User-Specific Integrated Circuit
UTQFP	Ultra-Thin Quad Flat-Pack
V	Volt
VAD	VApour phase Deposition
V_{CBO}	Collector-Base Voltage
V_{CC}	Supply voltage
V_{CEO}	Collector-Emitter Voltage
$V_{CE (sat)}$	Collector-Emitter Saturation Voltage
VCM	Virtual Channel Memory
VCO	Voltage Controlled Oscillator
VCSEL	Vertical Cavity Surface-Emitting Laser
VDSL	Very high bit-rate Digital Subscriber Line
VDSM	Very Deep Sub-Micron
V_{ECO}	Emitter-Collector Voltage
VESA	Video Electronics Standards Association
VFD	Vacuum Fluorescent Display
VHSIC	Very High Speed Integrated Circuit
VISC	Video Instruction Set Computing
VLD	Visible Laser Diode
VLIW	Very Long Instruction Word
VLSI	Very Large Scale Integration
VOA	Variable Optical Attenuator

VPE	Vapour Phase Epitaxy
VPM	Video Port Manager
Vpp	Volts peak-to-peak
VRAM	Video RAM
VRM	Voltage Regulator Module
VRP	Voice Recognition Processors
VSI	Virtual Socket Interface
VSPD	Variable Sensitivity Photodetector
VSP	Voice Signal Processor
VSWR	Voltage Standing Wave Ratio
W	Watt
WCDMA	Wideband Code Division Multiple Access
WDM	Wavelength Division Multiplex
WDT	WatchDog Timer
W/hr	Wafers per Hour
WLAN	Wireless Local Area Network
WRAM	Window RAM
WSI	Wafer Scale Integration
XIP	eXecute In Place
ZBT	Zero Bus Turnaround
ZIF	Zero Insertion Force
ZVP	Zoomed Video Port